Practical Electron Microscopy
for Biologists

PRACTICAL
Electron Microscopy
FOR BIOLOGISTS
2nd Edition

Geoffrey A. Meek
Senior Lecturer in Electron Microscopy
Department of Human Biology and Anatomy
The University
Sheffield

A Wiley–Interscience Publication

JOHN WILEY & SONS
London · New York · Sydney · Toronto

Library of Congress Cataloging in Publication Data:

Meek, Geoffrey A.
  Practical electron microscopy for biologists.
  Bibliography: p.
  Includes index.
  1. Electron microscope. 2. Electron microscopy—
  Technique. I. Title.
QH212.E4M43   1976   578′.4′5   75–4955

ISBN 0 471 59031 2 (Cloth)
ISBN 0 471 99592 4 (Paper)

Printed in Great Britain by
Unwin Brothers Limited
Old Woking, Surrey

TO MY WIFE

Preface to the First Edition

Morphological research in biology is at present undergoing an explosive renaissance which only the most farsighted could have predicted as little as twenty years ago. This is almost entirely due to the application of one particular instrument—the electron microscope. Possession of, or access to, an electron microscope is now virtually essential for all laboratories engaged on most kinds of work on cell or molecular biology. Unfortunately, electron microscopes are both expensive and complex, and are far more susceptible to costly damage than even the most complicated light microscope. It requires several years of continual practice before a reasonably competent research worker can be sure that he understands the nature of the problems involved in biological electron microscopy and the techniques most likely to solve them. Only then will he be on his way to turning out electron micrographs of similar technical standard to the masters of the art. For electron microscopy is an art—an applied science, but still in the end a graphic art. Photographic techniques play as great a part as those of specimen preparation, high vacuum and electronics.

Biological electron microscopy is a combination of applied physics on the one hand, and a specialized kind of biology on the other. When electron microscopes first became generally available to biologists ten or fifteen years ago, it was considered essential to buy a graduate physicist at the same time as the instrument, since nobody else could be expected to run it. Many of these physicists turned biologist, bringing a novel outlook with them, and some have made outstanding contributions to biology. The supply of electron microscopes, however, has long outstripped the supply of willing physicists, and now the biologist must fend for himself.

Few biologists will have had the opportunity to learn anything about electron microscopy during their undergraduate careers. When the time comes for the research biologist to choose and buy one of these expensive instruments, where can he turn for advice? There are several good textbooks dealing with biological techniques in electron microscopy, and several dealing with electron optics and the design of electron microscopes. This book is intended to fill the gap between instrument design on the one hand and specimen preparation on the other. It is intended to explain and expand the instructions given in the operating manuals provided by the electron microscope manufacturers, which generally seem to be written by physicists for physicists, and which experience has shown to confound and bewilder the beginner rather than to instruct him. It is also intended to survey the requirements of the various biological workers who might wish to acquire an electron microscope, and to suggest which of the many instruments at present available might best fulfil a given requirement for the least expenditure of funds. If the book saves one person from having to live with an expensive white elephant that once seemed perfect, it will have achieved its object.

The content of this book is based mainly on experience gained from organising and teaching several courses in practical electron microscopy to biologists under the auspices of the Royal Microscopical Society; and also from teaching the rudiments of electron microscopy to research and undergraduate students over a period of years in the Department of Human Biology and Anatomy at the University of Sheffield. Much effort has been expended in pruning theory to the irreducible minimum consistent with clarity. No rigorous mathematical treatments or proofs of the phenomena described in the book are offered. They are all available in the standard textbooks, many of which are cited in appropriate places. Biologists are constitutionally more familiar with descriptive rather than mathematical treatments, and in the author's opinion it is no more necessary for a biologist to be able to design an electron microscope from first principles than it is for him to be able to compute the lenses on the light microscope he uses every day.

For those who wish to become top-class microscopists, there is no substitute for an apprenticeship spent in a top-class electron microscope laboratory. Techniques cannot really be learned from reading books. But there is a dearth of practical information on electron microscopy in the books which are at present available; it is hoped that this book will go some of the way towards filling this gap. It is hoped that the book may enable any reasonably intelligent person, graduate or non-graduate, academic or technician, to make the best use of an electron microscope; to diagnose and repair common faults; and to make and interpret electron micrographs of a high standard of technical excellence.

Preface to the Second Edition

Since the first edition of this book was written five years ago, much progress has been made in the technology of electron microprobe instruments, to which family the conventional transmission electron microscope belongs. In particular, the use of high-brightness electron guns and the scanning mode of image formation in conjunction with electron spectrometry and computer-based techniques of image analysis have greatly improved the detection efficiency of the lower atomic number elements such as carbon, which are of special interest to biologists. The resulting increase in image contrast has made possible high resolution micrographs of native, unstained specimens of great biological importance, such as DNA molecules. Similar techniques are improving the sensitivity and accuracy of X-ray microanalysers, which are now able to quantitate and pin-point the position of as little as 10^{-18} g of certain elements, some of which are of great biological importance. In addition, the use of an accelerating voltage of a million or more in a conventional electron microscope, with the attendant advantages of increased specimen penetration and reduced specimen damage, is attracting the attention of more biologists, many of whom now have access to one of these very expensive instruments.

The potential usefulness of electron beam technology in biology is immense. Advances in instrument technology are now far outstripping their biological applications. Accordingly, this second edition has been brought up to date by the addition of new material introducing these novel physical microanalytical methods to the biologist. The opportunity has also been taken to update the specifications of those commercial transmission instruments which remain on the market; to correct a few misprints and mis-statements which regrettably slipped through the proof reading of the first edition; to improve the cross-referencing in the text; and to increase the scope of the index.

Acknowledgements

I would like especially to express my indebtedness to Professor Sir Hans Krebs, who first suggested that I should take up the art of electron microscopy; to Professor Robert Barer for his unstinting help and encouragement; to the Wellcome Foundation and the Nuffield Foundation for so generously providing the necessary very expensive hardware; to Professors Keith Porter and George Palade for their kindness and hospitality in allowing me to learn the basic essentials of the art of electron microscopy in their laboratory at the Rockefeller Institute in the early days; and not least to my wife and family for their forebearance, understanding and help during the time I was writing this book.

I would like to record my appreciation of the many helpful discussions I have had with my colleagues, especially Dr. Dan Goldstein and Dr. Ian Carr, and for their critical reading of parts of the manuscript. Many of their suggestions have been incorporated; any errors are my responsibility entirely.

I wish also to thank all the people, publishers and institutions who have generously allowed me to make use of their illustrations in this book. Acknowledgement is made with each illustration used. My thanks are also gratefully recorded to Mr. Peter Garlick, for his invaluable technical assistance, and to Mr. Michael Turton for photography. My grateful thanks are also due to Mrs. Lesley Conlin, who typed the manuscript.

G. A. MEEK
Sheffield
1970

Contents

xi

PART 2: USING THE ELECTRON MICROSCOPE

PART 3: SPECIMEN PREPARATION

PART 1

THE ELECTRON MICROSCOPE

CHAPTER 1

Some Basic Principles of Optics

1.1 Introduction

Very tiny things have always had a remarkable fascination for people, especially the very tiny things that can be seen. This immediately poses problems. How small is the smallest thing that can be seen? If we can see it, how much can we find out about it? Problems such as these have exercised microscopists ever since man discovered how to assist his eye to see smaller and ever smaller objects. The art and science of seeing the very small is the definition of microscopy, and the construction of ever better devices to enable him to see ever smaller things is the microscopist's chief preoccupation. To this end, man invented first the simple light microscope and then the compound light microscope. The latter instrument, with which earnest research scientists are frequently pictured in glossy advertisements, has been brought to a state of perfection which has possibly not been equalled in any other scientific instrument. Why, then, do we need that highly complicated, expensive and inconvenient device, the electron microscope?

The electron microscope has one basic advantage which the optical microscope, however perfect it may be, cannot possibly have. It can make clear, sharp images of objects up to one thousand times smaller than the light microscope can. This is the sole advantage which the electron microscope possesses over the light microscope. In nearly all other respects—for instance, simplicity, price, ease of specimen preparation, the ability to examine living cells, maintenance—the light microscope has the advantage over the electron microscope. But this one enormous advantage —the ability to resolve objects down to the dimensions of large organic molecules—has opened up a new world to the biologist. The electron microscope, in spite of its many disadvantages, is the sole key to this world of 'ultrastructure'.

We will examine the reasons why the electron microscope has this

1

fundamental advantage in much greater detail in the first section of this book. However, it will be useful first to summarize just what the electron microscope is all about. The key word is RESOLUTION. Most people, when asked what is the main advantage to be gained by using the electron microscope, will answer: 'It magnifies much more than a light microscope'. Such a statement is perfectly true, but this reason is only secondary, stemming from the primary reason—resolution. Resolution (or, strictly speaking, 'resolving power' if we are referring to the instrument and not to the finished picture) simply means the ability to discriminate fine detail. The light microscope can discriminate fine detail, but only down to a certain limit. This limit is imposed, not by any defect in the instrument, but by the fundamental nature of light itself. For the purposes of microscopy, light may be regarded as a train of waves belonging to a family of waves known as 'electromagnetic radiation'. This family embraces radio waves, heat or infra-red waves, visible light waves, ultraviolet waves, X-rays, gamma-rays from atomic disintegrations and cosmic rays from outer space. All these waves have one particular property in common: they travel at the same velocity (the speed of light) in a vacuum. They differ from one another only by the distance from wave crest to the next, known as the 'wavelength'. Radio waves may be from several miles long down to a few millimetres. Infra-red wavelengths are measured in hundredths of a millimetre; visible light waves in thousandths of a millimetre. All the other forms of radiation we have mentioned have wavelengths shorter than visible light.

Because our eyes are sensitive only to visible light, we have to use it to convey visual information to our brains. But the amount of information it can convey is limited by its wavelength. We can appreciate this by using as an analogy the waves of the sea. If we look down from a cliff on to a seashore on to which waves are breaking, we can see clearly from the pattern of the waves the presence of large boulders, but we cannot get any information about the size of the pebbles of the beach. This is because the pebbles are so much smaller than the wavelength of the breakers that the information about their size and shape is, so to speak, 'lost' in the wave. Water waves can carry information about the nature and size of objects larger than their wavelength, but can tell us little or nothing about the size of objects much smaller than about half a wavelength. In an exactly analogous fashion, light waves can only carry spatial information about objects which are significantly larger than one-half-wavelength. The resolving power of the light microscope, however perfect its construction, is therefore limited to objects larger than one-half-wavelength of visible light: about a quarter of a micrometre.

Although this fine structural information is all the time being carried into the eye by the tiny light waves entering it, everyone knows that the unaided eye cannot appreciate this fine detail. The reason is simple: the light-sensitive receptor of the eye, the retina, cannot itself resolve an image smaller than a certain size. It is made up of cells, the rods and cones, and a certain number of these must be stimulated before the presence of a discrete object can be appreciated. This limits the resolving power of the unaided eye to an external object about one-quarter of a millimetre across, held at the closest comfortable viewing distance under reasonable conditions of visibility. Obviously, this limiting size is only an approximation; it varies from person to person, with age, contrast, lighting, colour and so forth.

The eye must therefore be assisted by magnifying the image falling on the retina. This is the function of the microscope. The light microscope must ideally provide distortion-free magnification to a degree determined simply by the ratio of the resolving power of the eye to the resolving power of visible light, which is one-quarter of a millimetre divided by one-quarter of a micrometre: 1000 times. The struggle to provide this degree of magnification with minimum image distortion is the story of the development of the light microscope over the past 400 years or so. There is no instrumental difficulty in providing sheer magnification far in excess of this; a light micrograph projected on a cincma screen may easily be magnified to a total of a million times or more. But as the viewer approaches the screen, hoping to see more detail, all he sees is a larger and larger blur. There is no more spatial information to be gained about the image, and magnification much in excess of 1000 times is often referred to as 'empty magnification'.

How then can we increase resolving power to get more spatial information? There is only one solution to the problem: to use a form of radiation having a shorter wavelength. Unfortunately, as we go down the electromagnetic spectrum to the shorter wavelengths, the problems of forming a sufficiently magnified image become insuperable. Firstly, the eye is insensitive to them; secondly, waves shorter than ultraviolet cannot as yet be bent or refracted to form images; thirdly, the increasing energy associated with the radiation becomes lethal to living things and could have only limited biological application.

Until about 40 years ago, this problem appeared to be without a direct solution. How it was overcome is the story of the electron microscope and will be told in Chapter 2. To cut a long story short, a hitherto unknown, very short-wave radiation associated with fast-moving particles was postulated on theoretical grounds in 1924, its existence was confirmed in 1927, and it was applied to the formation of high-resolution images in 1933. The most convenient particles which may be used to generate this radiation for microscopy are electrons, since electron beams are easily refracted to form magnified images. The exact nature of this matter-associated, non-electromagnetic radiation, sometimes termed 'matter waves' or 'probability waves', is not fully understood but, after all, neither is the force of gravity. Just as we use gravity without fully comprehending its physical nature, so we may in practice make 'electron-optical' images without fully understanding the theoretical nature of the radiation involved.

The wavelength associated with the high-voltage (30–100,000 volt) electrons generally used in biological electron microscopy, which travel at about one-third the velocity of light, is approximately 100,000 times smaller than the wavelength of visible light. Theoretically, therefore, the resolving power of the electron microscope should be better than that of the light microscope by this huge factor. Unfortunately, the difficulties involved in forming magnified electron images in practice are such that the theoretical resolving power has not yet even been approached. The best result so far is an improvement of about 1,000 times over the light microscope. This is still a very large factor indeed by any standards, and has led to the world of biological and metallurgical ultrastructure undreamed of only a generation ago.

In spite of the complex nature and spectacular performance of the modern electron microscope, its basic optical principles are virtually identical with those of the light microscope. A thorough understanding of the electron microscope requires an equally thorough understanding of basic light optics. Sufficient information for this purpose will, it is hoped, be presented in the remainder of this chapter.

1.2 Magnification, Resolution and Contrast

Everyone knows that the simplest way to find out more about the fine detailed structure of an object is to 'magnify' it, and then to present the magnified image to the eye in such a way that the eye can appreciate the greater detail thus revealed. We say that we have been able to 'resolve' greater detail in the object, and that the magnified image has enabled us to improve upon the resolution of the unaided eye. 'Resolving power' or 'resolution', which is the ability to distinguish fine detail, is clearly bound up with magnification. Magnification is also a function of the distance of an object from the eye. The further away it is, the smaller it seems to be.

This affects resolving power; we cannot with the unaided eye resolve a tennis ball a kilometre away, but we can resolve it at a distance of 200 m if it stands out clearly against its background. This brings in another concept of great importance in microscopy—the concept of 'contrast'. A white tennis ball at 100 m is easily seen if it is held against a black background on a sunny day; we say that it has a high 'contrast' and thus stands out sharply against its background. On the other hand, a white tennis ball held against a white sheet at 100 m on a dull day would not be visible. This is not because the eye cannot resolve it; it is simply because the eye cannot distinguish it against its background. Clearly, the best resolving power in the world will be of no avail if the object to be resolved cannot be perceived.

As an object is brought closer and closer to the eye, more and more of the detailed structure of its surface can be resolved. The reason for this is that the image of the object on the retina of the eye is becoming larger and larger; we are receiving a magnified image which is capable of higher resolution by the eye. Magnification can therefore be achieved simply by bringing an object closer to the eye. Is there any limit to the closeness to which we can bring an object up to the eye? Everyone knows by experience that there is a limit, which is about 10 cm. It is a great strain on the muscles of the eye to focus an object so close, so the normal close working distance of the eye, called the 'near point', is generally accepted as being 25 cm. We can measure the resolving power or 'visual acuity' of the eye at this distance by simple tests. Under the best conditions, we find that the eye can resolve an angle of about 1 minute of arc. This angle corresponds to a separation of about 3 cm at a distance of 100 m. At the near point, the eye can separate the individual lines on a finely ruled grating under ideal conditions if they are 0·075 mm apart. It cannot separate smaller detail, because of the cellular structure of the retina, which is made up of rod and cone cells of finite width. Since this resolution is taxing the performance of the eye to the limit, and requires ideal conditions of contrast for its attainment, it is more usual to accept a lower, more comfortable figure of 0·25 mm at 25 cm as representing a reasonable value under normal conditions.

The question now is, how can the eye be assisted to separate two fine lines closer together than this minimum? The answer is to magnify the lines, so that the image of the lines impressed on the retina of the eye is made larger. The device we interpose between the object and the eye in order to achieve magnification is a microscope. It may be a simple microscope (which is the simple hand lens or 'magnifying glass'), a compound microscope or an electron microscope. The light microscope will give us a magnified image direct upon the retina of the eye; the electron micro-scope requires a further 'image translation system' between it and the eye before the eye can interpret the image it produces. The light microscope bends or refracts the rays of light in such a way as to enable the object to be held closer to the eye. In the case of the simple microscope, this is the actual process; in the case of the compound microscope, this is effectively what is done. Why, then, can we not just continue to magnify with the light microscope and thus discriminate more and more detail? The answer lies in the physical nature of light; the fact that light is an electromagnetic wave motion, and the waves are of finite length. The wavelength of light used determines the ultimate resolving power of any optical instrument.

Later in this chapter, we will see why no microscope, light or electron, can resolve two points which are closer together than about one-half-wavelength of the radiation used to image them. If the two points are closer together than the resolving power of the microscope, they will be imaged as one point. One-half-wavelength is the ultimate resolving power of either class of instrument. The resolution limit was reached for the light microscope in about 1890. The theoretical limit in the case of the electron microscope has not yet been approached by a factor of 100. Light lenses took more than three centuries to develop; electron lenses have only been under development for some 40 years. They are still in the compara-tively crude state of design of seventeenth-century light lenses. This does not imply any disrespect to the designers of electron lenses; it merely emphasizes the immensely greater difficulties of their problem.

How much magnification do we need? This is very readily calculated; it is simply the ratio of the resolving power of the eye to the resolving power of the microscope. This enlarges the smallest resolvable detail up to the point where the eye can readily distinguish it. In the case of the light microscope, the resolving power is one-half-wavelength of light, which is approximately $2 \cdot 5 \times 10^{-4}$ mm. The unaided eye, we have stated, can resolve 0·25 mm. The figure for maximum magnification is simply the resolving power of the eye divided by the resolving power of the micro-scope, which gives us 1,000 times. Magnification in excess of this figure gives us no more information—it merely magnifies a blur. Excess magnifi-cation is called 'empty magnification'.

The resolving power of the electron microscope is about one thousand times better than that of the light microscope; it can 'see' objects one thousand times smaller. Maximum magnification must therefore be one thousand times as great as that for the light microscope, which gives us a figure of one million times. Any figure in excess of this is once again empty magnification.

1.3 Units of Length in Microscopy

We have said that the resolving power of a light microscope is about one-half of the wavelength of light. How big an object does this in fact represent? The wavelength of green light, to which the eye is most sensitive, is 0·005 mm. The light microscope can therefore resolve a distance of 0·0025 mm under ideal conditions. The electron microscope, because it uses electromagnetic radiation very much shorter than that of light, can resolve distances of 0·00000025 mm. It is clearly necessary to have a unit smaller than the millimetre.

The unit used in light microscopy is the 'micrometre', symbolized 'μm', which is one-millionth part of the standard length of one metre. It used to be called the 'micron'. One micrometre is therefore one thousandth of a millimetre; 1,000 μm = 1 mm. This unit is too large for electron microscopy, so it is again divided into 1,000 parts called 'nanometres'. 1,000 nm = 1 μm. A more commonly used unit in electron microscopy is the Ångstrom unit, symbolized Å, which is one-tenth of a nanometre. 10 Å = 1 nm. The Ångstrom unit is sometimes referred to in the modern International System (S.I.) of length as the 'decinanometre', dnm. 1 Å = 1 dnm. Using these units, the resolving power of the light microscope is therefore about 0·25 μm or 2,500 Å; that of the electron microscope is 2·5 Å, which is one thousand times as great.

1.4 Geometrical Optics

The electron microscope is a completely logical development of the familiar compound optical microscope or 'light microscope' as it will subsequently be called throughout this book. The informed and intelligent user of both the light and the electron microscope is distinguished from the mere knob-twiddler by his or her grasp of the fundamental principles of optics which govern the design and operation of both instruments. The basic optical principles involving the use of refractile elements or 'lenses' in order to form magnified images are identical in both the light and the electron microscope, and anyone having a thorough grasp of the principles of the light microscope should have no difficulty in understanding the electron microscope, which differs from the light microscope only in the radiation it uses and in the way in which the radiation is bent or refracted. Both instruments are optical instruments; hence the use of the term 'light microscope' rather than 'optical microscope'.

It is necessary to have a reasonable understanding of elementary light optical theory in order to make the best use of either instrument. The remainder of this chapter attempts to deal in as simple and practical a way as possible with some of the fundamental principles of optics, and will endeavour to emphasize the inherent similarities between the two instruments, and to show the differences where they exist.

The study of optics can be divided into two branches. The more familiar branch is called 'geometrical optics', and is concerned with the study of the paths followed by 'rays' of light or electrons through lenses and apertures, and the geometrical constructions used to find the relative positions and sizes of objects and their images. A 'ray' of light or electrons is defined as an infinitely thin pencil or beam; physical optics shows that it is an abstraction and cannot exist physically because of an effect which we will meet later in this chapter called 'diffraction'. However, geometrical ray paths and diagrams are of great assistance in understanding the working of all forms of optical apparatus, and will be used as far as possible in this book, since they have the great merit of being easily understood.

The other branch is called 'physical optics', and is the more fundamental branch of the two. It deals with the wave motion of light and electrons. All the results obtained in geometrical optics can be derived from the principles of physical optics, together with other important phenomena such as interference and diffraction, which are not explicable in simple geometrical terms.

1.5 Glass and Electron Lenses

Lenses are used to bend rays of light or electrons so that they are deflected in a predictable way from their original paths. Light lenses do this because the transparent materials of which they are made cause a slowing-down of the velocity of light during its passage through the lens. If a beam of parallel rays of light strikes a perfect convergent lens in a direction perpendicular to its central plane (Fig. 1.1), the emergent beam will converge to a 'focal point' and then diverge again. All rays parallel to the lens axis will pass through this point. The distance along the axis of the focal point from the central plane of the lens is called the 'focal length' of the lens. A strong lens has a short focal length; a weak lens a long focal length. If a source of light is placed infinitely far from the lens, the rays from it which strike the lens will be parallel, and therefore a screen placed behind the lens in the focal plane will reveal a 'real' image of the source as a point of light on the lens axis. This type of lens is variously called a 'positive', 'convergent', 'convex', 'condensing' or 'magnifying' lens. If the glass surface is oppositely curved (Fig. 1.2), then parallel rays

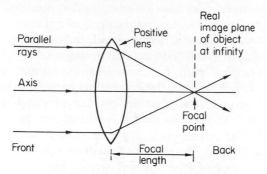

Figure 1.1 The action of a magnifying lens. Rays parallel to the axis are converged to a point at the focus of the lens. All magnetic electron lenses are of this type

striking the lens in a direction parallel to the axis will be caused to diverge, and will not be brought to a focus. This type of lens, various called 'negative', 'concave' or 'diminishing', cannot form a real image. Both types of lens are used in light microscopes, but only positive lenses are used in magnetic electron microscopes. It is not possible to construct a negative magnetic lens. Therefore, since the great majority of electron microscopes use magnetic lenses only, the divergent lens will not be discussed further.

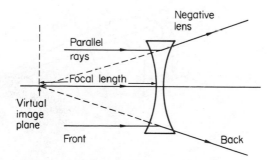

Figure 1.2 The action of a diverging lens. Rays parallel to the axis diverge and never form a real focus. If the paths of the divergent rays are projected to the front of the lens, they meet at a 'virtual focus', which is taken as the focal point of the lens. Magnetic lenses cannot be made like this, but electrostatic ones can

Magnetic electron lenses have a further fundamental way in which they differ from light lenses. The electron beam does not change in velocity as it passes through a magnetic field; instead, the deviation is caused by a force acting on each individual electron causing it to be bent or refracted away from its original path. In a light lens, the refraction of rays takes place at the interface between the lens and its immersion medium, which is usually glass and air. In an electron lens, the diverting force acts on the electron the whole of the time that it is within the lines of force of the magnetic field. The refraction is therefore continuous, and there is no sharp interface between the refracting medium (the magnetic field) and the immersion medium (the vacuum within the electron microscope column).

Fig. 1.3 shows the action of an axially symmetrical, curved magnetic field on a divergent electron beam. The curved field of force acts as a positive, condensing lens, and forms an image of the source on the opposite side of the lens in a manner which is geometrically analogous to the

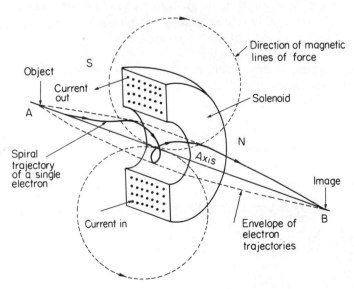

Figure 1.3   The action of a solenoid on an electron beam. An electric current passing through the coil produces an axial magnetic field. This is the refracting medium for the electrons. An electron starting at a point on the axis A and at an angle to it follows a helical path, returning to the axis at the point B. The action is basically similar to that of the converging light lens shown in Fig. 1.1

formation of a real image by a positive light lens. There is no sharp interface, and the lens can only converge the electron beam. If, however, the magnetic field is short compared with the distance of the object and image along the axis of the lens, then the magnetic field can be treated as a short, thin lens as shown in Fig. 1.4 and the ray diagrams used to construct image and object positions through thin-glass lenses can be used to construct electron images also. In the ray diagrams shown in the remainder of this book, electron lenses will be represented as having curved surfaces and interfaces analogous to light lenses. It must be borne in mind that this is a fiction dictated by convenience, and is not intended to represent the actual situation in an electron lens.

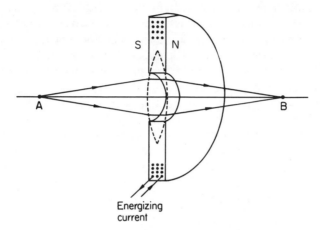

Figure 1.4   The action of a short, thin solenoid wound like a flat 'pancake coil'. A divergent beam of electrons leaving the point A will be converged to the point B by the magnetic field within the coil, which acts like a thin glass lens (shown dotted) on a divergent beam of light. Magnetic lenses can most easily be represented in this way, which will be used in the remainder of the diagrams in this Chapter

Another fact which will be ignored in the construction of electron microscope ray diagrams is the spiral path followed by electrons as they pass through a solenoid. The electron suffers a sideways force as it passes through an axial magnetic field, which causes it to follow a helical path as shown in Fig. 1.3. The pitch of the helix becomes shorter as the strength of the magnetic field increases towards the centre of the lens. However, a section through the envelope enclosing all the spiral paths of the electron trajectories gives the two rays shown in Fig. 1.4. These rays can be treated just as though the electrons followed straight paths through the lens, and the spiralling of the electrons can, as a first approximation, be ignored.

1.6 Ray Diagrams

Geometrical optics is based on the study of ray diagrams. The method of construction of ray diagrams is based upon two simple premises. These are: (a) all rays entering the lens parallel to the axis are brought to a common point on the axis, the focus; and (b) all rays passing through the geometrical centre of the lens (the point where the axis meets the central plane) are undeviated and pass straight on, no matter from which direction they come. The second premise assumes that the lens is 'thin', i.e. all the refraction takes place at one single plane, which does not apply in practice; electron lenses are of course thick lenses. However, the ray paths constructed from these two principles are sufficiently accurate to show how the electron microscope works.

The usual 'sign convention' of light optics will be followed in this book. Rays are always considered to travel from left to right in a diagram having a horizontal axis, and from top to bottom in a diagram having a vertical axis. The object is in the space to the left or above the lens; the image is in the space to the right or below. To avoid confusion, the object space will always be described as being 'in front' or 'before' the lens; the image space as 'after' or 'behind' (see Fig. 1.5).

1.7 Real Images

Fig. 1.5 is a ray diagram showing the paths of the parallel ray A and the chief or principal ray B through a thin positive lens L. The construction of the diagram is as follows. The paths of two rays are traced; ray A is the parallel ray, travelling from the point O on the object parallel with the lens axis. It strikes the central plane of the lens at right angles, is deviated through the back focal point F_1 of the lens and passes on. Ray B travels from the same point on the object through the geometrical centre C of the lens, and is undeviated. These two rays meet at a point I behind the lens. This is the position of the real image I of the object point O. A parallel ray from each point on the object can be traced through, and will be found to meet its corresponding central ray on the image plane. If a screen of ground glass is placed in the plane of I, it would reveal a 'real image' of the object O. Note that the real image is inverted. If O

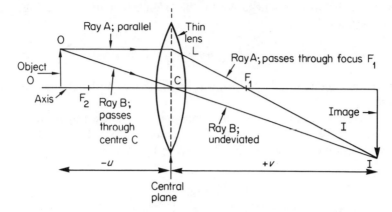

Figure 1.5 The geometrical form of construction used to find the position and relative size of an image I formed of an object O by a thin lens L (see text). A magnified image will be formed if the object is closer to the lens than a distance equal to twice the focal length. As the object approaches the focal distance F_2, the magnification becomes greater. At F_2, the magnification is theoretically infinite, but the image would lie at an infinite distance away from the lens. By convention, rays in ray diagrams are always considered to travel from left to right. The space to the left of the lens is the 'front'; to the right is the 'back'. F_2 is thus the front focal point, and F_1 the back focal point. Distances in front of the lens are generally considered to be negative, and those behind the lens to be positive. A positive lens therefore always has a positive focal length, and a negative lens a negative focal length (see Fig. 1.2). The 'sign convention' may be reversed in some accounts of geometrical optics

were a point source of light placed off the axis of the point O, then the image would be a point at the position I, provided the lens were perfect.

The positions of corresponding object and image points are called 'conjugate points', and the planes of object and image are called 'conjugate planes'. The image is only situated at the true or 'principal' focus when the object is situated at infinity.

The distance u of the object plane in front of the lens, the distance v of the image plane behind the lens, and the distance f of the focal plane from the central plane of the lens can be shown by simple geometry to be related by the expression:

$$\frac{1}{v} - \frac{1}{u} = \frac{1}{f}$$

Since the distance u is negative by the 'sign convention' (see Fig. 1.5), this becomes

$$\frac{1}{v} + \frac{1}{u} = \frac{1}{f}$$

This is the fundamental lens formula, applicable to both light and electron lenses.

The relative sizes of object and image can be shown again by simple geometry to be in direct ratio to their distances from the lens plane. The magnification of the lens is therefore given by the expression:

$$M = \frac{v}{u}$$

It will be seen that if

$$u = v = 2f$$

then $M = 1$

and the image is the same size as the object. If u is greater than $2f$, then M is less than 1 and the image will be smaller than the object. When the object is located between $2f$ and f, the focal distance M becomes greater than 1 and the lens magnifies. As magnification increases, the distance v becomes correspondingly larger and larger. Very high magnification can thus in theory be obtained with a single positive lens, but only at the expense of a very large and inconvenient image distance.

1.8 Virtual Images

When the object O is moved to a plane behind the front focal point F_2 of a positive lens as shown in Fig. 1.6, a different kind of image is formed. Since the rays A and B leaving the back of the lens are now divergent, no real image can be formed behind the lens. If, however, another optical system capable of forming real images (such as the eye) is placed behind the lens, then the rays from the object point O will appear to diverge from the point I, which is now situated in front of the lens. If a screen were placed in the plane of I, no real image would be formed. This apparent image is therefore called a 'virtual' image. It differs from a real image in that it is upright, not inverted. It is important to understand the formation of virtual images, because the intermediate lens in some electron microscopes, which is used to control the magnification, is used in this way in conjunction with the objective lens in order to reduce the size of the final image.

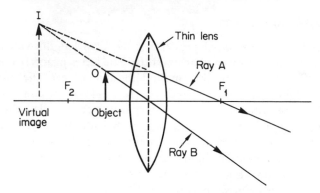

Figure 1.6   The formation of a 'virtual image' by a positive lens. If the object lies closer than a distance equal to the focal length in front of a lens, the rays coming from the back of the lens will diverge, and it is impossible for the lens to form a real image. If an optical system such as the eye is placed behind the lens, the divergent rays can be made to converge and form a real image. The intermediate lens of an electron microscope is sometimes used in this way, a real final image being formed by the projector lens system behind it. As the lens is changed from 'real image working' (Fig. 1.5) to 'virtual image working', the final image is inverted in direction, as will be seen

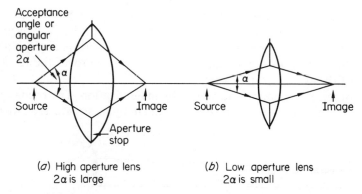

Figure 1.7   The angular aperture of a lens. The angle 2α is the acceptance angle of the lens, and the larger it can be made, the more information can the lens transmit. A large lens of high aperture can therefore tell us more about an object than a small lens of low aperture

1.9 Angular Aperture

A lens has two principal properties: its focal length, which we have already defined, and its angular aperture. The latter property controls the ability of the lens to gather information about an object. The information-gathering ability of a lens depends on the angle of the cone of rays it is able to accept from an object. This parameter is called the 'angular aperture'. Fig. 1.7 shows two lenses, each with the same focal length, forming an image I on the axis of a point source O, also on the axis. One lens is of large diameter or 'high aperture', and thus is capable of gathering more information about the object than the second lens, which is of smaller diameter or 'low aperture'.

1.10 The Function of the Microscope

The eye is a simple optical instrument, and the function of any microscope, be it simple, compound or electron, is to enlarge detail too small to be perceptible to the human eye until it is large enough for the eye to perceive clearly and thus to resolve. The eye is therefore the final arbiter of all microscopical results. It is an imperfect optical instrument, with important limitations in performance. One of these is its small angular aperture, which limits the angle of the cone or 'pencil' of light rays which can enter it. Bringing an object nearer to the eye increases the aperture, as is shown in Fig. 1.8, but there is a limit to the closeness to which an object can be brought to the eye. This near-point distance is normally accepted as 25 cm. At the near point, the pupil of the dark-adapted eye subtends a maximum angle at the object of about 8°. A typical oil-immersion light-microscope objective lens has an aperture angle of about 175°. This is one way in which the microscope improves the resolving power of the eye.

The other, more important, way is by magnification. The retina of the eye is limited in its ability to resolve fine detail by its structure, as we have

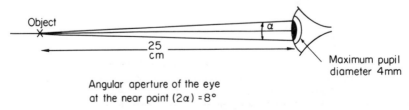

Angular aperture of the eye
at the near point (2α) = 8°

Figure 1.8   The angular aperture of the eye is relatively small

already noted. The central or foveal region is made up of a mosaic of cone cells, each of which has a diameter of about 2·5 µm. The precise mechanism whereby the retina translates light signals into nervous impulses is very imperfectly understood, and it is not known precisely how many cone cells must be stimulated in order to give rise to a sensation of light. However, it seems reasonable to assume that the more the image on the retina is enlarged, the greater will be the amount of information about the object that will be sent to the brain.

1.11 The Simple Microscope

It can be seen from the geometry of Fig. 1.8 that bringing an object closer to the eye will increase the acceptance angle α. Since this cannot be done physically, we must deceive the eye into thinking that an object is at the minimum distance for sharp focus when it is in fact nearer. This is done very simply by using a positive lens to form a virtual image as has already been described, and then using the lens of the eye to form a real image of the magnified virtual image on the retina. The lens is held close to the eye as shown in Fig. 1.9, and the ray diagram shows how the eye

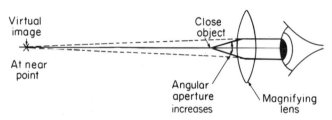

Figure 1.9   The angular aperture of the eye can be increased by the use of a lens. The lens in this case simply allows the object to be held closer to the eye, which is thereby enabled to gather more information

is deceived into thinking that the object is at the near point. Since the object is in fact held closer to the eye, it appears to be magnified. A lens used in this way is called a 'simple microscope', and is frequently used by electron microscopists to examine electron micrographs closely in order to see detail below the resolving power of the unaided eye. Like all other microscopes, it increases resolving power in direct ratio to magnification. It is not possible to obtain very high magnifications with a simple microscope, for the reason that there is a limit to the closeness to which a lens and an object can be held to the eye. Such microscopes are not to be despised; the distinguished eighteenth-century optician and observer

Antony van Leeuwenhoek, who has been called the 'father of micro-scopy', brought the use of simple microscopes to the highest art, and was able to obtain magnifications of 250 or more times. Since the resolving power of Leeuwenhoek's eyes must have been 250 μm, he could therefore resolve objects down to 1 μm in size. We know he was able to do this, for he was the discoverer of bacteria and protozoa.

1.12 The Compound Microscope

Although it is possible in theory to obtain a real image of any desired degree of magnification from a single positive lens, a microscope forming real images and using such a system would be very cumbersome because of the long lens-image distance. Also, the aperture of the lens, and hence its information-gathering power, would be very small. In practice, this distance is shortened by interposing a second lens behind the first as shown in Fig. 1.10. The second lens introduces a further stage of magnification by forming a secondary image I_2 of the primary image I_1 formed by the first or 'objective' lens. The total magnification is the product of the magnifications of the two lenses. This procedure can be repeated

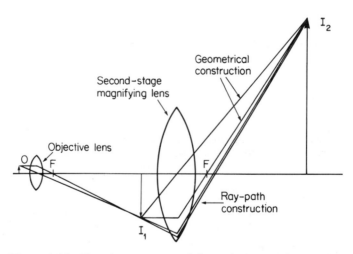

Figure 1.10   Two lenses are used in series to project a high magnification final image with a very much shorter object–final-image distance than can be achieved with one lens. This procedure can also be used with three lenses in series. Light microscopes can obtain sufficient magnification with two lenses; electron microscopes require three. The methods used in both geometrical and ray-path construction are shown

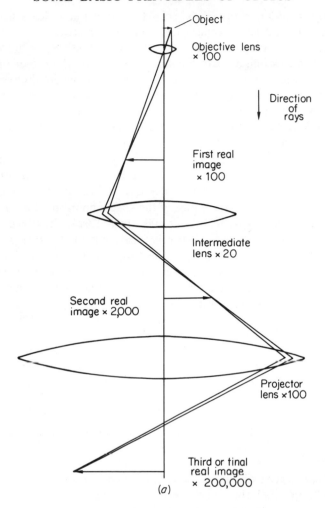

Object

Objective lens
× 100

Direction
of
rays

First real
image
× 100

Intermediate
lens × 20

Second real
image × 2,000

Projector
lens ×100

Third or final
real image
× 200,000

(a)

Figure 1.11(a)   A ray-path diagram showing how the final image is formed in a three-magnifying-lens system such as is found in most electron microscopes. This 3-real-image system is used for the high magnification range on a 3-lens electron optical system. The range covered is generally from about × 10,000 to × 200,000 overall magnification. Final magnification is changed by varying the strength of the intermediate lens from about × 1 to about × 20. This alters the position of the object plane of the intermediate lens, and consequently the current in the objective must also be changed to compensate when magnification is changed. This operation is called 'focusing'. On some designs, the strength of the projector lens is also changed in order to increase the magnification range covered

indefinitely in order to obtain higher and higher magnifications. In the light microscope, it is not usual to have more than two stages, since the resolution of the image is limited by the wavelength of light, and the required degree of magnification can be obtained by the use of only two stages of magnification. In the electron microscope, where the resolution of the image is limited only by the imperfections of the lenses and not by the wavelength of the electrons used, the high magnifications necessary to bring the resolving power of the instrument up to the resolving power of the eye have to be achieved by using three stages of magnification in order to keep the length of the microscope tube down to a reasonable size. The lenses used and their functions are shown in Fig. 1.11(a). The objective lens is a very strong lens of short focal length, values of 1·5 to 5 mm being common. It forms a primary image of the specimen in the primary image plane. This image, magnified about 100 times, forms the object for the intermediate lens, which is a relatively weak lens of variable power (or focal length). It is used to control the total magnification of the final image, and can usually be set to give magnifications of between zero and × 20 simply by varying the current flowing through the winding. This is the function of the magnification control on an electron microscope. The intermediate lens forms a secondary image lying in the object plane of the third or final projector lens, which is another strong lens. The magnification of the projector lens may be fixed or variable; it generally has a maximum magnification of around × 100. The total magnification of the three lenses combined when used at maximum power is the product of the individual magnification of each lens used singly; in the case shown in Fig. 1.11(a) this will be $100 \times 20 \times 100 = 200,000$. If a micrograph is made on a photographic plate of an ideal specimen with a high-resolution instrument in perfect adjustment, the plate will contain detail which is still too small for the eye to resolve. The plate can now be magnified in an optical enlarger by a further factor of 5, giving a final print magnification of 10^6, on which the unaided eye can resolve a distance of 2 Å. This is, at present, the useful limit of magnification.

This '3-real-image' optical system cannot be used to give electron optical magnifications below about × 10,000 because of lens imperfections. We shall discuss the reasons for this in Chapter 4. A different system, shown in Fig. 1.11(b), is used for low magnifications in some instruments. The objective lens is weakened so as to form an image behind the plane of the intermediate lens, which is then made very weak in order to form a diminished real image of the image which would have been formed by the objective in the absence of the intermediate. This image is shown dotted, and can be considered to be a virtual image formed by the

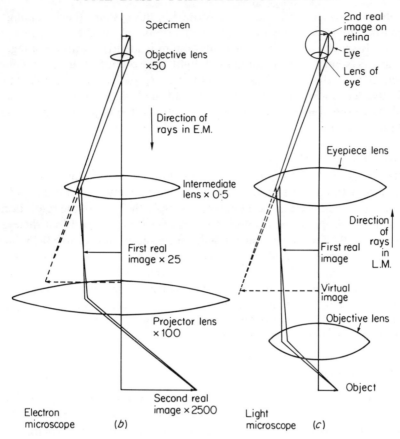

Specimen

Objective lens
×50

Direction of
rays in E.M.

Intermediate
lens × 0·5

First real
image × 25

Projector lens
× 100

Electron
microscope   (b)

Second real
image × 2500

2nd real
image on
retina

Eye

Lens of
eye

Eyepiece lens

Direction
of
rays
in
L.M.

First real
image

Virtual
image

Objective lens

Object

Light
microscope   (c)

Figures 1.11(b) and 1.11(c)   Ray-path diagrams showing how the optical system of a 3-lens electron microscope is used to give a range of lower magnifications by forming 2 instead of 3 real images. The objective lens is made weaker so as to form its primary image, shown dotted, beyond the intermediate lens instead of in front of it as shown in the previous diagram. The intermediate lens is now made very much weaker, so as to form a diminished real image of the object in the front conjugate plane of the projector lens. This system provides a low magnification range from about × 1,000 to × 10,000. Notice particularly how the final image is reversed in direction compared with the 3-real-image system shown in Fig. 1.11(a). The dotted primary image which would be formed by the objective lens in the absence of the intermediate lens can be considered to be a 'virtual' image. The correspondence between this system and the formation of a virtual image using a simple microscope shown in Fig. 1.6 can easily be seen if the second real image in this diagram is considered as the object and the ray paths are reversed (Fig. 1.11(c))

objective. Only two real images are formed in this system; the final image therefore appears to be inverted when compared to the 3-real-image system. Inspection will show that this system is identical with the image-forming system when a compound light microscope is used in conjunction with the lens of the eye, but the ray paths are reversed. The objective lens in Fig. 1.11(b) corresponds to the lens of the eye; the intermediate to the eyepiece lens and the projector to the objective lens of the light microscope.

1.12.1 Maximum Magnification

It is necessary here to emphasize the point that increase in magnification is useless without increase in resolution. The maximum overall magnification which can usefully be employed is simply the ratio of the resolution of the microscope to the resolution of the eye. If it is agreed that the eye can resolve 0·25 mm and that the light microscope can resolve 0·25 μm, then the maximum magnification that can usefully be used with the light microscope is × 1,000. In the case of the electron microscope, if an instrument is capable of resolving 2·5 Å, then clearly the maximum magnification is × 1,000,000. This is a thousand-fold greater than in the light microscope, and is one of the many reasons why electron microscopes are more complex than light microscopes.

1.13 Physical Optics

Geometrical optics gives us, as we have seen, a satisfactory explanation of how lenses and microscopes work, simply by following the paths of the rays through the optical components. It does not offer any explanation for three extremely important facts. These are:

(1) the resolving power of a microscope is effectively limited to one half of the wavelength of the illumination used;
(2) that light or any other form of radiation does not necessarily travel in straight lines, but can in certain circumstances 'bend around corners'; and
(3) that in certain circumstances, and especially in the electron microscope, objects are surrounded by haloes called 'Fresnel fringes'.

Since the explanation of Fresnel fringes and the limitation of resolving power lies in the ability of light to bend around corners, we will first examine how this curious phenomenon comes about. Let us first consider a point source of light suspended in space. Fig. 1.12 represents the way in which the radiation it gives out forms a spherical envelope of waves with the point source at the centre. This envelope travels at the velocity of light (3×10^{10} cm/s) out into the vacuum of space and the radius of the wavefront becomes increasingly greater as the distance from the source increases. If the source is infinitely far away, the wavefront is parallel.

1.14 Interference

If we try to work with spherical wavefronts, life becomes impossibly complicated. We therefore have to work with sections through spherical wavefronts, which are, of course, circles. A section through a spherical wavefront resembles the ripple pattern on the surface of a pond after a small pebble has been thrown in. The ripples travel outwards from the centre, forming circles of increasingly greater radius, which are a section through a spherical wavefront. This is the pattern represented by Fig. 1.12.

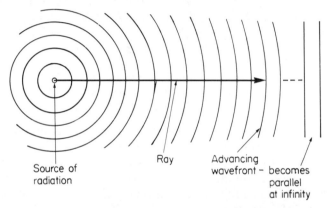

Source of radiation

Ray

Advancing wavefront – becomes parallel at infinity

Figure 1.12   This represents a source of radiation or waves surrounded by a spherical wavefront. The diagram is an instantaneous section through the wavefront in two dimensions only. The wavefront travels outwards from the point source at 3×10^{10} cm/s in the case of light; electrons travel at about one-tenth the velocity of light. The wavefront becomes parallel at infinity. A 'ray' is the locus of a point on the wavefront. It always travels perpendicular to the wavefront

We can now make a further simplification. We can take a section of the two-dimensional section, and represent a ray of light or electrons by a sine wave travelling out from the source, as shown in Fig. 1.13(a). The physical characteristics are defined by the wavelength or crest-to-crest distance from one wave to the next, and the amplitude, or crest-to-trough height of the waves. Another relationship also comes in—that of time. If a second identical wave is placed beside the first (Fig. 1.13(*b*)), but shifted in space so that a crest from one wave is exactly opposite a trough from the other, then the second wave must have left the source exactly one half-wavelength in time after the first. It is said to be exactly 180° out of phase with the first wave, and if the two waves coincide exactly both in

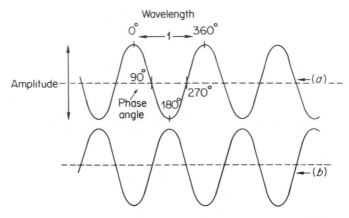

Figure 1.13   The time relationships or 'phase' of a wave. The two waves (*a*) and (*b*) are exactly out of phase; they will cancel one another out if superimposed

space and in time, then the effect of one will exactly nullify the effect of the other, and the result will be a cancellation of the light. The sum of the amplitudes of the two waves will be zero, and there will be a redistribution of energy in space. If, on the other hand, two identical waves arrive at the same place at the same time exactly in phase, then the one will reinforce the other, and the net effect will be the sum of the two amplitudes separately. The waves may be out of phase by any value between 0° and 180°, and the resulting light intensity will vary between the square of the arithmetical sum of the amplitudes of the waves and zero. This superposition of the two wave trains in space and time causes the modification of one wave by the other, and gives rise to the phenomenon of 'interference'.

1.15 Diffraction

Let us now suppose that a parallel wavefront strikes an opaque barrier (Fig. 1.14). What now happens is that the part of the wavefront which clears the barrier continues unmolested in a straight line, but the part which hits the edge goes 'round the corner'. It does this by starting up a new spherical wavefront at the barrier edge. This explanation was put forward by Huygens some 300 years ago, when he proposed that 'each point on a wave front may be regarded as a new source of waves'. This is a very difficult concept to grasp, but the fact that waves do in fact go around an obstacle in this way can easily be demonstrated by the water

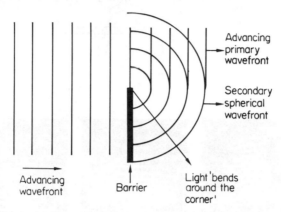

Advancing primary wavefront

Secondary spherical wavefront

Advancing wavefront

Barrier

Light 'bends around the corner'

Figure 1.14   When a wavefront strikes a barrier, it can 'bend around the corner' by giving rise to a secondary wavefront at the edge, since each point on a wavefront can give rise to a new source of waves. This phenomenon is called 'diffraction'

ripple analogy. Take a large circular bowl of water and support two glass plates vertically in it leaving a slit opening about 2 cm wide between them (Fig. 1.15(a)). When the water surface is quite still, observe it by light reflected from the sky and then allow a drop of water to fall behind the plates. The ripples from the drop will be seen to pass through the slit, and to form a new *circular* pattern in front of the plates. This effect is called 'diffraction'.

The effect can be demonstrated very beautifully by making the ripples on the surface of the water more clearly visible in an apparatus called a 'ripple tank'. All serious students of optics should study wave motion in a ripple tank. Ripple tank photographs (Figs. 1.15(b), (c) and (d)) show how the effect of diffraction becomes less as the width of the slit is increased; in other words, diffraction effects are only noticeable in general when the size of the object is of the same order as the wavelength of the radiation used to illuminate it.

That diffraction also occurs with light can be demonstrated by a simple experiment. Take two photographic plates, expose them to light, develop, fix and dry them. Now scribe across each plate a very fine line with the

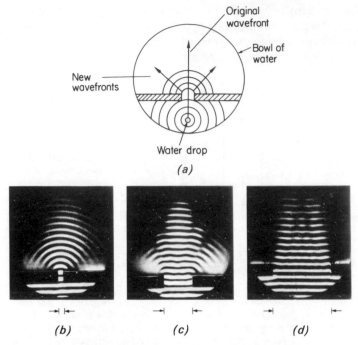

Figure 1.15   (a) Diffraction simply demonstrated on the surface of
a bowl of water (see text). The effect is difficult to see, but can be
demonstrated very clearly on the surface of a ripple tank, as shown
in the three succeeding ripple tank photographs

(b) The width of the slit is the same as the wavelength of the parallel
wavefront advancing from below. A new circular wavefront spreads
out beyond the slit, showing quite clearly how the wavefront 'bends
around the corner'

(c) The slit is several times wider than the wavelength of the ripples.
A certain amount of spreading out takes place, but less than in the
preceding case.

(d) The slit is very wide compared with the ripple wavelength.
There is practically no sideways spread beyond the slit

Ripple tank photographs (b), (c) and (d) by courtesy of Dr. W.
Llowarch and Oxford University Press

corner of a sharp razor blade. The finer the lines, the clearer will be the
effect to be described. Set one plate up close to a small, powerful lamp—
a car headlamp bulb is ideal. Switch on the lamp, hold the second plate
close to the eye, and look through the second (eye) slit at the lamp filament
seen through the first (source) slit. Instead of an image of the fine slit, a

wide pattern of coloured bands is seen, spread over several centimetres. The appearance of the pattern is shown diagrammatically in Fig. 1.16.

This experiment demonstrates two facts. Firstly, the diffraction of light —the spreading of a slit by the 'bending-round-the-corner' effect—occurs

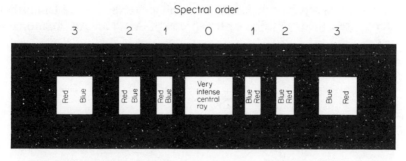

Figure 1.16   A diagram of the appearance of a slit brightly illuminated with white light when viewed through a single dispersing slit held close to the eye. The first three orders of spectra are shown on either side of the intense central image of the source slit. If the slits are very narrow, the spectral colours are very pure and beautiful. The successive spectra and dark spaces merge into one another and are not sharply defined as shown on the diagram. The pattern is very difficult to photograph, because the intensity of successive spectra falls off greatly as the order increases. The fourth order is hardly visible to the eye in the simple experiment described

quite independently of any lenses, since we have used no lenses to demonstrate it. Secondly, the effect has caused a severe loss of resolving power. The image of our source slit, perhaps 0·1 mm wide, has been broadened to an apparent 5 cm or more—a loss in resolution of 5,000 times.

How can this effect be explained? An explanation is provided by Huygen's principle, combined with the principle of the reinforcement and cancellation of light by interference. Our source slit near the lamp provides us with a single wavefront, selected from the incoherent jumble of wavefronts emanating from the lamp filament. Such a single wavefront is called a 'coherent source'; the concept is a very important one in electron microscopy. This coherent wavefront advances towards the eye slit. When it strikes the latter, part of the wavefront passes straight through the slit, forming the broad, bright band at the centre of the pattern (Fig. 1.16). But two secondary spherical wavelet patterns are set up close to the two edges of the slit. These two patterns are coherent with one another, since the source wavefront was coherent. Therefore, parts of the pattern will be in phase and reinforce, and parts will be out of phase and cancel. In particular, the points on the wavefront marked A, A′ in Fig. 1.17 are in phase and will reinforce; so also will B, B′ and C, C′. But A is in phase not only with A′ but also with B′, C′ and D′. If we join up the reinforcing parts of the wavefront by drawing tangents to them, then AA′ reinforce to give the powerful central part of the pattern; AB′ and A′B reinforce to give the two separate patterns on either side of the central band (these patterns are called the 'first order spectra'); AC′ and A′C reinforce to

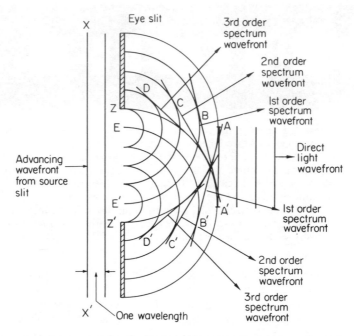

Figure 1.17 A very much simplified diagram to show how the three spectral bands shown in the previous figure are formed. Consider an advancing parallel wavefront XX′ travelling towards a slit ZZ′. Any two points such as EE′ on the part of the wavefront entering the slit can be considered by Huygen's principle to be the sources of two new wavefronts, shown as a series of circles in the diagram. These new wavefronts are coherent and will therefore interfere. This gives rise to the spectral bands. Interference between two successive wavefronts AB′ and A′B gives rise to the strong first order spectra; the second order spectra are due to interference between wavefronts separated by two wavelengths AC′ and A′C, and are consequently weaker; the third order spectra arises from interference between wavefronts separated by three wavelengths, AD′ and A′D, and so on. The higher order spectra become successively less intense

give the next pair of spectra—the 'second order'—and so on. The higher orders of spectra become progressively dimmer as the displacement from the central band increases.

What does this mean in terms of resolving power? Let us first consider the results in terms of aperture. Our very fine slit caused tremendous broadening of the image. Will a wider slit give more or less broadening?

Scribe a broader slit on the second plate and repeat the experiment. The coloured bands will now be seen to be far narrower; resolving power has therefore been improved. This is because the angular aperture of the system has been increased by increasing the secondary slit width. The effect of secondary wavelet formation has been overwhelmed by the effect of the central band. The effect is shown in Fig. 15(b), (c) and (d).

Let us now consider the result in terms of the coloured bands or spectra. Examination of the image pattern shows clearly that the width of the pattern as seen with *blue* light is considerably narrower than the width seen with *red* light. The narrower the band, the better the resolving power. Resolving power by blue light is better than by red light. Blue light is of a shorter wavelength than red light. The experiment demonstrates that the shorter the wavelength, the better the resolving power.

The results of our diffraction experiment can therefore be stated: resolving power increases as angular aperture increases, but decreases as wavelength increases.

1.16 Fresnel Fringes

Our experiments so far have been concerned with the effects of diffraction at the edges of a single slit aperture. What happens if we remove one half of the slit? Will diffraction still take place, blurring the image of the single edge we are now using? We can show very easily that blurring does take place simply by illuminating a straight edge from a coherent source, and examining the edge of its shadow cast on to a screen. We can use the simple apparatus described on p. 25 to demonstrate this effect. Using the illuminated slit as a source, place a white screen some distance away so that a dim rectangular patch of light from the source slit falls on it. It is necessary to carry out this experiment in a darkened room. Now place a razor blade so that its edge intercepts the light propagated from the slit. Close inspection of the screen will show that the shadow of the edge of the razor blade is not sharp. It is blurred by the formation of a series of fringes, which are sharpened if monochromatic light is used. If a photographic plate is suspended on the screen, the fringes can be photographed. Fig. 1.18(a) is a print from a plate exposed to blue light in this way. If a microdensitometer tracing is made of the plate, the curve shown in Fig. 1.18(*b*) is obtained. The fringes are formed by interference, due to difference in path length between the main wavefront and the secondary wavefront formed at the edge in accordance with Huygen's principle. The

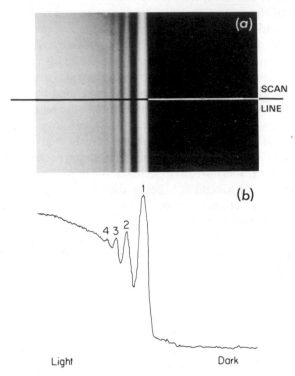

Figure 1.18(*a*)   A photograph of the shadow cast by the edge of a razor blade illuminated by mono-chromatic (blue) light rendered coherent by passing through a narrow slit. The photograph was taken by holding a plate on the wall where the shadow was cast. The photograph has been enlarged × 2·5

(*b*) A microdensitometer trace of the negative from which the photograph above was made. Four Fresnel fringes can be distinguished

way in which they are formed is shown in Fig. 1.19. The mathematical derivation of the fringe spacing in this general case of diffraction is more complex than in the simpler case of diffraction at a slit. These fringes are called 'Fresnel fringes' (pronounced FRE-NELL), after the French mathematical physicist Augustin Fresnel (1788–1827) who first investigated them mathematically.

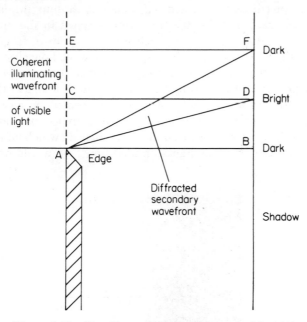

Figure 1.19   How Fresnel fringes are formed with visible light. The sharp edge A is illuminated by a coherent wavefront. Diffraction at the edge gives rise to a secondary wavefront spreading into the illuminating wavefront, which interferes with the illuminating (primary) wavefront in the following way. If the path length AD differs from the path length AB by one full wavelength, the wavefronts reinforce and a bright fringe results at D. However, where the path length differs by one-half wavelength as at F, the wavefronts cancel and a dark fringe results. Multiples of whole and half wavelengths cause a series of fringes to be formed, of which the first is the brightest. No fringes are formed unless the illuminating wavefront is coherent. Note that the real fringes are formed in the illuminated side of the shadow of the edge

1.17   Electron Fresnel Fringes

Fresnel fringes are also formed when a beam of electrons strikes an opaque or semi-opaque edge. Fig. 1.20(a) is an electron micrograph at very high magnification of a small part of the circumference of a hole in a carbon film. The film is black, and the hole white in the photograph. A system of fringes very like that formed by light at a sharp edge (Fig. 1.18(*a*)) can be seen, but comparison of the two photographs will show that the electron fringes are formed in the image of the film (the dark area), whereas in the case of light, the fringes are formed in the bright area alongside the shadow. A second, much weaker system of fringes corresponding to the light system is also formed inside the hole (Fig. 1.20(*c*)), but is generally too weak to be seen on the screen of an electron microscope. The presence of the strong fringes in the dark image of the film is explained as follows. Two systems of fringes are formed, because the edge in this case is not completely opaque, and interference also takes place between the diffracted wavefronts of reduced intensity which pass through the film. From our previous light considerations, we would expect this

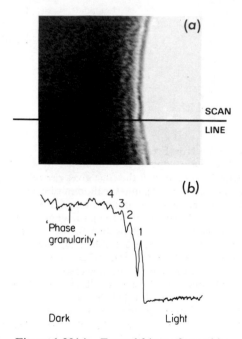

Figure 1.20(*a*)   Fresnel fringes formed by electrons. These fringes are formed outside the edge of a hole (white) in a carbon film (black)

(*b*) A microdensitometer tracing of the fringe system; the pattern is identical with the Fresnel fringe system formed by visible light shown in Fig. 1.18(*b*)

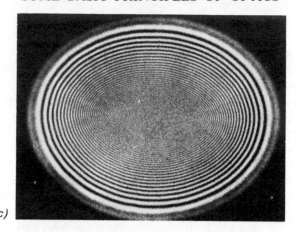

(c)

Figure 1.20(c)   An underfocused image
of a hole in a film, showing a complete
system of about 40 Fresnel fringes inside
the hole. Such micrographs are only
possible with a highly coherent electron
source of very small aperture, using a
pointed filament and fully defocused
condenser system. The exposure was
6 minutes and demonstrates that specimen drift must have been less than
0·05 Å/sec. Taken on Siemens Elmiskop
101; micrograph courtesy Siemens AG

system to be weaker than the system formed by interference with the direct
wavefront, which is the one we see in the case of light Fresnel fringes. It
is in practice very much stronger; the reasons for this have not yet been
satisfactorily explained in terms of electron optics. However, the strong
fringe system gives a microdensitometer trace (Fig. 1.20(b)) which is
almost indistinguishable from the trace given by light fringes (Fig. 1.18(b)).
The scale, however, is quite different. The distance apart of the fringes at
the object plane is of the order of tens of Ångstrom units; the fringes in
Fig. 1.20(a) were magnified 250,000 times by the optical system of the
electron microscope, followed by a further 5 times in an enlarger. By
contrast, the magnification of the light fringes shown in Fig. 1.18(a) is
only 5. This difference in scale is due to the fact that the wavelength of
electrons, as we shall see in the following chapter, is about 100,000 times
shorter than the wavelength of visible light.

Four fringes are shown in the figure; the first is very strong. The others
can only be seen (Fig. 1.20(c)) using an electron source of very high

coherence. In general, only the first fringe can be seen on the fluorescent screen. This first fringe is of the greatest importance in aligning and operating the electron microscope, as will be shown later in this book.

The appearance of the single 'first bright fringe' and its mode of formation are shown in Figs. 1.21(a) to (d). An edge is illuminated by an electron beam, resulting in the formation of a real, bright fringe behind the edge. Close to the edge, the fringe is narrow and sharp; further out from the edge the fringe becomes wider, dimmer and less well defined. The edge is imaged on the electron microscope screen by the objective lens (plus any other image forming lenses which are also in use). Due to the presence of the fringe, the appearance of the image of the edge of the hole will change as the focal length of the objective lens is changed, that is, as the microscope is focused. If the focal length of the objective is made smaller (Fig. 1.21(a)) the lens will be focused on the plane AA', and it will 'see' the out-of-focus edge plus the first bright fringe behind it. In this condition, the lens is said to be 'overfocused', since too much current is passing through it. If the amount of overfocus is relatively large, with the lens focused on a plane several microns behind the edge, the fringe will be wide, blurred and well removed from the image of the edge (Fig. 1.21(a)). If the overfocus is small (Fig. 1.21(b)), the fringe will be narrow, bright and close to the edge. At exact focus, no fringe can be imaged (Fig. 1.21(c)), because the lens is now focused in the exact plane CC' of the edge. The fine detail of the structure of the edge can now be seen. If the current in the lens is reduced still further, the focal length becomes too great, and the lens is said to be 'underfocused'. It can now no longer image the system of real fringes behind the edge, but instead the system of 'virtual' fringes in the space in front of the edge is imaged. This system is shown by broken lines (Fig. 1.21(d)). The apparent image of the edge is displaced towards the hole, and is surrounded by a bright line. The fine detail of the edge is again lost, since it is out of focus, but the visibility of the edge is greatly enhanced due to the outlining by the bright line. The underfocus bright line therefore gives rise to a considerable increase in apparent image contrast, but it must be emphasized that the increase in contrast is gained at the expense of loss of resolution, because the object is no longer in exact focus. Experienced electron microscopists frequently use this trick of underfocusing to enhance contrast, especially at lower magnifications where resolution is not of paramount importance. The use of the method will be discussed in Chapter 12.

The appearance of the Fresnel fringe is therefore a very sensitive indicator of the exact point of focus of an electron microscope. The sensitivity can be judged from the fact that a clear fringe can be seen at

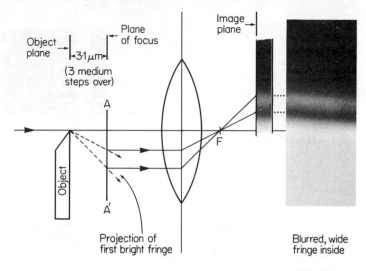

(a) Objective very overfocused – lens far too strong

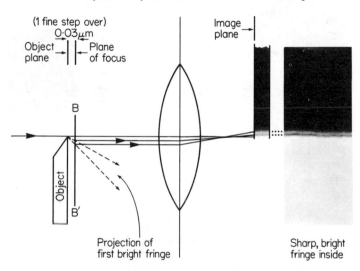

(b) Objective slightly overfocused – lens a little too strong

Figure 1.21(a) to (d)   Ray diagrams showing how the first bright Fresnel fringe is imaged in the electron microscope at four different focal settings of the objective lens. The micrographs show the appearance of the fringe on the screen at each focal setting

(a) Objective lens very overfocused (far too strong—+3·1 μm). The edge is very ill-defined and cannot be located accurately

(b) Objective lens slightly overfocused (slightly too strong—+0·03 μm). The edge is sharper but is still inaccurately located

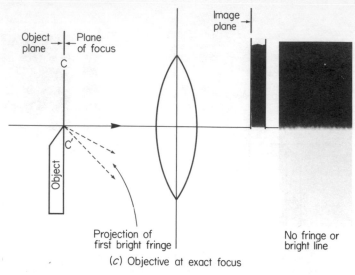

(c) Objective at exact focus

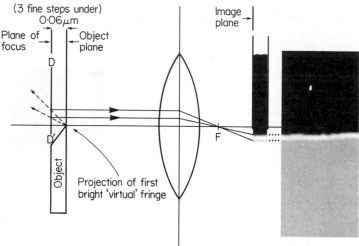

(d) Objective underfocused – lens too weak

(c) Objective lens at exact focal point. The fine structure of the edge is now apparent

(d) Objective lens slightly underfocused (too weak——−0·06 μm). The edge is now better defined but displaced by the width of the bright line surrounding it. This bright line enhances the edge and the eye tends to select the resulting apparent increased contrast as the point of true focus. It is permissible to use this underfocus effect to enhance contrast at relatively low magnifications provided the width of the bright line does not exceed the required resolution

Micrographs of the edge of a hole in a thick carbon film. AEI–EM6B, instrumental magnification × 250,000, total 1 × 10⁶

high magnification for a change in focus of 1 part in 20,000 or less of the objective lens. The fringe also has other very important applications. Since the spacing is so critically dependent on focal length, it can be used to determine if the lens is radially symmetrical. If this is not so, the lens is said to be 'astigmatic' (see Chapter 4), and the fringe spacing is used to correct the defect. Watching for random changes in fringe spacing at high magnification is the most sensitive test for general instrumental stability, both mechanical and electrical. The fringe spacing is also used to measure resolving power. The fringe is, in fact, the perfect test specimen—the 'selfless object'. A thorough understanding of its mode of formation is therefore of the greatest importance to the electron microscopist.

1.18 The Airy Disc

So far, in all our discussions of interference and diffraction, we have only considered the interaction of light waves with sharp edges. No optical systems have been involved, except for the examination of the fringe systems produced at the edges. Does diffraction affect the resolving power of optical systems, and if so, how?

Let us consider the simplest possible optical system. Fig. 1.22 shows a convergent lens forming an image of a perfect, infinitely small point source of light O in an image plane I. The aperture of the lens is defined by a circular hole, shown in section at AB. If no physical aperture were present, the aperture would still be defined by the circular edge of the lens. Will the image I be a perfect replica of the source O? We know already that this cannot be, because the lens, even if it were infinitely large, cannot collect more than half the information emanating from the object. The image must therefore be imperfect, due to the presence of the lens itself; imperfections arise by diffraction at the aperture edges AB, which gives rise to interference fringes. The point source is imaged as a bright disc of finite diameter, surrounded by alternating bright and dark rings. This pattern, which is circular if the lens aperture is circular, is called an 'Airy disc', after the nineteenth-century Astronomer Royal, Sir George Airy, who first showed how it is formed. The appearance of the Airy disc is shown in Fig. 1.23.

If an Airy disc is photographed and a positive image is scanned across the centre with a microdensitometer, a curve such as is shown in Fig. 1.24(a) will be obtained. It can be shown that 84 % of the incident energy is contained in the central peak, the remainder being spread into the first order, second order etc. diffraction bands surrounding it.

The diameter of the Airy disc, which is usually given as the radius of

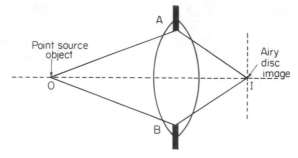

Figure 1.22   A perfect point source O cannot be imaged by the lens L as a perfect point image, due to the presence of the aperture AB (which may be the edge of the lens). Diffraction at this aperture gives rise to a series of fringes, which surround the image formed of the point source. The pattern produced is called an 'Airy disc'

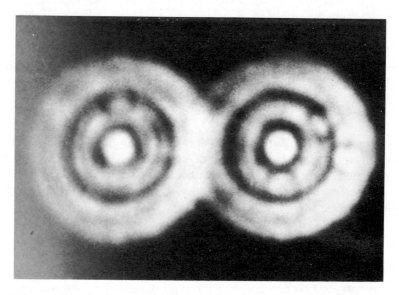

Figure 1.23   A photograph taken with a light microscope, using an objective lens with a relatively low aperture, of two pinholes in a layer of aluminium evaporated on to a glass surface. The pinholes appear as discs surrounded by diffraction bands. The two Airy discs shown here just overlap, but the two central discs are well separated and can easily be resolved

the first dark ring, varies with the same two parameters as the resolution of the slit in our previous experiment—the aperture of the system used to image it, and the wavelength of the illumination. It also depends on the refractive index of the medium between the lens and the object, because this affects the velocity and hence the wavelength of the light. The radius of the first dark ring is given by the expression:

$$r = \frac{0 \cdot 612\lambda}{n \sin \alpha}$$

where λ is the wavelength of the light in vacuum; the angle α is half the angular aperture; and n is the refractive index of the medium between the lens and the object. Strictly speaking, this expression applies only to infinitely small self-luminous objects. It was derived originally from work on astronomical telescopes, where the objects are stars which fulfil this criterion almost perfectly.

1.19 Aperture Stops and Field Stops

Strictly speaking, the 'aperture' of a lens is defined as half the acceptance angle. The acceptance angle itself is generally defined by the diameter of a circular hole, called the 'aperture stop', placed at or near the lens centre (see Fig. 1.22). The aperture stop may also be placed at the front or back focal plane of the lens. In some electron microscope objective lenses, the aperture stop has to be placed as close to the back focal plane as possible, because the focal length of a very powerful magnetic lens is so short that the specimen holder has to be at the lens centre. A characteristic of an aperture stop is that it is never in focus with the specimen.

The field of view is sharply defined by a second class of stop, a 'field stop', which is in sharp focus with the specimen. In a light microscope, it is generally placed in the eyepiece. In an electron microscope, it is generally placed in the final projector lens (which corresponds to the eyepiece), and is adjusted so as to cover the final screen or photographic plate completely. In most instruments, it cannot be altered from outside the column.

1.20 Resolving Power

The term 'resolving power' expresses the ability of an optical system to discriminate between two objects. It is given as the minimum separation at which two objects can be seen as separate entities and not blurred

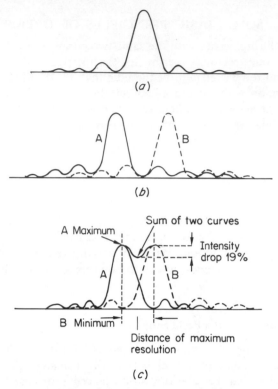

(a)

(b)

(c)

Figure 1.24(a)   The form of a microdensitometer
tracing across an Airy disc. The central peak contains 84 % of the incident energy received from the
source, the remainder being spread into the diffraction rings surrounding it

(b) When two Airy discs approach one another, the
energy spread can be calculated by summing the two
curves. Provided the two images are separated by
the width of the central peak, the two sources will
be completely resolved

(c) There is a minimum distance of separation between the centres of the two central peaks for
resolution into two to be possible. If the two peaks
approach more closely, it is impossible to say if the
image is of one elongated object, or of two separate
objects. The minimum distance of separation for
resolution into two to be possible was suggested by
Lord Rayleigh as the distance between the centre of
the central peak and the centre of the first dark
fringe. Summation of the two Airy disc curves then
gives an intensity drop between the two central
peaks of about 19 %

together as one. The smaller the spacing, the greater is the resolving power.

Lord Rayleigh in 1896 proposed to define resolving power as follows. Let us consider two very small self-luminous objects of equal brightness. Each will be imaged as an Airy disc. As the two sources approach each other, the two Airy discs will overlap. Lord Rayleigh proposed that the criterion for resolving two Airy discs should be when the separation between the centres is the same as the radius of the first dark ring. When two Airy discs are superimposed in this way, the intensity pattern shown in Fig. 1.24(c) is obtained. It can be shown that the intensity drop between the two bright peaks A and B is then 19 %. This is a purely arbitrary figure; it was suggested as a visual criterion. Photographic techniques can resolve two Airy discs which are closer together than this. Rayleigh's criterion is however a good approximation of resolving power.

The great German theoretical optician Ernst Abbe derived an expression for the resolving power of a light microscope from a consideration of the light diffracted at the specimen. He realized that a lens of large aperture would collect more of the diffracted light (i.e. more of the spectral orders shown in Fig. 1.17) and would therefore give more information about the object. He introduced the concept of the 'numerical aperture' (N.A.) of a lens, which he defined as:

$$\text{N.A.} = n \sin \alpha$$

where n is the refractive index of the medium and α is half the lens acceptance angle. He defined the resolving power of the objective lens of a light microscope for two 'artificial star' objects (minute pinholes in a thin metal film) by the expression:

$$\text{R.P.} = \frac{0\cdot61\lambda}{\text{N.A.}}$$

which is virtually the same expression as that derived by Rayleigh.

The value of the constant 0·61 is somewhat controversial. It depends on the coherence of the light and criteria of visibility. It is also affected by the fact that normal microscopical objects are not small self-luminous points, but are semi-transparent objects of finite size which themselves introduce diffraction effects in addition to those due to the lens aperture.

It is of interest to calculate the resolving power of a light microscope from this formula. The maximum reported numerical aperture is 1·5, for an objective lens using monobromonaphthalene of refractive index 1·66 as the immersion medium between lens and object. If green light of wavelength 0·5 μm is used, the resolving power will be:

$$\text{R.P.} = \frac{0\cdot6 \times 0\cdot5}{1\cdot5} = 0\cdot2 \ \mu m$$

Resolving power in the light microscope is therefore limited by diffraction effects to about one-half the wavelength of the illumination used. This distance cannot, as far as can be seen, be made significantly smaller by any improvement in lens design, since no lens can be made with an acceptance angle exceeding 180°, and suitable immersion media of very high refractive index are as yet unknown. Ernst Abbe himself was fully aware of the limitations imposed by diffraction when he remarked: 'It is poor comfort to hope that human ingenuity will find ways and means to overcome this limit'.

How, then, can resolving power be increased? There is only one way out of the dilemma—to reduce the wavelength. How this came to be done is the story of the electron microscope, and we will discuss it in the following chapter.

CHAPTER 2

The Development of the Electron Microscope

2.1 Electromagnetic Radiation

We have seen in the preceding chapter that the only way in which we can increase resolving power and thus make smaller objects visible is to reduce the wavelength of the light used to illuminate them. Let us see how this is done.

Light is a form of electromagnetic radiation to which the eye is sensitive. There exists a continuous spectrum of electromagnetic radiation shown in Fig. 2.1 ranging from radio waves at the one end to gamma-radiation from nuclear distintegrations at the other. All these radiations travel through space with the same velocity (approximately 300,000 km/s) and differ only in wavelength. Radio waves range from several hundred metres in length down to the so-called 'centimetric' waves used in radar. As the wavelength decreases, the radiation becomes heat or infrared waves, which are about 1,000 μm long. Visible light ranges in wavelength from 0·7 μm to 0·4 μm. Differences in wavelength in the visible region of the spectrum are interpreted by the eye as differences in colour; blue being approximately 0·47 μm, green 0·52 μm, yellow 0·58 μm, and red 0·65 μm. Wavelengths longer than 0·7 μm are designated infrared; shorter than 0·4 μm are called ultraviolet. Ultraviolet radiation is followed by X-rays; these are followed by gamma-radiation and cosmic radiation. Electro-magnetic waves of different wavelengths vary in their mode of interaction with matter. Radio waves and X-rays pass through matter with little absorption or refraction, but infrared, visible and ultraviolet radiation is either absorbed or refracted by matter. Some forms of matter such as glass do not absorb visible light, but reduce its velocity. This reduction in velocity is called 'refraction'; it gives rise to the bending of oblique rays of light and the action of lenses.

Because of the wave nature of electromagnetic radiation, there is clearly a limit to the size of an object which will be able to impress itself upon a wave, so that the wave can carry away information about it. Quite a large

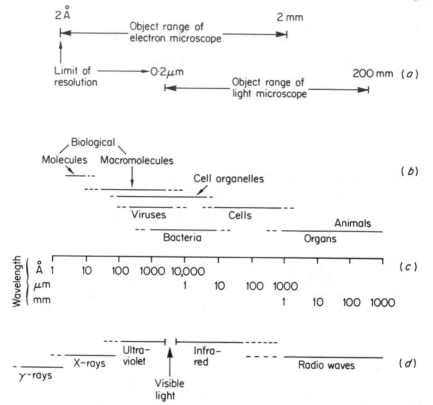

Fig. 2.1   A composite table showing:

(a) The range of sizes of objects which can be examined with the light and electron microscopes. In each case, resolving power determines the lower limit; instrument design the upper limit. The upper range of the light microscope can be extended by using an optical enlarger as a photographic microscope

(b) The range of biological objects covered by each instrument. The light microscope ranges from large organs down to some cell organelles; the electron microscope ranges from tissues or very small organs down to small macromolecules

(c) and (d) The wavelengths and nomenclature of the electromagnetic spectrum corresponding to these biological objects

boulder has little effect on an ocean wave, but quite a small pebble can interfere with a pattern of ripples. In general, an object can only impress itself upon a wave if its size is of the same order as or larger than the wavelength of the radiation. Visible light waves can resolve objects as small as bacteria, but cannot detect most viruses. We have seen in the previous chapter that the resolving power of a perfect optical system is about one-half the wavelength of the radiation it uses. The ultimate resolving power of a microscope using visible light is approximately 0·25 μm. How, then, do we obtain increased resolving power?

There is only one answer—by using electromagnetic radiation of shorter wavelength, to which the eye is unfortunately not sensitive. Fig. 1.1 shows us that ultraviolet radiation could be used in order to give some improvement. However, glass is opaque to ultraviolet radiations at wavelengths shorter than about 0.3 μm, and as the wavelength is reduced, the oxygen of the air begins to absorb significantly. A shortwave ultraviolet microscope has to use curved mirrors to form the image and must be operated in a vacuum, and it can in any case only increase resolving power by a factor of 5 at most. This is a very small gain for such a great increase in complexity.

2.2 The X-ray Microscope

X-rays would be the obvious answer, but unfortunately no substance is known which will bend or refract X-rays sufficiently to enable an image to be formed. Experiments have been made in using mirrors to form X-ray images, but little progress has been made due to the immense practical difficulties that must be overcome.

A more practical form of X-ray microscope is the 'point projection' or 'shadow' microscope. This uses the simple geometrical principle that an object illuminated by a point source of light or X-rays will cast a sharp shadow on a screen. If the screen-to-object distance is greater than the object-to-source distance, the shadow will be enlarged, as shown in Fig. 2.2. Since X-rays can penetrate material which is opaque to light, the internal detail in the object will also be enlarged in the shadow image. The resolving power of this method is limited by the size of the source. Since a perfect point source cannot be achieved, the sharpness of the 'shadowgraph' depends on the relative sizes of the source and the object. The edges of the image will be blurred by a penumbra, which limits resolution. X-ray sources can be made much smaller and more intense than light sources, so this principle can be used to construct a simple X-ray microscope without lenses. Such instruments have been experimented with and resolutions of the order of 0·1 μm have been obtained by using an *X*-ray source 0·1 μm in diameter with a magnification of about × 1,000. Such tiny sources are inevitably very weak *X*-ray emitters, and the shadow-graphs can barely be seen on a fluorescent screen. Photographic exposure times of the order of hours or even days must be used to record the image.

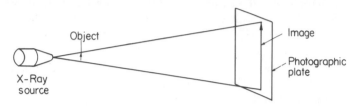

Figure 2.2 The principle of the point-projection or 'shadow' *X*-ray microscope. Resolution is limited by the diameter of the *X*-ray source; magnification by the minimum intensity of the image which can be detected

Image intensifiers (Chapter 17) might make the point projection *X*-ray microscope into a practicable proposition, but much more work remains to be done on these instruments.

If a practicable *X*-ray microscope with the resolving power of present-day electron microscopes could be devised, the electron microscope would probably be rendered virtually obsolete in biology, since it might be possible to examine living cells lying deep within tissues; this is something the electron microscope can never hope to do.

At the beginning of this century, Abbe himself declared that microscopy had come to halt for the want of a suitable short wavelength radiation to overcome the resolution limit imposed by diffraction. But it is a curious irony of fate that even before Abbe had perfected his optical theories, physicists were working on the very principles which were to form the basis of the electron microscope. Around the middle of the last century, great interest was being shown in the spectacular and beautiful discharges produced when high-voltage electricity from electrostatic generators and induction coils was passed through gases at low pressure contained in glass tubes. In the 1850's, Geissler, an instrument maker of Bonn, dis-covered how to seal metal electrodes into glass tubes with vacuum-tight joints. Using his 'Geissler tubes', the phenomena of electrical discharges in gases were closely investigated by Plücker, Hittorf, Goldstein and Hertz in Germany, and by Crookes, Perrin and J. J. Thomson in England. These workers discovered that the so-called 'cathode rays', as they were termed by Goldstein, were propagated in a linear fashion, were negatively charged and could be deflected by electrostatic and magnetic fields. In particular, Wiechert found in 1899 that they could be 'concentrated' or focused by an axially symmetrical magnetic field formed when a current was passed through a coil of wire wound around the discharge tube. Such axial coils are termed 'solenoids'. His apparatus is illustrated in Fig. 2.3. He relied

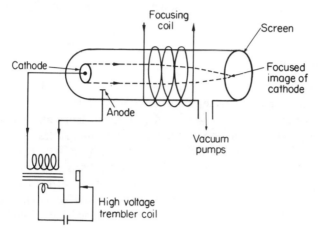

Figure 2.3   An experiment dating back to the late nineteenth century, demonstrating how 'cathode rays' (beams of electrons) could be 'concentrated' or focused by an axial magnetic field from a solenoid. Electrons are drawn from the cold metal plate cathode by a high voltage generated by a trembler coil, and are focused on to the end of the Geissler tube by the action of an electric current passing through the coil. The apparatus contains all the essentials of a single electron lens

on the natural fluorescence of the glass walls of the tube to demonstrate the rays, although the fluorescent screen was first introduced by Braun in 1897. This experiment demonstrates the basic principle of the electron microscope. It contains all the basic essentials: an evacuated tube; an electron source; a high voltage electron accelerator; an electromagnetic electron lens; and a device for converting electron irradiation into visible light. J. J. Thomson refined this apparatus and was able to show that the cathode rays were negatively charged corpuscles with a mass of about one thousandth of that of the hydrogen atom. The corpuscles were termed 'electrons' by G. Johnstone Stoney; Thomson himself never used the term. Although theories on the wave nature of cathode rays had been current before Thomson's experiments, his evidence in favour of the corpuscular theory was so overwhelming that the wave theory was completely dropped, and thus any ideas which may have been in people's minds for using Wiechert's device as a microscope were discarded. Thus, although the mechanics, so to speak, of the electron microscope were known during Abbe's lifetime (he died in 1905), the fundamental discovery linking electrons with short wavelength radiation was still a quarter of a century off.

2.3 Electron Waves

The two great universal principles which were to link fast-moving electrons with short wavelength radiation were the quantum theory of Max Planck and the theory of relativity of Albert Einstein, both published in the early 1900's. The consequences of combining these two principles were worked out by the French mathematical physicist de Broglie in 1924. His calculations showed that any fast-moving ('fast' here means close to the velocity of light) particle would have a new form of radiation associated with it. The wavelength λ corresponding to a particle of mass m travelling at a velocity v is given by the de Broglie relationship:

$$\lambda = \frac{h}{mv}$$

where h is the universal quantum number or Planck's constant. Such waves cannot be detected unless the particle has a very small mass and is travelling faster than about one-tenth the speed of light. If the particles are electrons and the velocity is about one-third the speed of light, the wavelength works out at about 0·05 Å. This is 100,000 times shorter than the wavelength of green light, and if the potential increase in resolution could be fully utilized, the resolving power of a microscope using this radiation would be 100,000 times better than that of the best light microscope. When de Broglie first published his theoretical analysis, the experimental physicists saw that the extremely short wavelengths predicted could best be measured by observing diffraction phenomena, and the predicted result was confirmed independently in 1927 by Davisson and Germer in the United States, and by G. P. Thomson, the son of J. J. Thomson, in England.

The precise nature of these 'electron waves' or 'matter waves' is very difficult to understand or describe in material terms. They are not electromagnetic radiation of the kind to which light, X-rays and radio waves belong. They constitute a sort of quantum or 'packet' of radiation which accompanies each individual electron, following its path and not radiating outwards from it. The characteristics of the electron wave depend on the exact position in space of a given electron at a given instant in time, and express the probability of finding the electron at that time in that place. Electron waves are sometimes described as 'waves of probability', which although possibly a more accurate description, conveys little to anyone other than a quantum mathematician. The electron waves must not be confused with the radiation which is emitted when an energetic electron interacts with matter, and loses energy by being slowed down (see p. 68). These waves are true electromagnetic radiation, generally lie in the X-ray part of the spectrum, and are progagated in space in the same way as other electromagnetic waves. Electron waves are part of an electron, and remain with it.

2.4 The Idea of the Electron Microscope

Looking back with hindsight, it is very curious that even after the experimental demonstration of the existence of focusable radiation of extremely short wavelength, the idea of using this new radiation in an ultra-high-resolution microscope did not immediately occur to anyone. To quote Mulvey: 'The postulate of the wave nature of electrons put forward by de Broglie (1924) had no influence on the early history of the electron microscope; the idea of the exclusively corpuscular nature of the cathode rays was by now so deeply ingrained that wave properties were considered to be purely a mathematical formality'.

It is not clear from the published literature who it was had the original idea of using electron beams to form images of objects at resolutions higher than that of the light microscope. Gabor has stated that speculation on the possibility of an electron microscope had been going on in Berlin as far back as 1928, but the fact remains that the early development of practical electron optics was aimed at the perfection of the high-speed cathode-ray oscillograph, urgently wanted at that time for the investigation of high-voltage surges in electrical power distribution lines.

Electron optics had been very thoroughly investigated by the German experimental physicist Busch, who in 1926 published the results of his experiments on the trajectories of electrons in magnetic fields. He showed that axial magnetic fields refract electron beams in a way geometrically analogous to the refraction of light beams by glass lenses. He showed that the geometrical laws of optics were obeyed by electron systems, and so laid the foundations of the science of electron optics. Busch, however, did

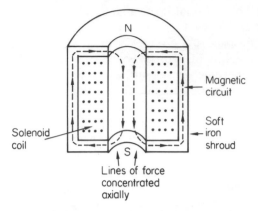

Figure 2.4   The axial magnetic field formed by a solenoid coil can be greatly increased by surrounding the outer part of the coil with a shroud made of a permeable material such as soft iron. This increases the strength of the magnetic lens for the same current flowing in the coil. This is the basic design of weak magnetic lenses used in electron microscopes

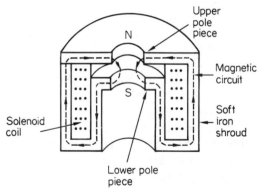

Figure 2.5   An even greater axial concentration and hence a stronger lens can be made by extending the shroud internally, and only allowing the magnetic field to emerge from the iron shroud at 'pole pieces' set a short distance apart. The curved magnetic field between the polepieces is the actual lens. This is the basic design of strong magnetic lenses used in electron microscopes

not combine his results with de Broglie's predictions, nor did he suggest the possibility of an electron microscope.

The practical application of Busch's results to high-speed oscillographs was undertaken by a team of workers under Matthias in the laboratories of the Berlin Technische Hochschule between 1928 and 1934. The team was headed by Max Knoll, who divided the research between Ruska (electron lenses); von Borries (image recording); Knoblauch and Freundlich (electron sources); and Lubszynski (electromagnetic shielding). This team in fact invented and developed the electron microscope, although the first record of its fundamental concept does not appear in the literature until 1931, when Reinhold Ruedenberg filed a German patent (which was not at first granted) on behalf of the instrument-making division of Siemens. Ruedenberg is therefore in patent law the inventor of the electron microscope, although he played no part in its development.

2.5 The Development of the Magnetic Lens

Although the Berlin team experimented with electron lenses using electrical fields, Ruska's main contribution to the design of the electron microscope was the development of the magnetic electron lens. The simple solenoid (Fig. 1.3) wastes a great deal of the magnetic field, since the lines of force are not concentrated along the axis. Gabor in 1926 discovered that magnetic lenses could be improved by shrouding the solenoid coil in an external soft iron case. Fig. 2.4 shows how the magnetically permeable iron case forms a magnetic circuit around the energizing coil, and concentrates the lines of force in an axial direction. In this way, a much more powerful axial magnetic field can be obtained for the same amount of current flowing in the solenoid winding. Gabor used this type of 'concentrating coil' in his high-speed cathode-ray oscillograph. Ruska carried the shrouding one stage further, and enclosed the solenoid coil inside the axis as well as externally, as shown in Fig. 2.5. This improvement completes the magnetic circuit around the solenoid, leaving only a short gap between the circular polepieces of the shroud. This mode of construction forms a short, thin lens such as that obtainable from a torus (Fig. 1.4), but with very much greater efficiency. The magnetic field so obtained can be treated in simple ray diagrams as a short axial lens instead of a diffuse refracting space. This basic type of lens construction is now universally used in all modern magnetic electron microscopes.

Ruska applied his improved electron lenses to Gabor's oscillograph, and reported in 1931 the construction of what was in effect a two-lens

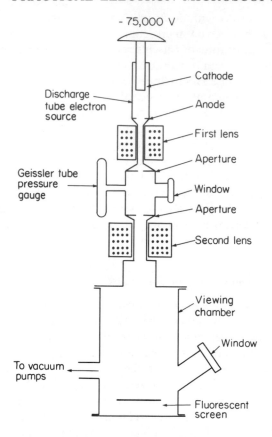

Figure 2.6   A diagram of Ruska's 1931 electron source for a high-speed oscillograph. This instrument could image the apertures placed in the electron optical system, but it was not a true microscope, since there was no provision for inserting a specimen

electron source for a high-speed recording oscillograph. This instrument is shown in Fig. 2.8(*a*) and diagrammatically in Fig. 2.6. Electrons were formed at the cold metal plate cathode by the bombardment of its surface with positive ions formed in the space between the anode and the cathode. Such cold-cathode electron sources are not very suitable for electron microscopes, because they will only work in a poor vacuum (otherwise positive ions cannot be obtained) and are unstable and pressure-dependent. The electrons so formed were accelerated by the potential (in this case,

75 kV) between cathode and anode plate, and some of the electrons passed through a hole in the anode into the optical part of the instrument. This roughly-collimated electron beam was then focused by either or both of two solenoid lenses placed around the brass tube enclosing the electron path and forming the vacuum chamber. An image could be formed on a

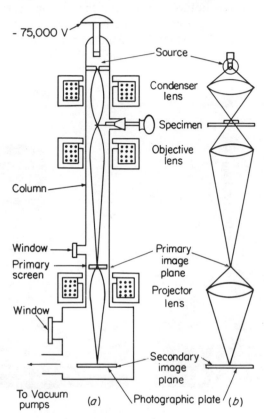

Figure 2.7 A diagram of the first true electron microscope, together with its light microscope analogue. This instrument was developed by Ruska and his colleagues between 1931 and 1933 from the electron source shown in the previous diagram. The basic optics of electron microscopes have remained unchanged since this time. Modern improvements consist largely in the addition of more lenses to increase the magnification range and flexibility of the instrument

fluorescent screen placed at the bottom of the vacuum tube or 'column', and could be observed through a stout glass window in the side of the observation chamber. The object that was imaged was the electron source, i.e. the hole in the anode aperture, or else a second aperture introduced behind the first lens. The device was in fact an electron-optical bench; transmission specimens were not introduced into the vacuum. It is the counterpart of the illuminating system of a modern double-condenser electron microscope; the upper lens is the first condenser (C_1) and the lower lens the second condenser lens (C_2). The maximum magnification reported was × 17.

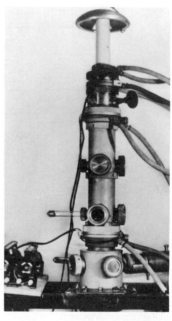

(a)   (b)

Figure 2.8   A series of photographs showing the development of the modern Siemens Elmiskop I electron microscope from the basic oscillograph electron source shown in Fig. 2.6

(a) The 1931 high-speed oscillograph electron source. The two lenses, discharge tube pressure gauge, observation chamber, part of the diffusion pump, and the water-cooling pipes for the cold-cathode electron source can be seen

(b) 1933. The first true electron microscope, shown diagrammatically in Fig. 2.7. The water-cooling pipes for the 3 lenses and the cold cathode source can be seen, together with the intermediate and final observation windows, specimen insertion knob and discharge tube pressure gauge. The sliding rheostats control lens currents

2.6 The Development of the Electron Microscope

The sharpness of the images formed with this apparatus encouraged Ruska to construct the first true electron microscope. This instrument is shown in Fig. 2.8(*b*), and diagrammatically in Fig. 2.7 together with its light optical analogue. Its remarkable basic similarity to the electron microscopes which were developed from it can be seen by comparing it with the instruments shown in Fig. 2.8(*c*) to (*f*) and in Chapter 9.

Ruska followed the analogy between light and electron optics, and based his instrument on the design of the compound light microscope. A vacuum-tight metal tube (the column) about 1 metre long was supported vertically with a cold-cathode discharge electron source at the upper end,

(c)   *(d)*

Figure 2.8(*c*) The prototype Siemens and Halske instrument of 1938, constructed by Ruska and von Borries. The large tank at the top contains the filament battery for the hot source and the high voltage generator. The electron optics are fundamentally the same as the 1933 instrument, but operating facilities are greatly improved

(*d*) The 'First-series' electron microscope, which was the 1939 version of the 1938 prototype and differs only in detail. It was the first commercially available electron microscope; some 40 instruments were built

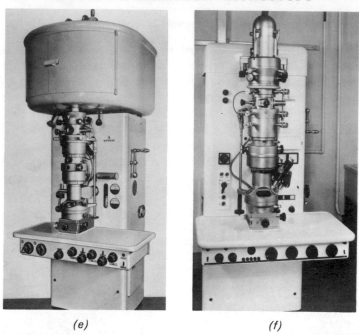

(e)   *(f)*

Figure 2.8(*e*)   The first post-war Siemens instrument, the 1950 ÜM-100. It was virtually a cleaned-up First-series instrument with improved operating facilities

(*f*) The famous Elmiskop I, first available in 1954. This greatly improved instrument had 2 condenser and 3 imaging lenses. This basic design remained, with detailed improvements, for 20 years.

and a fluorescent screen was placed at the lower end. The screen could be observed through a stout glass window, or it could be replaced by a photographic plate for recording the image. Three electron lenses of his polepiece type were arranged about the tube. The electron source was still of the cold cathode type operating at 75 kV. The electron beam was collimated as before by a small hole drilled in the anode plate, which was maintained at earth potential to avoid the risk of electric shock to the operator. The resulting beam was focused by the condenser lens on to the specimen, which was mounted on a holder just beneath it. The specimen holder could be rotated from outside the vacuum, so that the specimen could be brought in and out of the beam. A primary, magnified image of the specimen was formed by the magnetic field of the objective lens in the primary image plane just below the centre of the column. A small fluorescent screen with a hole in the centre was placed here, and could be observed through a central window. The electrons passing through the hole in this screen could be imaged by a second lens, the projector, which formed the final image on the fluorescent screen at the bottom of the tube. The apparatus was evacuated through a metal tube joined to the observation chamber. The lenses were cooled with water in order to dissipate the considerable heat from the large amount of power used to excite them; the cooling water pipes can clearly be seen in Fig. 2.8(b).

Images were obtained of the edges of pieces of metal foil and of cotton fibres, although the latter were carbonized by the intense electron bombardment. This destruction of the specimen seemed at the time to be an insuperable difficulty. These discouragements caused Knoll to leave the team for the field of television, but Ruska and von Borries persevered. In spite of poor vacuum and specimen damage, they were able in 1934 to demonstrate a resolution of 500 Å, an improvement of a factor of 5 over the optical microscope. The very complicated apparatus needed to obtain this rather trivial improvement in resolving power, and the fact that it did not seem possible to achieve even this small improvement without destroying the specimen, suggested that the electron microscope, while of considerable academic interest, had no practical application. Ruska was not able to obtain any further financial support to continue his experiments and was forced to abandon his development work and join Knoll in television.

At the same time as Knoll and Ruska were developing the electromagnetic electron microscope, a similar instrument using electrostatic lenses was being developed by Brüche in the laboratories of A.E.G., also in Berlin. Berlin must therefore be regarded as the birthplace of the electron microscope.

The pioneer work of Ruska and Knoll aroused great interest in other countries, because the soundness of the principle had been established and it was apparent that all that was lacking was the intensive development work needed to overcome lens defects and specimen damage. This was taken up by Marton in Brussels, who improved the instrument by adding the hot filament electron source. He was able to publish late in 1934 the first biological electron micrograph, taken of a plant leaf impregnated with osmium. At about the same time Driest and Müller in Berlin modified Ruska's 1933 instrument, and were able to publish in 1935 electron micrographs of the leg and wing of a fly showing 400 Å resolution. These encouraging results enabled Ruska to obtain financial support from the Berlin electrical firm of Siemens and Halske, and the Jena optical firm of Carl Zeiss. He set up a developmental laboratory in the Siemens factory and together with his former colleague von Borries started on the design of a greatly improved instrument with commercial series production in mind. This instrument, shown in Fig. 2.8(c), had a resolving power of around 100 Å when fully developed in 1938, and led to the final design of the first series-production electron microscopes (Fig. 2.8(d)), the first of which was delivered in 1939. The first-series Siemens electron microscopes had one condenser and two imaging lenses and gave a resolving power of better than 100 Å. About 40 were built; several were brought to Britain after the war. One was still in use, very little modified, in London, over 30 years after it was built. An improved version of the first-series microscope was brought out by Siemens in 1950, designated the ÜM–100 (Fig. 2.8(e)); this was followed in 1954 by the famous Siemens Elmiskop I, a greatly improved instrument with two condenser and three imaging lenses (Fig. 2.8(f)), capable of a resolution of better than 10Å. Many hundreds of this model—one of the most commonly encountered electron microscopes—have since been sold. Further improvements and refinements led to the Elmiskop IA, basically almost identical with the Elmiskop I but with a guaranteed resolution of 4 Å. This instrument, brought out in 1961, has now been superseded by the latest model, the Elmiskop 102, which is described in Appendix I.

Although the Siemens has the longest production history, commercial instruments were available in America and Britain in the 1940's. The first electron microscope to be built in North America was designed and constructed by James Hillier and Albert Prebus at the University of Toronto in Canada, and was operating in 1939. The first series production commercial instrument marketed in the United States was developed by the Radio Corporation of America. It was developed by a team of electronic engineers and physicists led by Marton (who had left Belgium in 1936). Their efforts culminated in the first commercial American electron microscope, the RCA–EM–B, first manufactured in 1940 and having a resolving power of about 50 Å. Several of these instruments were shipped to Britain under the lease-lend agreements.

Research on electron microscopes in Britain began as early as 1935 when Professor Martin of Imperial College commissioned the electrical engineering firm of Metropolitan-Vickers to build an instrument in his department. This instrument, christened the MV–EM–1, formed the basis for the design of the first British series production electron microscope, the MV–EM–2, first manufactured in 1946. This instrument, like the Siemens and RCA, led to a family of electron microscopes, culminating in the present GEC–AEI Corinth (Metropolitan-Vickers is now part of Associated Electrical Industries). The AEI–EM–6B was the first instrument designed especially for biologists, and has a resolving power of about 3 Å.

In Holland, le Poole designed and constructed an electron microscope in the University of Delft in 1944. This led to the design of the well-known Philips EM–100, introduced in 1950, the EM–75, EM–200 and EM–300 family of Philips instruments. The Japanese firm of Hitachi was also early in the field, having an electron optical bench under development in 1941, which led to the HU series of electron microscopes. Electron microscopes are now also manufactured in Switzerland, Russia, East Germany, Czechoslovakia and Jugoslavia. It is estimated at the time of writing that at least 20,000 instruments are in use throughout the world.

CHAPTER 3

The Classification of Electron Microscopes

3.1 Introduction

The unqualified term 'electron microscope' is generally used to describe the conventional transmission type of instrument, the development of which was described in the previous chapter, just as the term 'light microscope' refers to the familiar transmission-type of compound light microscope. The transmission electron microscope, or 'TEM' as we shall refer to it from now on, was the first of the new family of so-called 'electron probe' instruments to be developed. The very rapid pace of development of electron probe instruments over the past decade or so has made available several closely related instruments, most of which can be classified as specialized types of electron microscope. These newer instruments give different kinds of information about a specimen which supplement or complement the information obtained from the straightforward transmission image. The significance and capabilities of these new instruments should be appreciated and understood by the biologist, because they can give additional information about a specimen which would be difficult or impossible to obtain from a conventional TEM. For example, the three-dimensional structure of surfaces, the shape and configuration of biological molecules in their native state, and the distribution and amounts of specific atomic species in a specimen are now all susceptible to investigation at high resolution.

To understand these newer types of electron probe instruments, it is necessary to examine in detail the various ways in which a beam of energetic electrons interacts with a piece of matter, the specimen. The most important of these effects are shown diagrammatically in Fig. 3.1. Six major phenomena are involved:

1. The specimen transmits electrons;
2. The specimen reflects electrons;

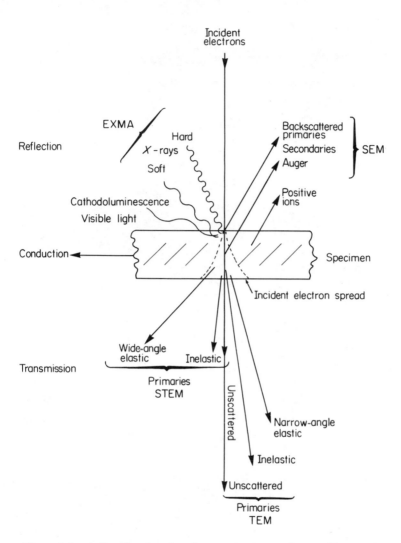

Figure 3.1   A diagram showing the most important forms of interaction between a beam of high-energy incident electrons and a piece of matter, the specimen

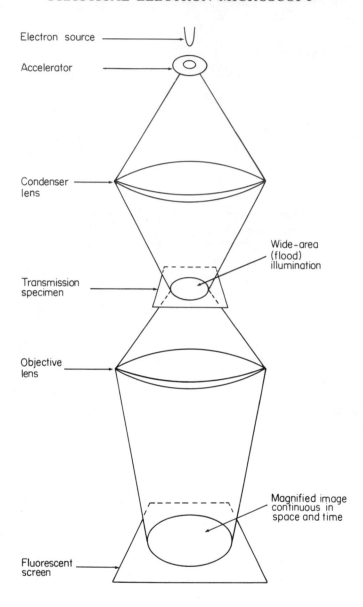

Figure 3.2   A diagram showing how a conventional transmission
electron microscope (TEM) uses a wide-area or 'flood' beam of
illumination to form by means of imaging lenses a magnified
image which is continuous both in space and in time

3. The specimen absorbs electrons;
4. The specimen emits electrons;
5. The specimen emits electromagnetic radiation;
6. The specimen emits positively charged ions.

In addition, there are two quite distinct modes by which a beam of electrons may be made to form an image of a specimen. Firstly, a homogeneous beam of comparatively wide area (between 1 μm and 1 mm dia) may be used to illuminate a correspondingly wide area of the specimen.

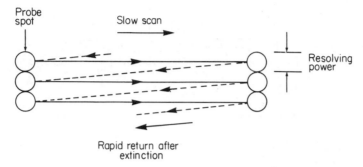

Figure 3.3   A diagram showing how a specimen is 'scanned' with a small probe spot which passes over each specimen point in turn, one scan line at a time, forming a 'raster'

A magnified image is then formed by a train of electron lenses, the imperfections of which limit resolving power. This method is called variously a 'fixed-beam', 'flood-beam' or 'conventional' operation (Fig. 3.2). Secondly, a very narrow, very intense beam may be focused to form a very small spot of illumination (between 2 Å and 200 Å dia) on the specimen. This fine spot is then moved sideways by deflecting the beam so that a very narrow ribbon of specimen, of width corresponding to the diameter of the spot, is traversed across the specimen (see Fig. 3.3). The spot is then returned very rapidly (in general, the beam is first extinguished) to a point on the specimen one spot diameter above or below the original starting point, and the adjacent ribbon of specimen is then traversed. The process is repeated until the whole area of the specimen has been covered by the scanning spot. The pattern of lines generated is called a 'raster', and is familiar to anyone who has ever looked closely at the screen of a television set. A visual image corresponding to the signal produced by the interaction between the beam spot and the specimen at each point along each scan line is simultaneously built up on the face of a cathode ray tube in exactly the same way as a television picture is generated. Magnification is simply

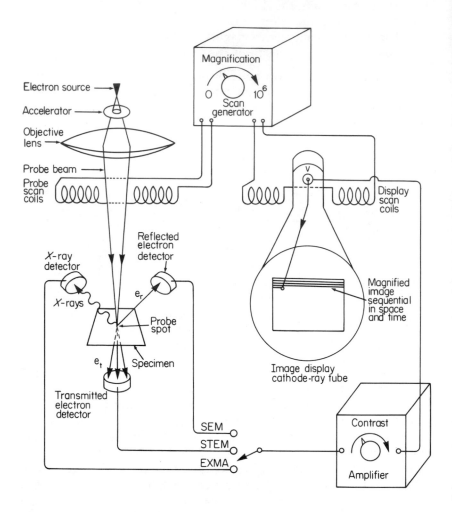

Figure 3.4 A diagram showing how a number of different sets of image information (e.g. SEM, STEM and EXMA) corresponding to specific interactions between the incident electron probe raster and the specimen may be formed. The
interaction products (electrons, X-rays or positive ions) are captured in suitable
detectors and a sequence of pulses corresponding to the arrival of each electron,
photon or ion modulates the brightness of the synchronous display spot. A
magnified image, which is sequential in space and time, is formed without the
use of any imaging lenses

the ratio of the length of the scan line on the specimen to the length of the scan line on the display tube. No imaging lenses are used; resolving power is determined by the size of the probe spot. This method (see Fig. 3.4) is called 'scanning' or the 'scan mode'.

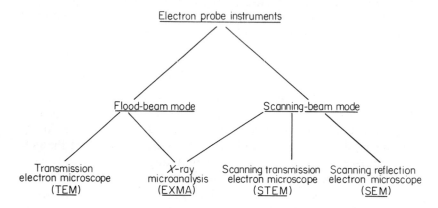

Figure 3.5   A diagram showing the relationships between the most important members of the family of electron microprobe analytical instruments

In theory, it would be possible to use both conventional and scanning modes to examine and record all six interaction phenomena listed above, thus giving rise to twelve different instrumental types. In practice, however, only particular combinations have practical advantages, therefore the number of types of instrument is smaller. The most important types (see Fig. 3.5) are as follows:

1. The conventional flood-beam transmission electron microscope (TEM);
2. The scanning beam transmission electron microscope (STEM);
3. The scanning beam reflection electron microscope (SEM);
4. The electron probe X-ray microanalyser (EXMA).

3.2 Nomenclature

Since the descriptive terms for these instruments, although accurate, are lengthy and clumsy in the extreme, most workers are now using acronyms. The exact form of initials used has not settled down yet; the TEM is sometimes referred to as the CTEM (C for conventional) or the FBEM (fixed- or flood-beam EM); the scanning reflection instrument as the SREM. The electron microprobe X-ray microanalyser is sometimes referred to as EMMA, but this delightful acronym is the registered trade name of the particular instrument marketed by GEC–AEI. Philips are endeavouring to popularize the acronym TEAM for their transmission electron analytical microscope, which is a form of the EXMA. With the present popularity and convenience of this type of nomenclature, others will doubtless be devised. The acronyms TEM, STEM, SEM and EXMA will be used henceforth in the remainder of this book.

3.3 The Flood-beam Transmission Electron Microscope (TEM)

The TEM can only image a specimen thin enough to transmit a substantial proportion (50–90 %) of the electrons incident on it. The general principles of the TEM, the first to be developed, were discussed in the previous chapter. The optics are basically the same as those of the light microscope (see Fig. 3.2). An electron illuminating system is placed on one side of the transmission specimen, and the electrons transmitted by the specimen are focused to form a magnified image on a fluorescent screen (or photographic film or plate) by means of a train of up to four magnetic lenses on the opposite side. Resolving power is limited by the imperfections or aberrations of the lenses (see Chapter 4). Intensive development of these lenses has yielded a resolving power of about 2 Å on a perfect specimen under ideal conditions, but further improvements appear to involve almost insuperable difficulties. Significant improvements in resolving power, e.g. to enable individual atoms of lighter elements to be imaged, seem unlikely in the near future. However, the TEM has the immense advantage of providing a spatially continuous image of a specimen which the eye and brain can appreciate immediately. For general morphological information at medium resolution (down to about 10 Å) the TEM is unrivalled, and this will probably continue into the forseeable future.

3.4 The Flood-beam Reflection Electron Microscope (TEM/R)

It is possible to reflect a wide beam of electrons at a glancing angle from the surface of a specimen, and then to form an image of these reflected primary electrons using a conventional train of lenses. Although the depth of field (see p. 89) of electron microscope lenses is greater than that of the light microscope, chromatic aberration in the reflected beam (see p. 87) is very great, and the reflection image is of very low resolution. The method has now been completely supplanted by the scanning reflection electron microscope (SEM) which will be described in Section 17.3.

3.5 The Scanning Transmission Electron Microscope (STEM)

In the STEM, a probe spot of finite size is scanned in a raster across a thin transmission specimen identical to a TEM specimen. The transmitted electrons, instead of passing into an imaging lens, are simply captured by a detector. As the spot scans across the specimen, the current signal from the detector is a measure of the product of the mass and the thickness (mass-thickness) of the area of the specimen immediately below the spot. This current is amplified and is used to modulate the brightness of a synchronous recording spot moving across the face of a display cathode ray tube. Resolving power is determined primarily by the physical size of the electron probe spot; magnification by the ratio of specimen to display scan line lengths. The STEM is discussed in greater detail in Section 17.2, page 376.

3.6 The Scanning Electron Microscope (SEM)

Fig. 3.1 shows that when high-velocity electrons strike the surface of a solid object in a vacuum, electrons are either reflected or emitted from the surface. Reflected electrons are the 'back-scattered primaries' deflected through large angles by interaction with atomic nuclei (see Fig. 5.1). Secondary electrons are of very much lower energy, and originate in the volume of the solid penetrated by the incident beam. They are formed by ionization of the specimen atoms by the incident primary electrons. As the electron probe spot is moved in a raster across the specimen, a picture of the surface may be built up on a synchronous display tube (see Fig. 3.4) in the 'light' either of back-scattered primaries or of emitted secondaries. The SEM is discussed in greater detail in Section 17.3, page 384.

3.7 The Electron Micro-probe X-ray Microanalyser (EXMA)

The X-ray microanalyser is not, strictly speaking, an electron microscope. It is a method for making a chemical analysis of the atomic species which make up a normal TEM, SEM or STEM specimen. The X-ray analysis is made during the normal course of examination of a specimen in any of these forms of electron microscope. The EXMA apparatus is therefore found attached to electron microscopes, and so the principle of the method will be briefly discussed in this section. The apparatus used is discussed in greater detail in Section 17.4 on page 389.

When high-velocity electrons interact with matter (see Fig. 3.1), electromagnetic radiation (X-rays and visible light) is also emitted. The wavelength λ and the energy E of all electromagnetic radiations are inversely related according to the fundamental expression $\lambda = hc/E$. Visible light is long-wave, low-energy electromagnetic radiation; 'hard' X-rays are short wave, high-energy radiation. 'Soft' X-rays are of intermediate wavelength and energy.

There are, broadly speaking, two sources of electron-induced electromagnetic radiation. The first is due to loss of energy from the incident electron to the outermost electron shells of the specimen atoms as it is slowed down during its passage through the specimen. This reappears as quanta of soft X-radiation. Since the slowing process is continuous and not stepwise, a continuous or 'white' spectrum of soft X-rays is emitted, which are sometimes called 'Bremsstrahlungen' or 'braking rays'. The minimum wavelength of the white spectrum is determined by the velocity (energy) of the incident electron; the higher the velocity, the shorter the minimum wavelength. The range is between about 100 Å and 1000 Å.

As the energy of the incident electrons increases above about 1 kV, they are able to interact with the inner shells of specimen atoms. This causes an orbital electron to jump from a lower energy shell to a higher energy shell. When it drops back, a quantum of energy in the form of a 'hard' or short-wave X-ray is emitted (see Fig. 5.1). Electron transitions from one shell to another involve very precise energy quanta, and so the emitted X-ray is of a very precise wavelength or energy. Because the transition energies differ for each individual atomic species, the emitted wavelengths are characteristic of the atoms forming the specimen. Therefore, by measuring either the wavelength or the energy of each of the emitted X-rays with an X-ray spectrometer, a qualitative analysis of the atoms composing the specimen may be made. Also, if the relative intensity of the particular X-ray corresponding to a particular atomic species is measured, a quantitative analysis of the specimen may also be made. This is the fundamental principle of the electron microprobe X-ray microanalyser (EXMA).

3.8 Cathodoluminescence

When incident electrons interact with the outer electron shells of certain mid-Z-range elements, the electromagnetic quanta emitted are in the visible spectrum. This emission of visible light is called 'cathodoluminescence'. It differs from fluorescence, which is the immediate reradiation of visible light after stimulation with shorter wavelength light, especially ultraviolet. Phosphorescence is the delayed emission of light after light stimulation. Strictly speaking, the so-called 'fluorescent' screen of an EM should be termed a 'cathodoluminescent' screen. The phenomenon has been widely studied in order to develop suitable phosphors for coating the luminescent screens of colour television sets and fluorescent lamps. Biological materials are very seldom cathodoluminescent, but may become so after exposure to certain reagents. For example, it is reported that the surfaces of red blood cells become intensely cathodoluminescent after exposure to paraformaldehyde vapour. It is possible that this technique may prove useful in the ultrastructural localization of antibodies on cell surfaces. Certain dyestuffs are cathodoluminescent, and it is possible that a technique similar to fluorescence microscopy might be developed.

3.9 Conduction Contrast

It is possible to use the electrons absorbed by a specimen to form an image in a scanning EM, the charge on the specimen being passed to earth via a high-value resistor and the resulting voltage being used to modulate the brightness of a display tube spot. This method has been of value in investigating the structure of certain metals and semiconductors, notably of the planar transistors used in integrated microcircuits. As far as is known, there have been no biological applications of the technique presumably because biological tissues and preparations are, in general, non-conductors.

3.10 Electron Emission Microscopes

If the specimen to be examined is a metal or semi-conductor containing an abundance of free electrons, an electron image may be obtained by causing electrons to leave the surface of the specimen. Some form of energy of excitation must be supplied. This may be thermal energy applied by heating the specimen (thermionic emission); or it may be transferred to the specimen by bombarding it with an electron beam (as in the SEM) or an ion beam, in which case we have kinetic emission; or with ultraviolet light or X-rays, in which case we have photo-emission. Alternatively, the electrons may be dragged out of the surface of the specimen with a very strong electric field, in which case we have field emission. In any event a form of electron emission microscope is possible. In addition, positive ions may also be induced to leave the surface under any of the above stimuli, and an image of the surface may be formed using these ions. This type of instrument is known as an 'ion analyser', and is an ionic and not an electron microscope.

The most familiar application is that of the thermionic emission electron microscope. When a normal TEM is focused so as to image the under-heated filament on the screen (see p. 230), it is acting as a thermo-emission EM, the specimen being the surface of the tungsten filament, as shown in Fig. 11.1(a). Resolving power is very poor owing to the high degree of thermal agitation in the specimen, and the method is not used in practice other than as a convenient way of lining-up a conventional TEM.

Kinetic emission of secondary electrons following bombardment by primary electrons is used to form the image in the SEM (see p. 385). Resolving power is of the order of the diameter of the scanning spot. If secondary emission image were to be formed by using a primary flood beam together with conventional electromagnetic imaging lenses, re-solving power would be very poor due to the uncorrectable chromatic aberration in the imaging lenses. This is why the flood-beam reflection EM was a failure (see above). The problem does not arise in the SEM, because no imaging lenses are used.

3.11 The Field Emission Electron Microscope

Electron emission images can be formed from a metal surface without the use of imaging lenses in a very simple manner. Electrons are simply dragged out of a very fine tip of a high-melting-point metal such as tungsten, using the apparatus shown diagramatically in Fig. 3.6. The emitted electrons travel through the vacuum within the glass bulb, accelerated by the voltage on the anode ring, and strike the fluorescent screen coated on the inside of the bulb. The presence of a regular lattice of atoms at the surface of the emitting tip gives rise to an emission pattern, certain parts of the lattice emitting electrons more copiously than others. The metal lattice pattern is thus reproduced on the screen at a magnifica-tion which is simply the ratio of the radius of the emitting tip to the tip–screen distance. Magnifications of the order of millions can easily be obtained without the use of imaging lenses, since the tips of hard metals, notably tungsten, can be etched to radii of only a few hundreds of Ang-strom units. The application of a comparatively low accelerating voltage of only 1,000 V or so to the anode ring gives rise to a colossal electric field at the tip of the order of hundreds of millions of volts per centimetre, and a copious emission of electrons is obtained with the cathode tip at a much lower temperature than that of a thermionic source such as the conventional heated tungsten hairpin. A reasonably high resolving power of the order of 10–15 Å is obtainable with this simple apparatus, enabling micrographs of metal surfaces down to almost atomic dimensions to be obtained. To work satisfactorily, an electron emission microscope needs a very much higher vacuum than can be obtained with the use of conventional diffusion pumps. The apparatus forms the field emission electron source which is used with the high-resolution STEM (see p. 378).

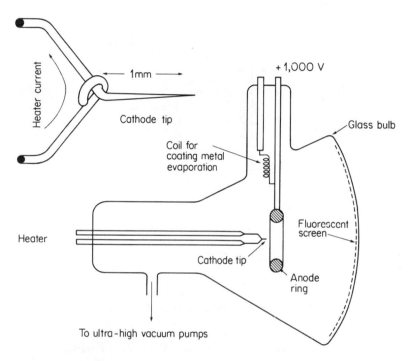

Figure 3.6   A schematic diagram of a field-emission electron microscope. Redrawn from Good and Müller, *Handbuch der Physik*, Vol. 21, Springer-Verlag, Berlin

3.12 The Field Ion Microscope

Some of the drawbacks of the field emission electron microscope have been overcome by the application of a brilliant idea due to Professor Müller of Pennsylvania. This is to form the image by using heavy ions and not with electrons. The microscope now becomes a 'field-ion microscope', and is not an electron microscope at all. The principle is to

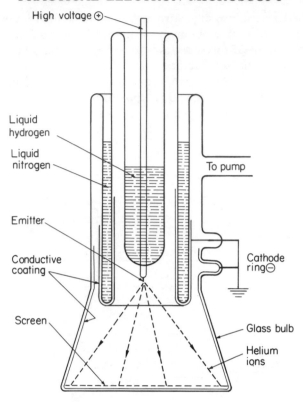

Figure 3.7 A schematic diagram of a field-ion microscope. The apparatus is basically similar to the field-emission electron microscope, but the polarities of the electrodes are reversed and the pattern of the cooled tip is formed by helium ions. Redrawn from Müller, 1960, Advances in Electronics and Electron Physics, 13:83

introduce an inert gas (generally helium) into the bulb of a field emission electron microscope (this overcomes the disadvantage of the necessity for an ultra-high vacuum), and to reverse the connections to the electrodes. The finely-etched metal point now becomes the anode, and attracts electrons. When a neutral atom of helium drifts close to the point, the intense electric field drags an electron out of it into the metal of the tip. The helium ion is now repelled at very high velocity from the positively charged tip, and travels in a straight line to the screen. Since the ionization of helium atoms is strongest at points of highest field intensity on the tip, corresponding to the pattern of the atomic lattice, the helium ions form a corresponding pattern on the screen at an equally high magnification to the field emission electron microscope. The resolving power can now be made much greater, since the ions, being very much more massive than electrons, have a very much shorter de Broglie wavelength. Also, the tip does not need to be hot in order to emit electrons; on the contrary, it can be cooled almost to absolute zero by means of a liquid hydrogen bath. The lattice pattern is therefore almost stationary, and resolutions of better than 1 Å can be obtained. This is the highest resolution yet obtained in any microscope because of the complete absence of any lenses. The pattern on the screen is very faint indeed, and an image intensifier (see Chapter 17) is necessary to make the image visible. This type of instrument has the great advantage that emitters of low melting point can be used, so it is not limited to metals such as tungsten. It might conceivably have biological applications, but is at the present time confined to the study of metals and alloys.

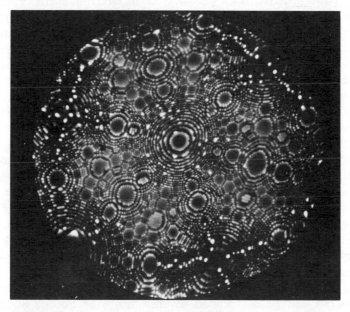

Figure 3.8 An image of a pure iridium tip at the temperature of liquid nitrogen formed in the field-ion microscope by helium ions. The pattern is formed by bright spots, each of which corresponds to the position of a single iridium atom. Approx. × 1,000,000; resolution about 1 Å. (Courtesy Drs. Ralph and Page, Cambridge University)

3.13 The Photo-emission Electron Microscope

When a very intense beam of ultraviolet light is focused on the surface of a metal or alloy in a vacuum, a sufficiently copious and monochromatic electron emission is obtained to enable an enlarged image of the surface to be formed using a conventional magnifying train of electron lenses. The vacuum must be very high (10^{-7} Torr or better). Resolution of the order of 150–300 Å is possible, enabling micrographs at magnifications up to $\times$ 10,000 to be made. This form of electron microscope is becoming increasingly important in metallurgy and materials science, especially for examining integrated circuits. A photo-emission instrument, the Metioskope KE3, is manufactured by Balzers of Liechtenstein.

3.14 The Ion Microprobe Analyser

This is not an electron microscope, any more than the electron microprobe X-ray microanalyser (EXMA) is an electron microscope. It is, like EXMA, a method for obtaining chemical information from a specimen, but in this case the positive ions liberated by sputtering from the surface under bombardment (either with electrons or with a primary exciting positive ion species) are analysed and quantitated in a mass spectrometer. This apparatus sorts the emitted ions into species by mass, and charge, just as the EXMA X-ray spectrometer sorts the emitted X-rays into species by energy (see p. 389). Either a scanning or stationary microprobe technique can be used. Ions of a given species are collected on a photomultiplier, and a scanning image in the 'light' of any given ionic species can be formed on the face of a synchronous cathode ray tube.

The ion microprobe analyser has a number of advantages over the X-ray microprobe analyser. Firstly, it is equally sensitive to the whole elemental range from hydrogen to uranium; the sensitivity of EXMA falls drastically in the range of mass numbers lighter than sodium, which is the range of greatest interest to biologists. Secondly, the sensitivity is higher by a factor of 100 or so. Thirdly, isotopic abundance ratios can be measured, which is not possible with EXMA.

It is therefore ideally suited to geological research, in which the age of minerals can be determined by measuring the isotope abundance ratios. It has also been applied to the measurement of impurities in the ultra-pure semiconductor materials used in modern transistor and integrated circuit technology. It seems unlikely that this instrument has any application to normal biological tissues. The instrument is marketed by GEC-AEI and the French firm, Cameca.

CHAPTER 4

Some Properties of Magnetic Electron Lenses

4.1 Introduction

An electron beam can be focused by the action either of magnetic fields or electrostatic fields. Lenses may be constructed using either principle, or a combination of both. With the one important exception of the lens used to shape the emergent electron beam in the electron gun, electrostatic lenses are not used in modern electron microscopes. Although they have some advantages over magnetic lenses in image formation, electrostatic lenses require very powerful electrostatic fields, which can lead to electrical breakdown or 'arcing-over' inside the column, especially under conditions of poor vacuum. For this reason, electrostatic lenses cannot be made with focal lengths as short as magnetic lenses, and therefore, as we shall see in this chapter, they cannot be as well corrected for spherical aberration. Image-forming electrostatic lenses will not be discussed in this book.

4.2 Magnetic Lenses

Basically, these are simple solenoids, the action of which is shown in Fig. 1.3. When a current is passed through an axially wound coil, it behaves like a hollow bar magnet. Part of the magnetic field runs inside the coil parallel to the axis; part is radiated out into space outside the coil. Since the only useful part of the magnetic field as far as an electron lens is concerned is the inner part parallel to the axis, the external field from a simple solenoid is wasted.

The efficiency of the simple solenoid lens was improved by enclosing the energizing coil in a sheath of soft iron, as shown in Fig. 4.1. Soft iron has the property of concentrating the lines of force in a magnetic field, and thus becoming itself magnetized by induction. No other material has this 'ferromagnetic' property developed to such a high degree. As the current in the energizing coil is increased, so the induced magnetism in the iron sheath increases, and the electron lens becomes stronger, i.e. the focal

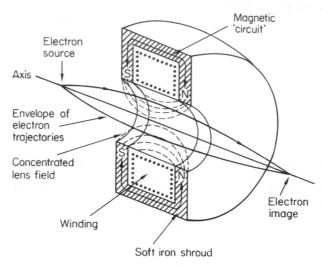

Figure 4.1 The axially symmetrical magnetic field inside a simple shrouded solenoid. The lines of force are shown dotted

length decreases. Unfortunately, focal length and current are not linearly related. Fig. 4.2 shows how the strength of a lens varies with the current in the exciting coil. If the lens starts at the point O, completely de-magnetized, the strength increases in a slightly sigmoid fashion as current increases, but after a while, more and more current is needed to give a corresponding increase in lens strength. Finally, at the point A, no further increase in strength can be obtained, and the lens is said to be 'saturated'. Since instrumental magnification depends on the strength of the inter-mediate lens, it is important to realize that magnification is not linearly related to lens current. Fig. 4.3 shows a typical curve of magnification versus intermediate lens current, taken from a Philips EM–200.

4.3 Hysteresis

A further property of a magnetic lens is that its strength cannot be calculated exactly from the current flowing in it; the strength depends to some extent on the previous magnetic history of the lens, as can be seen from Fig. 4.2. When the current in a saturated lens is reduced, the lens strength does not retrace the path AO. Magnetized soft iron retains some of its magnetism, a property known as 'remanence'. As the exciting current is reduced, the lens strength follows the curve AB. Even when the

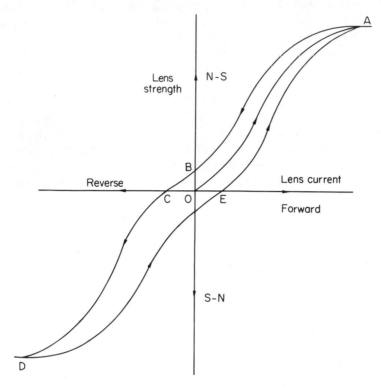

Figure 4.2   Curves showing how the magnetization of soft iron (lens strength) is related to the magnetizing force (lens current). An unmagnetized lens starts from the point O and follows the path OA as lens current increases. At the point A, further increase in lens current produces no further increase in lens strength; the lens is said to be 'saturated'. When lens current is reduced, the path OA is not retraced; a different path AB is followed. This displacement is called 'hysteresis'. At zero current (point B), some residual lens strength remains; this is called 'remanence'. To bring the lens back to zero strength, a reverse current OC must be applied. Lens strength then increases with increasing reverse current, following the path CD. The polarity of the lens changes, but this does not affect its focusing power; only the spiral electron path is reversed. Because of hysteresis, it is not possible to calibrate a lens current meter accurately in terms of lens strength or magnification

current is zero, the lens retains some magnetism, which can only be removed if the current is reversed and taken to the point C. If the reversed current is then increased, the lens will become resaturated with opposite polarity, at the point D. This does not affect the image-forming property of the lens except inasfar as image rotation is concerned (see p. 85). When the current is reduced, the lens strength will follow the curve DE, passing again through the remanence point. On increasing the current again in a forward direction, the lens strength will follow the path EA until the saturation point is reached once more. This phenomenon is called 'hysteresis'. The consequence of hysteresis is that lens strength for a given excitation current may lie anywhere between two extreme values, depending on the immediate past magnetic history of the lens. Hysteresis can cause errors in magnification of $\pm 10\%$ or more. It can be eliminated by taking a lens to saturation and then following the curve down to the required magnification. This is done by increasing lens current to the maximum and then going straight to the required value without

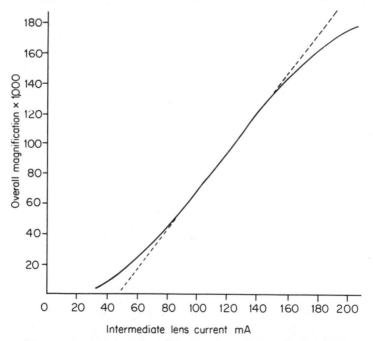

Figure 4.3   An actual curve of magnification (as a function of inter-mediate lens strength) versus lens current. The non-linear shape of the curve can be seen by reference to the dotted straight line. (Data from Philips 200, courtesy Philips, Eindhoven)

overshooting and returning. Some instruments have a button which enables this to be done rapidly and accurately. Remanent magnetism can be removed from a lens by rapidly reversing the lens current while steadily reducing it. However it is done, hysteresis must always be allowed for when making accurate measurements involving the calibration of magnification with an electron microscope.

4.4 The Polepiece Lens

Lenses of the simple shrouded type just described are suitable for relatively weak lenses, such as the second condenser and the intermediate lenses in an electron microscope. For strong lenses of short focal length, the magnetic field must be concentrated into a shorter distance along the axis. This is done, as shown in Fig. 4.4, by carrying the soft iron shroud

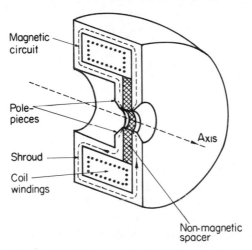

Figure 4.4   The construction of a polepiece magnetic electron lens. The non-magnetic spacer is generally made of brass

along the lens axis, and by adding cylindrical, tapering iron polepieces so as to bring the effective magnetic field as close to the axis as possible. This form of construction enables the whole of the external magnetic field to be concentrated into a very short distance (a few mm) along the lens axis. This highly concentrated magnetic field enables lenses to be designed with focal lengths as short as 1 mm before the iron saturation limit is reached. The axial distribution of field for a solenoid, a shrouded lens and a polepiece lens are shown in Fig. 4.5.

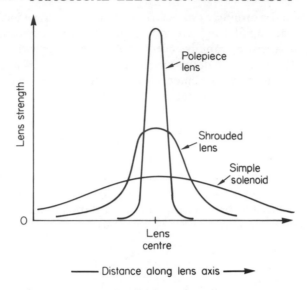

Figure 4.5 The distribution of magnetic field along the axis of a single solenoid, a shrouded lens and a polepiece lens, for the same current in the solenoid winding

The focal length of a magnetic lens can be changed very easily and conveniently simply by varying the current flowing in the windings of the exciting coil. This means that magnification, focus and illumination are very easily controlled by the operator, who has only to rotate the correct lens current controls. Focal length can also be changed, although less conveniently, by altering the diameter of the bore of the lens polepiece. This method has certain advantages in reducing image distortion at low magnifications, and is fitted to the projector lens in some electron microscopes.

4.5 Lens Aberrations

No lens, light or electron, is able to form a perfect point image of a perfect point source, quite regardless of the limitations imposed by diffraction. A lens can only intercept part of the wavefront emanating from an object or source, and so cannot possibly image all the information which it sends out.

Lenses suffer from a number of specific defects called 'aberrations', which can be classified and analysed so that their ill effects can be minimized. The two major lens defects are called 'spherical aberration' and 'chromatic aberration'. The first is due to the inability of a lens to bring all the rays which enter it to a common focus; the second arises from the fact that neither light nor electron beams are composed of radiation of one single wavelength. The designer of glass lenses is able to apply corrections for minimizing the major aberrations, because he has at his disposal different glasses of different refractive indices and dispersions which he is able to combine using positive and negative lenses so as to minimize a defect in one element by introducing an equal and opposite defect in another. In this way, he is able to reduce image defects formed by lens aberrations to values less than those caused by diffraction. The electron lens designer is very limited in his attempts to reduce aberrations. He has only one refracting medium at his disposal—a magnetic field— and he is confined to using positive lenses. The only effective way to reduce most of the aberrations in electron lenses is by reducing the aperture, which, as we have seen, means reducing resolving power. The choice of the most suitable lens geometry also helps, but this art is still in its relative infancy.

Although the light microscope and the electron microscope work on broadly the same principle, the analogy must not be carried too far. One of the most important fundamental differences between them is the fact that in the light microscope, resolution is limited by diffraction effects; in the electron microscope on the other hand resolution is limited by lens defects. Over the four centuries or more during which the light microscope has been under development, physical opticians have learned how to perfect the image-forming qualities of optical lenses until the loss of resolution due to lens defects is less than that imposed by diffraction. If electron lenses could be corrected to the same degree, the resolving power of an electron microscope would also be one-half wavelength, which would be approximately 0.02 Å. The best result achieved so far is approximately 2 Å—100 times as great. In point of fact, since interatomic distances are of the order of 1–2 Å, one wonders whether improved resolving power would lead to further morphological discoveries. The reason for the discrepancy between theoretical and practical resolving power is simply that the only effective means at present known for reducing spherical aberration is by reducing aperture, which, as we have seen, reduces resolving power.

4.6 Spherical Aberration

For a lens to refract an oncoming wavefront to form a perfect converging image-forming wavefront, the curvature of the interface between the lens and the medium in which it is immersed must be correct. Unfortunately for the designers of glass lenses, this curvature is not a part of the surface of a sphere; it is a more complex surface which cannot be generated by a simple grinding process. Glass lens designers generally have to be content with a spherical surface; hence the term 'spherical aberration'.

The image formed by a lens bounded by a spherical surface will be defective. The defect arises from the fact that the peripheral rays are bent more than the rays close to the axis of the lens (Fig. 4.6). The image of a

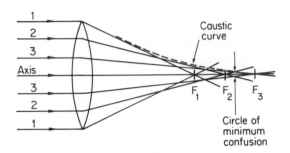

Figure 4.6   Spherical aberration arises because the peripheral rays are brought to a focus which is closer to the lens than rays which are nearer to the axis (see text)

point source will be spread out over a finite distance over the axis. The shape of the envelope of rays converging to the image is called a 'caustic curve'. The rays never converge to a point; they converge at one place on the axis to a disc of minimum diameter. This disc is called the 'circle of minimum confusion' and its position on the axis is taken as the practical focal point of the lens. Spherical aberration varies as the cube of the numerical aperture of the lens; it can therefore be greatly reduced by cutting out the peripheral rays or 'stopping-down' the lens. This reduces the information-gathering power and, as Abbe's formula shows, the resolving power is also reduced. All electron lenses so far constructed suffer from spherical aberration which can only be effectively minimized (not corrected) by stopping the lens down in this way. This is the primary reason why the resolving power of an electron microscope is so much less than would be expected from the very small wavelengths employed in image formation.

It can be shown that the resolving power of an electron lens, if limited solely by spherical aberration, is given by the expression:

$$RP = K.C_s.f.\alpha^3$$

where C_s (the 'spherical aberration coefficient') is a parameter dependent on lens design; f is the focal length and α is the aperture angle in radians. C_s is dependent in a complex fashion on focal length, as shown in Fig. 4.7 where C_s is shown plotted against lens strength (the inverse of focal length).

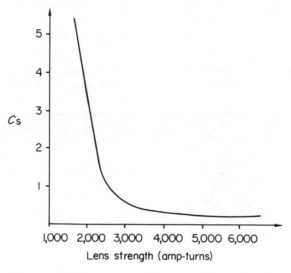

Figure 4.7   The amount of spherical aberration (C_s) in a magnetic electron lens decreases as the lens strength increases

It will be seen that C_s decreases rapidly as focal length is reduced, approaching an almost constant value at very high lens strengths. The minimum value of C_s at present attainable with modern objective lens designs having a focal length of about 1·5 mm and working with the iron polepieces close to saturation is about 0·03, which with an aperture angle of 10^{-3} radians (more than a thousand times narrower than the aperture of a light microscope) gives a theoretical resolving power of about 2 Å.

4.7 Distortion

Spherical aberration gives rise to another defect which can be very prominent in electron microscope images; this is called 'distortion'. In a lens suffering from spherical aberration, the peripheral rays are imaged by what is in effect a stronger lens than the central rays. The image formed by the peripheral rays will therefore lie in a plane different from the image plane of the axial rays, and it will therefore be at a different magnification. The magnification of the image will therefore increase or decrease with increasing distance from the axis. If the lens is of very small aperture, as in the case of an electron microscope lens, the depth of field will be large (see p. 89) and all the images will be in focus together. Magnification therefore varies with increasing distance from the lens axis, and a square object will not be imaged in the form of a square. It may have inwardly curved sides, resembling a pincushion; or it may have outwardly curved sides, resembling a barrel. These two forms of distortion are therefore called 'pincushion' and 'barrel' distortion respectively. An electron micrograph of a square grid free from distortion is shown in Fig. 4.8(a).

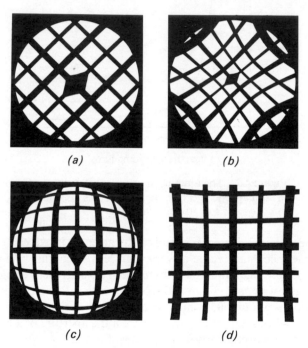

(a)   *(b)*

(c)   *(d)*

Figure 4.8   Four electron micrographs at very low magnification (approx. × 50) showing: (a) negligible distortion; (b) pincushion distortion; (c) barrel distortion; and (d) sigmoid distortion. The distortions were introduced deliberately by altering the focus of the intermediate lens, which was being used as a long-focus objective

The same grid is shown in Fig. 4.8(b) with pincushion distortion, and in Fig. 4.8(c) with barrel distortion. The amount of distortion is exaggerated for clarity in these illustrations, but distortion on this scale can unwittingly be obtained on some instruments by an inexperienced operator. A further effect of distortion is that intensity of illumination also varies across the image field if distortion is present, leading to unevenly exposed micrographs.

4.8 Image Rotation

A further property of a magnetic lens is that an electron does not follow a trajectory curved in one plane only. An electron moving through a magnetic field is subject to the same forces as any current-carrying conductor, and therefore Fleming's 'left-hand' or 'motor' rule will apply. The moving electron will suffer a force at right angles to the direction of its motion. It will therefore follow a spiral path through the field of a magnetic lens, as shown in Fig. 1.3. A section through the envelope of trajectories followed by electrons moving with the same initial velocity is in the shape of a ray diagram; it is therefore permissible to use ray diagrams in electron optics.

When the effect of rotation is combined with that of distortion, a straight line will be imaged as a sigmoid shape (like the letter 'S'). This effect is shown in Fig. 4.8(d). The effect is barely noticeable in well-designed instruments.

The possibility of distortion being present must always be borne in mind when accurate measurements for stereology (see p. 409) are required from electron micrographs. Errors of 5 % between centre and edge at low magnifications are accepted as reasonable by manufacturers. Such errors are quite unnoticeable until actual measurements on regular structures are made.

The effect of all three kinds of distortion is most noticeable at low magnifications, becoming increasingly less at high magnifications. Distortion can be reduced to almost negligible proportions by careful design of the two final image-forming lenses. The method used is to balance pincushion distortion in the one against barrel distortion in the other. For each value of magnification, there is an optimum current in each lens for minimum distortion. On modern instruments, these values are selected automatically as magnification is varied. On some instruments, however, distortion is under the control of the operator. If all the lens currents can be freely controlled, values can easily be chosen which, although they may appear to give the required magnification, also give unacceptable distortion.

4.9 Astigmatism

This defect arises because it is impossible in practice to make a perfectly symmetrical lens. All real lenses will have a focal length in one direction which is slightly different from the focal length in another direction, due to imperfections in grinding the lens bore and inhomogeneities in the polepiece material. This means, for example, that the north–south axis of

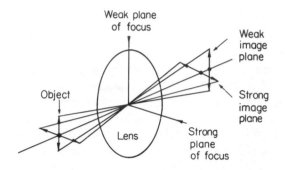

Figure 4.9   The cause of astigmatism. If a lens is not axially symmetrical in strength, it will be stronger in one plane and weaker in another plane perpendicular to the first. An object in the form of a cross will therefore be imaged with one arm in focus and the other arm blurred (see text)

a specimen will be imaged at a plane slightly different from the east–west axis. An object in the form of a cross shown in Fig. 4.9 will be imaged with one arm in one plane, and the other arm in another plane. A lens with this defect is said to be 'astigmatic'. The two astigmatic planes can be considered as a first approximation to be at right angles to one another. It is therefore possible to apply a first order correction by increasing the focal length of the astigmatic lens in one direction only. This can effectively be done by adding a cylindrical lens of the correct focal length and in the correct direction to bring the two out-of-focus specimen planes together. Astigmatism is in practice the most important defect limiting the resolving power of an electron lens, and some means for correcting it by applying a correcting field equivalent to a cylindrical lens must be provided. The device which does this is called the 'stigmator', and is always provided in the objective lens and sometimes also in the illuminating system. The use of the stigmator is discussed in detail in Chapter 12.

4.10 Chromatic Aberration

This defect arises in the light microscope because light of short wave-lengths is slowed more when passing through a medium of high refractive index than is light of a long wavelength. Blue light is therefore brought to a focus closer to the lens than red light. Chromatic aberration is not related to spherical aberration; the effect increases linearly with aperture, and not as the cube. It is therefore much less susceptible to reduction by reducing numerical aperture. Electron lenses suffer in the same way from this defect, but since electron wavelength is associated with velocity, the 'chromatic' effect in electron lenses is really a velocity effect. Fast, short-wavelength electrons are deviated less by an electron lens than are slow electrons of longer wavelength. The effect thus works in the direction opposite to that in the light microscope, as is shown in Fig. 4.10. The net effect, however, is a similar blurring of the image by out-of-focus effects. In both light and electron microscopes, chromatic aberration can be reduced by using monochromatic light to view the object. A 'mono-chromatic' beam of electrons is one in which all the electrons have identical

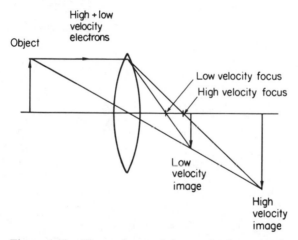

Figure 4.10   The analogue of chromatic aberration in an electron lens. Fast (short wavelength) electrons are brought to a focus at a point further from the lens than slow (long wavelength) electrons. The image formed by the high-velocity electrons will therefore be larger than that formed by the slow ones. An effect called 'chromatic change of magnification' is produced, which is aggravated by thick specimens and low accelerating voltages

velocities; it is, unfortunately, impossible to achieve owing to the random emission of an electron source. It can be almost eliminated by a very high degree of stabilization of the electron accelerating voltage. This is only partially effective, because the object itself introduces a chromatic defect by slowing some electrons which pass through it more than others. This effect will be discussed in the following chapter. Chromatic aberration can be reduced after the specimen by means of a 'velocity filter' (see Chapter 17).

4.11 Chromatic Change in Magnification

All the electrons which are slowed down in their passage through the specimen but still pass through the objective lens will be imaged at different planes. The effect will be similar to that given by distortion; the low velocity electron image will be at a different magnification to the high

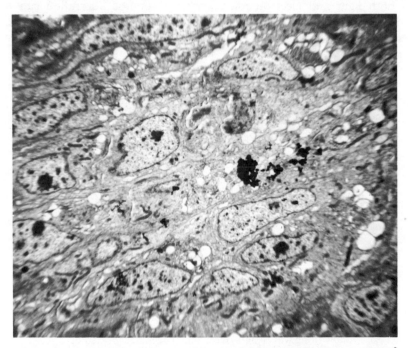

Figure 4.11   The effect of chromatic change of magnification. A 1,000 Å thick section of embedded tissue was imaged by a 30 kV beam. The central part of the micrograph is sharp, but the out-of-focus effect becomes increasingly noticeable further from the axis. The effect is particularly noticeable at low magnifications. × 5,000. See also Fig. 13.3, p. 307

velocity image. The net effect will be to blur the image; it will be greatest at the periphery and least at the axis. Fig. 4.10 shows diagrammatically how the effect arises; Fig. 4.11 is an electron micrograph showing the image blurring it causes.

The effect is worst with thick specimens and low accelerating voltages; it is further aggravated by distortion. The effect is least when the instrumental distortion is at a minimum, achieved by balancing the projector and intermediate lens currents, and when the specimen is thin and when a high accelerating voltage is used. The two last-mentioned requirements also lead unfortunately to reduced image contrast, as will be explained in the following chapter.

4.12 Depth of Field and Focus

We have seen that any lens, however perfect, can only image a point object as a disc, the diameter of which is the resolving power of the lens. This fact has important practical applications; it makes the task of focusing the image (which would otherwise be almost impossible at very high magnifications) a great deal easier. How it does this is shown diagrammatically in Fig. 4.12(a).

Suppose a lens brings a pencil of rays from a point O on the axis towards the image point I. As the pencil converges on the image side of the lens, it reaches a diameter equal to that of the resolving of the lens at a point on the axis RP_1. In theory, the pencil continues to converge, but in practice it cannot do so, because the finite resolving power (the disc of minimum confusion) prevents this. The pencil will not appear to diverge until it reaches once again the diameter of the resolving power of the lens at a point RP_2 on the axis. There is therefore a finite distance along the axis, D_{f_o}, where the image is equally sharp. This distance is called the 'depth of focus'. It means that if an image screen were placed anywhere between RP_1 and RP_2, the image would still be in as perfect a focus as it is possible to achieve. Depth of focus therefore simplifies the task of focusing.

Because of the existence of the depth of focus on the image side of the lens, there must exist an equivalent distance along the axis on the object side over which the object could be moved and still give a maximally sharp image, provided the position of the screen remained unchanged. This distance is called the 'depth of field' D_{f_i}, and is related to the depth of focus by simple geometry.

Depth of field and focus are an inverse function of the numerical aperture of the lens. Fig. 4.12(b) shows how a lens of the same resolving

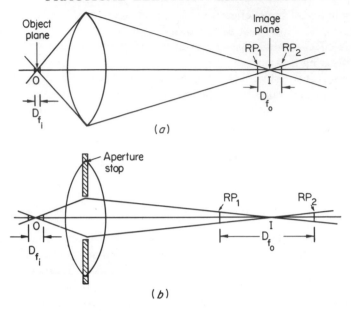

Figure 4.12   Depth of field and depth of focus, and the effect of lens aperture on them. As lens aperture is reduced, both depth of field and depth of focus are increased

power as the lens in Fig. 4.12(a) but having a smaller aperture gives rise to an increase in both depth of field and focus. For lenses of very small aperture, it can be shown that the depth of focus varies as the square of the magnification according to the expression:

$$D_{f_0} = \frac{M^2 . RP}{NA}$$

When applied to a compound magnifying system, the magnification M is the product of the magnification of all the lenses. A typical electron microscope has a lens NA of about 10^{-3}, a resolving power of 5 Å and a total magnification of 10^5. The formula above gives a depth of focus at the final image plane of 5,000 m, which is for all practical purposes infinite.

The value of this in practice is that the fluorescent screen or photographic plate or film can be placed anywhere beneath the projector lens; the final image will be equally sharp. A 35 mm film can therefore be placed immediately after the projector lens or a plate can be placed well after the viewing screen; all three will be in equally sharp focus, although the magnification will differ.

The depth of field of the objective lens will be correspondingly great; greater in general than the thickness of the specimen. Depth of field is given by the expression:

$$D_{f_i} = \frac{\lambda}{NA^2}$$

As we have seen, 50 kV electrons have a wavelength of 0·04 Å, and the NA of an electron objective lens is of the order of 10^{-3}. The depth of field is therefore 4×10^4 Å, or 4 μm. The thickness of a section is of the order of one hundredth of this distance. All the detail in the specimen will therefore be superimposed equally sharply in the image. For an average biological specimen embedded in an epoxy resin, the resolution is given very roughly by Cosslett's rule, which states that the resolution to be expected is approximately one-tenth the section thickness. This depends on the nature of the specimen and its orientation to the incident electron beam. If the required structure can be suitably oriented to the beam, very much higher resolutions can be obtained from a section. It is in this respect that the use of a tilting stage, allowing the section to be tilted through a large angle (up to ±45°), is very advantageous. This facility can now be obtained on most biological instruments, and will be discussed at greater length on p. 268.

CHAPTER 5

Contrast and Image Formation

5.1 Introduction

In this book, we are dealing mainly with transmission electron microscopy. Basically, this means that some form of radiation, bearing no relevant information, is made to interact with a specimen by passing through it. When the radiation reappears on the other side of the specimen, it carries information about the specimen, which is then processed in such a way that the eye can see and the brain can interpret it.

The eye is sensitive to changes in light intensity (amplitude contrast) and wavelength (colour contrast). The information given by the light microscope can be interpreted directly by the eye; that given by the electron microscope cannot. The information carried by the electrons can as yet only be converted into amplitude and phase contrast; only black-and-white images can be formed. It is not practicable to convert changes in electron wavelength into changes in colour; in any case, very little more information could be expected by doing so. All electron micrographs are therefore monochromatic, and are generally reproduced in black and white.

The terms 'amplitude contrast' or simply 'contrast', as used in this book, mean differences in light intensity which can be interpreted by the eye. The eye is insensitive to phase contrast. This amplitude contrast can be formed either on a fluorescent screen directly, or it can be recorded on a photographic plate carrying an emulsion sensitive to electrons which gives amplitude contrast after suitable processing. The image seen on the fluorescent screen is transient, and due to screen granularity contains less information than a permanent photographic record. After the exposed emulsion has been processed, the permanent image can be treated as a normal object for further magnification (or occasionally reduction) in an ordinary photographic enlarger. In this case, the developed plate will record the presence of electrons as dense areas. The enlarged print will of course reverse the density values on the plate; dark area on the print will record the absence of electrons just as dark areas on a light micrograph record the absence of light. The reversed, printed image is thus accepted by the eye as a conventional photograph, corresponding to the image originally seen on the fluorescent screen.

5.2 The Specimen

To form a transmission image, the electron beam must first pass through a specimen. Fast electrons are very readily slowed or stopped by matter. They can only traverse a few millimetres of air or a micron or so of water. The transmission specimen must therefore be very thin, and must be mounted if necessary upon an equally thin support. Specimen support films are generally made of carbon about 50–100 Å thick. Biological specimens are generally embedded in a hard plastic such as an epoxy resin, and are then sliced into sections about 500 Å thick with an ultramicrotome. The details of support film and specimen preparation will be dealt with in Chapters 13 and 18.

5.3 Image Forming Processes

In both light and electron microscopes, four fundamental physical processes take part in the formation of the image. These are absorption, interference, diffraction and scattering. In general, all four contribute together to image formation. Absorption gives rise to amplitude contrast, the differences in intensity to which the eye is sensitive. It is generally the most important factor in the formation of the conventional light microscope image. Interference gives rise to phase effects, to which the eye is completely insensitive. These must first be converted to amplitude contrast before the eye can appreciate them; the effect is then known as 'phase contrast'. Diffraction effects in general degrade the image by the formation of fringes and haloes which reduce resolution and which may give rise to spurious images. Diffraction effects can be used to enhance contrast in both light and electron microscopes, but only at the expense of a certain loss of resolution.

Scattering plays little part in the formation of the light microscope image, except in certain special circumstances; this effect is on the other hand the most important of all in the formation of the electron microscope image. We shall now consider in greater detail how these four processes give rise to the electron image.

5.4 Scattering

When a beam of electrons passes through a thin specimen, the atoms of the specimen will interact with the incident fast electrons in one or both of two ways. They can interact either with the nuclei of the specimen atoms, or else with the electrons in the cloud surrounding the nuclei (see Fig. 5.1). In a thick specimen, one incident electron may suffer both types of interaction. It must be remembered that matter is virtually completely empty space; the dimensions of nuclei and electrons are of the order of 10^{-5} to 10^{-6} Å, whereas interatomic distances are of the order of 1 Å or more. The trajectory of an electron must pass very close to a specimen nucleus or electron before it is deflected. The majority of incident electrons will therefore pass straight through a thin enough specimen (Fig. 5.1(a)).

The two forms of interaction have different effects on the transmitted electrons, due to the great difference in mass between an electron and a nucleus. If a fast electron passes close to a nucleus, it will be deflected through a large angle, but will suffer practically no energy loss, because of the great difference in mass between the two particles. The effect is similar to the dropping of a ball bearing on to a massive steel plate. The ball will rebound almost to the height from which it was dropped, losing very little energy. Similarly, an electron will rebound from a carbon nucleus, which is some 20,000 times its own mass, without loss in energy, but with considerable deviation in trajectory. This type of interaction is called 'elastic scattering' (Fig. 5.1(c)).

If, on the other hand, a fast electron collides with a slow electron orbiting around a nucleus, they will divide their combined energies between them, since their masses are the same. If a ball-bearing rolling on a steel plate collides with another one, they will share their velocities, obeying the laws of conservation of momentum. The high velocity imaging electron in this case suffers not only a change in direction, but also a change in velocity. This type of interaction is called 'inelastic scattering'. (Fig. 5.1(b)). Since there are far more electrons in a specimen than there are nuclei, inelastic scattering is the most important effect in the formation of the electron microscope image.

In addition, two other important effects take place during inelastic scattering which, however, do not contribute to transmission image formation. Firstly, X-rays are produced due partly to loss of energy in slowing down and partly to specimen electrons jumping between orbits.

This phenomenon is utilized in X-ray microanalysis (EXMA—see p. 389). Secondly, specimen electrons close to the incident surface may be ejected. This phenomenon is utilized in scanning electron microscopy (SEM—see p. 384).

Since the possibility arises that incident electrons may be slowed down but not appreciably deflected during their passage through a specimen,

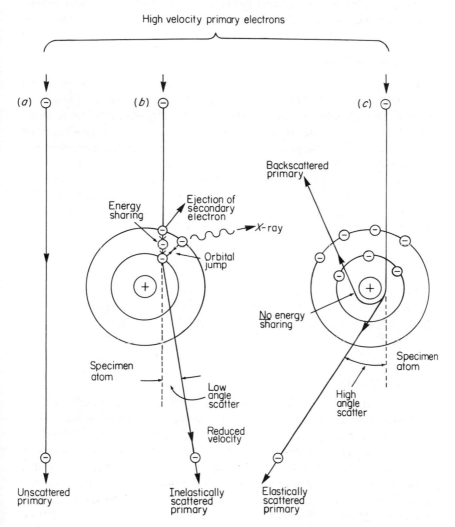

Figure 5.1   The principal modes of interaction of high-energy electrons with the atoms of matter constituting the specimen

the emerging electron beam may suffer chromatic aberration. Reduced velocity results in increased wavelength. The thicker the specimen, the more chromatic aberration will be introduced, however monochromatic the incident beam may have been. This means that some of the electrons reaching the final screen after being inelastically scattered will degrade the image by the effect of chromatic change in magnification (Figs. 4.10 and 4.11). The effect will be similar to 'glare' in the light microscope image, and will serve only to reduce the contrast of the final electron image.

5.5 Mass Thickness

The larger the number of atoms the imaging electrons encounter in their passage through the specimen, the greater will be the probability of scatter. The amount of scatter is therefore directly proportional to specimen thickness. Another factor, however, also comes into play. The larger the atomic nucleus, the greater is its positive charge and the more electrons are present in its surrounding electron cloud. The size of the atomic nucleus is directly related to its atomic number and hence, roughly, to its density. Therefore, the greater the specimen density, the greater is the probability of scatter. The total scatter is proportional not simply to thickness, but to the product of density and thickness. This is called the 'mass-thickness' of the specimen. Variations in mass thickness in various parts of the specimen give rise to differential scattering and thus to contrast in the image on the screen.

Mass thickness is measured in $\mu g/cm^2$. The density of carbon is about 2, so the mass thickness of a 100 Å (10^{-6} cm) thick carbon film is 2 $\mu g/cm^2$. The average mass thickness of a biological thin section is of the order of 10 $\mu g/cm^2$.

5.6 Absorption

For an electron to be absorbed by the specimen, it must suffer a whole sequence of inelastic collisions until its velocity is reduced sufficiently for it to be captured by the specimen. This will only take place if the mass-thickness is high. Absorption, which in a thin specimen accounts for very little image contrast, gives rise directly to amplitude contrast. In a thick conducting part of the specimen, such as the bars of the specimen support grid, absorption of the incident electrons can be total. Absorption involves considerable energy transfer to the specimen, which accordingly will become hot. This is undesirable, because differential heating causes differential thermal expansion in the specimen, resulting in specimen movement and image blurring. Any heavily absorbing object resting on a thin plastic section or support film, such as a tiny fragment of metal or dust, will heat up by absorption. The heat will be transferred to the section, which, being a poor heat conductor, may melt and tear.

5.7 Image Formation

The process of information transfer from the specimen to the homogeneous electron beam consists therefore in the removal of electrons from the beam, mainly by the process of inelastic scattering. At a plane a little below the specimen, the emergent beam will be deficient in electrons in places corresponding to areas in the specimen where mass-density is high. These 'holes' in the emergent electron beam are imaged by the magnifying system as dark areas in the final image. The more electron-transparent areas of specimen scatter fewer electrons, and therefore appear brighter.

5.8 The Objective Aperture

Where do the scattered electrons go to? If they are allowed to pass through the objective lens, they will be imaged at random on the final screen, contributing nothing to the information and degrading the image by reducing contrast. Any electrons which are scattered through an angle which is greater than the acceptance angle of the objective lens can readily be removed by making them strike a metal surface and be conducted back to earth. The metal surface used is a thin disc or thin foil of a non-magnetic metal such as platinum or molybdenum having a very small hole of the order of 20 to 50 μm bored in the centre. These discs are called 'objective apertures'. They serve both to define the lens acceptance angle and hence its numerical aperture, and also to collect up the scattered electrons and thus to improve contrast. The ideal position for the objective aperture is at the optical centre of the lens. This position is not possible in a modern short focal length immersion objective (see p. 126) because of the presence of the specimen holder. It is therefore placed as close as possible to the back focal plane, where it acts as an aperture stop at high magnifications (see p. 39). As magnification is reduced and focal length is consequently increased, it acts more and more as a field stop. For very low magnification working with the objective lens switched off, it must be withdrawn completely.

The aperture must be approximately circular, and must be centred very accurately about the lens axis. There is an optimum size of aperture for maximum lens performance. If the aperture is too small, it will limit the numerical aperture of the lens and thus reduce its ultimate resolving power. Contrast will be higher, but two other factors must be taken into consideration. Firstly, too small an aperture will mask part of the specimen area at low magnifications, since it is not generally possible to place it exactly at the lens centre. Secondly, and more important, the edges of the aperture become increasingly susceptible to an effect known as 'contamination', which will be discussed at greater length in Chapter 7. Contamination is caused by the deposition of a waxy, non-conducting substance on any surface in the vacuum system which is struck by electrons. These deposits tend to build up on certain spots on the aperture edge, rather than in a symmetrical fashion. Being non-conducting, they become electrically charged by collecting electrons, and act therefore as small,

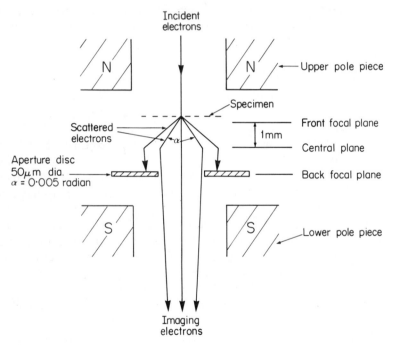

Figure 5.2   The action of the objective lens physical aperture in a short focal length immersion objective lens under high magnification conditions. Widely scattered, high chromatic aberration electrons which degrade the image are intercepted by the aperture disc. The smaller the hole, the more scattered electrons are intercepted and the higher the image contrast. The hole also serves to define the angular aperture of the lens α

randomly disposed electrostatic lenses perturbing the main electron beam. They will therefore introduce astigmatism. The smaller the aperture, the closer will the contamination deposits be to the imaging beam, and the more quickly will they build up. So, the smaller the aperture, the higher will the contrast of the final image be, but the more rapidly will the aperture become unusable due to the effects of contamination. Placing the aperture at the back focal plane of the lens also reduces the effect of contamination because the imaging pencil is narrowest at this position, and the edges of the aperture are consequently removed further from the beam.

The effect of the objective aperture on contrast can readily be demonstrated by obtaining a clear, well-focused image and suddenly withdrawing the aperture. The resulting contrast reduction is dramatic; the image almost disappears.

5.9 Defocus Contrast

So far, we have dealt with contrast only when the image is exactly in focus. There is, however, another source of contrast which can easily be demonstrated. If, after suddenly withdrawing the objective aperture as described above and noting the sudden reduction in contrast, the focus control (which alters the focal length of the objective lens) is swung rapidly back and forth on either side of the true focus point, a marked increase in contrast on either side of focus will be noted which is sometimes called 'blinking'. The increase is greatest at a point where the objective lens is slightly underfocused (Figs. 5.3 and 12.1). The point of minimum contrast is a very sensitive indication of true focus, and is used in practice as will be described in Chapter 11 for focusing at low magnifications.

This increase in contrast on either side of the point of true focus cannot be explained in terms of scatter. It can only be explained in terms of interference. The increased out-of-focus contrast is called 'defocus contrast', and is due to the formation of Fresnel fringes (see Chapter 1) about any parts of the specimen where there is a rapid change in mass-thickness. The fringes, as has been shown, serve to enhance such points or edges by delineating them with a bright line in the underfocused position, or with a dark line in the overfocused position. Fresnel fringes disappear completely at the point of exact focus; image contrast is therefore at a minimum when the image is exactly in focus. The enhancement of contrast by the use of Fresnel fringes is almost always taken advantage of in practice; very few electron micrographs of thin sections of biological specimens are taken at the point of exact focus. Care must be taken in using Fresnel fringes in this way, since resolution is bound to be lost. Judgment of the correct amount of underfocus to choose is a skill which the operator learns by experience. Fresnel fringes cause not only a loss in resolution, but may also cause spurious images to be 'resolved', especially if the specimen contains objects which have a close, regularly repeating spacing. The correct degree of underfocus for maximum contrast enhancement is such that the slight loss in resolution is undetectable by the unaided eye on the final print, but the enhancement of contrast is readily appreciated. The use of Fresnel fringes in practice will be discussed in Chapter 12.

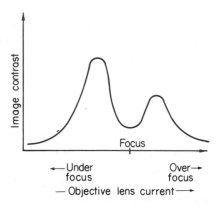

Figure 5.3   A curve showing the variation of image contrast with objective lens focus. Fresnel fringes produce a so-called 'defocus granularity' or spurious out-of-focus contrast which is maximal at a point just below focus, where detail is enhanced by the bright underfocus line. The effect is shown in the micrographs of Fig. 12.1

There is also a physiological accentuation of contrast at a sharp edge when perceived by the eye. This effect, known as the 'Mach effect', is ascribed to suppression of retinal response in the cells immediately surrounding a bright image. It naturally assists physical contrast in the perception of microscopic images.

5.10 Phase Contrast

The 'defocus contrast' effect described above is an interference phenomenon similar to the effect used in phase contrast light microscopy, which renders visible transparent objects which differ in refractive index from their surroundings. The phase contrast light microscope converts phase differences in light emerging from the specimen into black-and-white amplitude contrast which is appreciated by the eye. The 'phase contrast' of electron microscopy is generally considered to arise from interference effects due to the finite aperture angle and uncorrectable spherical aberration of the objective lens. Path length of rays passing through the periphery of the lens is greater than path length at the central axis, thus rays from the same point in the specimen can interfere to give differences in intensity at the image plane. The effect is only detectable close to the absolute limit of resolution of the lens, and in effect determines its ultimate resolving power. It causes an extremely fine image granularity which is only discernible very close to focus at the highest magnification. The term 'phase contrast' should not be used to describe defocus contrast.

5.11 Electron Noise

A further source of granularity in very high magnification electron micrographs is the inhomogeneity of the electron beam itself. This is due to the random emission of electrons from the hot filament. It is independent of the specimen, and can be photographed in the absence of a specimen. The degree of granularity is of the same order as that due to phase contrast. It is very difficult to separate electron 'noise' from photographic 'noise', which is the basic granularity of the photographic emulsion. Longer exposure times tend to reduce granularity caused by these two effects by collecting more electrons and more grains, but bring disadvantages of their own, as will be discussed in Chapter 14.

5.12 Contrast in Biological Specimens

Biological specimens—cells and tissues—are composed mainly of the lighter atoms of low atomic number, such as carbon, hydrogen, oxygen and nitrogen. These atoms have a relatively low electron scattering power. To make matters worse, biological specimens generally have to be embedded in an organic material such as an epoxy resin in order that they may be sliced into sufficiently thin sections. Since the constituent atoms of the embedding medium are the same as the atoms of the biological specimen, there will be little difference in scattering power between specimen and embedding medium. The inherent electron contrast of the specimen will be very low. The relative scattering power of a biological specimen must therefore be increased. This is done, as will be discussed in Chapter

18, by impregnating either the tissue or the section or both at some stage of the preparation with salts of heavy metals such as osmium, tungsten, uranium or lead. Different chemical components of cells and tissues take up these 'electron stains' in varying degrees. The membrane systems of cells, for example, take up osmium preferentially. The cell membranes in a thin section of osmium-stained tissue will therefore scatter electrons preferentially, and will appear black on the final positive electron micrograph.

5.13 Specimen Thickness and Accelerating Potential

A number of conflicting factors govern the choice of specimen thickness and accelerating potential. The electron microscopist is always seeking both maximum contrast and highest resolution. These two requirements are unfortunately mutually incompatible. High contrast demands a thick section. A thick section will have a mass of fine detail distributed throughout its thickness. All this detail will be sharply imaged by the magnifying system, due to the great depth of field. It will therefore be superimposed as a jumbled, unresolvable mass in the final image. So, for high resolution it is necessary to work with a thin specimen. This will give low contrast, which can be overcome by the use of a lower velocity, more easily scattered electron beam; in other words, a lower accelerating potential. But a low potential beam has less theoretical resolving power owing to its longer wavelength; also the brightness of the final image will decrease considerably, making it more difficult to focus accurately.

All these effects must be borne in mind when choosing the optimum conditions for contrast and resolution. Experience has shown that an accelerating potential of 50–60 kV is most suitable for thin sections prepared by modern techniques. Such sections are generally about 500–600 Å thick. If the detail is contrasty and not too closely packed, a resolution of about 15–25 Å should be attainable in the final image. A thinner section will give lower contrast, but this can be increased if a lower accelerating potential of 20–40 kV is used; no significant loss in resolution is found in practice due to the use of longer wavelength electrons. A thicker section of 1,000 Å or so will give higher contrast, but resolution will inevitably be poorer due to increased specimen chromatic aberration even if a higher accelerating voltage (80–100 kV) is used in order to increase penetration and reduce the effects of chromatic aberration. The full resolving power of the modern high-resolution instrument can seldom if ever be used on biological sections in the present state of the art; for resolutions of 5 Å or less the special specimen preparation techniques described in Chapter 13 generally have to be used.

CHAPTER 6

The Modern Transmission Electron Microscope

6.1 Introduction

The electron microscope is basically a system of electron lenses with an electron source at one end, a viewing and recording system at the other, and means for mounting the specimen to be examined somewhere in between them. The simplest instruments have only two image-forming lenses; the most complex have two condenser and four image-forming lenses.

Two other auxiliary systems are needed to operate the column. These are the vacuum system, which will be discussed in Chapter 7, and the electrical power supplies, which will be discussed in Chapter 8.

6.2 Lens Construction

The lenses are the most important components of the column. They are generally electromagnetic in modern instruments, although permanent magnetic lenses, with or without supplementary electrical windings, are occasionally used. Electromagnetic lenses as we have seen in Chapter 4 are hollow cylindrical shrouds of soft iron containing a coil of copper wire to energize the magnetic gap between the polepieces. This gap forms the actual lens itself. The lenses are generally 6″ to 8″ in external diameter, and weigh 25 lb or more apiece.

A longitudinal section through the axial plane of a strong magnetic lens is shown diagrammatically in Fig. 6.1. The detailed method of construction varies from manufacturer to manufacturer. The lens is a hollow cylindrical soft iron box B which forms the shroud around the energizing coil C. The inner part of the box forms the lower polepiece S–S. The upper polepiece N–N is formed in the lid L of the box. The box and lid together

103

form a soft iron magnetic circuit which is complete except for the gap G between upper and lower polepieces. The lines of force flow from the iron into this gap, and form the lens itself. The lid is made removable so that the energizing coil C, which is wrapped with tape and impregnated with epoxy resin for maximum insulation, can be slipped into the space between box and lid.

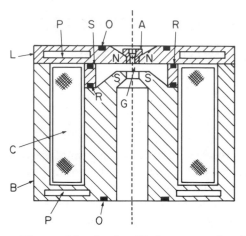

Figure 6.1 A simplified cross-sectional diagram of a typical design of strong magnetic lens with non-demountable pole-pieces such as is used for a first condenser or final projector

In the case of a strong lens, more heat is generated by the voltage drop across the energizing coil than can reasonably be dissipated at the lens surface. It is necessary to cool the lens by passing cold water through passages P which are machined into the shroud. It would be very inconvenient to have the energizing coil in the vacuum, for it would evolve gas, and the column would be almost impossible to evacuate sufficiently. A spacer S made of brass or other non-magnetic material and carrying rubber O-rings R is therefore placed between the box and the lid in order to confine the vacuum to the space between the polepieces. Further rubber O-rings O seal the top and bottom of the lens on to the mating components of the column, thus rendering the whole vacuum tight.

In order to reduce inherent lens astigmatism to the smallest possible amount, the lens must be as axially symmetrical as it is possible to make it. The shrouding boxes must be machined to the highest degree of accuracy from carefully chosen, magnetically homogeneous soft iron billets. The actual polepieces, especially in the objective lens, must be ground true to within a few microinches, and must then be matched in pairs. The position of minimum astigmatism is then found by rotating one polepiece with respect to the other. It is therefore very unwise to attempt to dismantle an objective lens other than simply to remove the polepieces (if they made separately from the shroud and lid), for its performance can so easily be ruined by incorrect reassembly.

6.3 Physical Lens Apertures

A physical aperture is placed either between the polepieces, or is inserted into one of them. This aperture can be fixed or movable. Fixed apertures (Fig. 5.1, A) inserted into polepieces are generally made of non-magnetic bronze or aluminium, and can be removed for cleaning. The exact centration of the aperture in some lenses, notably the second condenser and the objective, is very critical, since the aperture may be imaged close to the plane of focus of the objective lens, and would therefore obscure the field of view if it were off-centre. For the same reason, these two apertures are the most subject to the effects of contamination, and must be readily removable for cleaning. The objective and second condenser

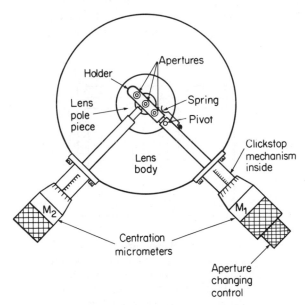

Figure 6.2   A plan view of a moveable lens aperture selector and centration mechanism

apertures are therefore mounted in a movable holder between the pole-pieces. They are usually discs of platinum or molybdenum, about 3 mm in diameter by 0·5 mm thick, and are drilled with holes ranging from 10 μm to 500 μm in diameter. Objective apertures range from 10–100 μm; condenser apertures from 100–500 μm. The apertures are slid into a pivoted holder which can be moved radially by means of the micrometer M_1 (Fig. 6.2). The right-hand micrometer is provided with a click-stop selector mechanism, so that any one of (generally) three apertures can be selected. The left-hand micrometer M_2 moves the holder at right angles to its long axis about a pivot. In this way, each of the three apertures provided can be selected and centred in turn. Apertures may be punched into a strip of very thin molybdenum foil. These have the advantage of being self-cleaning (see p. 162).

Lenses are frequently built in pairs of one strong and one weak lens for greater accuracy of alignment in manufacture. The pairs most frequently encountered are the two condenser lenses, the objective and intermediate, the diffraction and intermediate, and the intermediate and projector.

6.4 Double Lenses

It is possible to use a single excitation coil and shroud to form two lenses simply by providing a gap at either end of the iron shroud, as shown in Fig. 6.3. By suitably arranging the excitation and gap dimensions, a

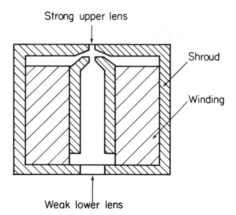

Strong upper lens

Shroud

Winding

Weak lower lens

Figure 6.3 A diagram of a section through an electromagnetic 'double lens' with strong upper and weak lower components. Such lenses are occasionally found as C_1–C_2 or objective–projector combinations

primary image can be formed between the two gaps by the upper lens. This can then be magnified or diminished further by the lower lens. In this way, two lenses can be made for the price of one. This double lens unit is applicable to a simple form of double condenser, or to an objective–projector combination for a simple form of basic microscope (see p. 194). The strengths of the two lenses cannot be changed independently unless some form of polepiece changing mechanism (a polepiece turret) is incorporated into one of the lens gaps.

6.5 The Column

The column can conveniently be divided for description into three main parts: (1) the illuminating system, consisting of the electron source plus the condenser lens or lenses; (2) the image-forming and specimen translating systems; and (3) the image viewing and recording systems. The general construction of the column will first be considered, followed by a more detailed discussion of the functions of each of the three main parts.

The column is generally mounted vertically (Figs. 6.4(a) and (b)) with the electron source at the top and the imaging system at the bottom, for reasons of mechanical rigidity. It must always be borne in mind that if an object is to be resolved to 2 Å, then the image must remain stationary to within less than ± 1 Å with respect to the photographic plate for the length of time needed to record it—1 to 5 seconds. This demands a very high degree of rigidity and freedom from vibration in the column. All high-resolution instruments have the column mounted vertically. Some older medium-resolution microscopes have the column mounted on its side with a transparent fluorescent forming the bottom of the tube, following the design of a cathode ray or television tube. Such a design requires elaborate mechanical support to obtain sufficient rigidity, whereas the free-standing vertical column requires the minimum of support, and has the added advantage that its own weight pressing down on the vacuum seals assists in keeping them leak-free.

The complexity of the column depends on the number of lenses provided, and the number of centration and tilt adjustments provided for each lens. Medium-performance TEMS have 4 or 5 lenses, high-performance TEMS have 6 (see p. 197). Simple 'basic' instruments can have as few as 2 lenses, dispensing with condenser lenses.

A simplified diagrammatic cross-section of a simple medium-performance 4-lens column is shown in Fig. 6.4(a). Above the specimen plane is the single-condenser illuminating system; below it the 3-lens imaging system. Then follow the viewing chamber with screen and window, followed by the

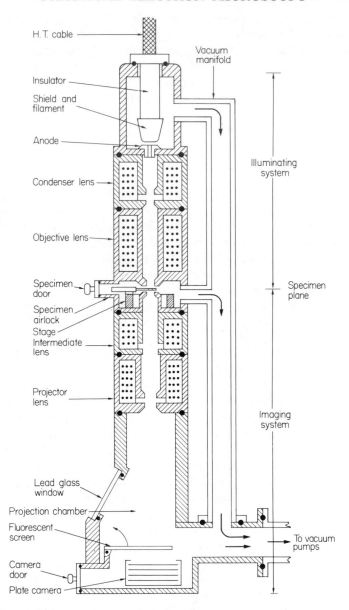

H.T. cable

Vacuum manifold

Insulator

Shield and filament

Anode

Condenser lens

Objective lens

Specimen door

Specimen airlock

Stage

Intermediate lens

Projector lens

Lead glass window

Projection chamber

Fluorescent screen

Camera door

Plate camera

Illuminating system

Specimen plane

Imaging system

To vacuum pumps

Figure 6.4(*a*) A simplified section through a simple single-condenser 4-lens medium performance electron microscope

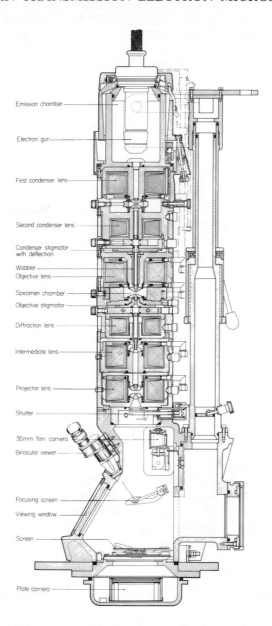

Emission chamber

Electron gun

First condenser lens

Second condenser lens

Condenser stigmator
with deflection

Wobbler

Objective lens

Specimen chamber

Objective stigmator

Diffraction lens

Intermediate lens

Projector lens

Shutter

35mm film camera

Binocular viewer

Focusing screen

Viewing window

Screen

Plate camera

Figure 6.4(*b*)   A simplified section through a complex double-condenser 6-lens high performance electron microscope, the Philips EM 200. (Courtesy Philips, Eindhoven)

camera mounted below the screen. The column is connected as directly as possible to an oil diffusion pump (not shown) by means of a vacuum manifold.

A section of a complex high-performance 6-lens column, that of the Philips EM 200, is shown in Fig. 6.4(*b*). It is identical in principle to the 4-lens column, but has a double-condenser illuminating system above the specimen and a 4-lens imaging system below it. Only 3 of the imaging lenses are normally used for transmission microscopy; the objective, diffraction and projector for low magnifications, and the objective, intermediate and projector for high magnifications. All four can be used together for the very highest magnifications. The column construction is further complicated in the later EM 300 and EM 301 by the inclusion of vacuum airlocks for the gun chamber, viewing chamber and cameras. It is constructed in units which can be separated by built-in lifting jacks, and each lens unit can be separately aligned from outside the vacuum while the instrument is working.

The image viewing and recording systems are similar in both simple and complex instruments, except that the simpler instrument has one camera and one screen, while the complex one has 3 independent cameras and two viewing screens. The screens are contained in a projection or viewing chamber, which is provided with large viewing windows so that the screen can be seen clearly by the operator and several other observers. The screen may be tilted upwards for more convenient viewing (this, of course, distorts the image by elongation). In some instruments, the screen may be tilted so that the image can be brought into the object plane of an optical magnifier for critical focusing. The latter is generally a long working distance binocular microscope (or short working distance telescope) with very large aperture objective lenses to gather as much light as possible. A separate focusing screen may be used.

The cameras may be mounted above or below the final viewing screen; depth of field maintains focus. The screen must be made to tilt out of the electron beam, so that the image can be projected on to a photographic plate or film lying below it. In simpler instruments, the tilting screen can also act as the photographic shutter. Because of the presence of photographic material in the projection chamber and the consequent 'outgassing' (see Chapter 7), it is necessary to connect it as directly as possible to the vacuum pumps through a wide bore (3″–4″ dia.) tube to ensure maximum pumping speed. It is also necessary to connect the gun chamber and the specimen chamber separately to the pumps via a vacuum manifold to ensure an adequate vacuum in these components.

The current through all the lenses can be varied by turning controls mounted conveniently for the operator to use for controlling illumination intensity, magnification and focus. The specimen can be moved approximately 2 mm in two mutually perpendicular directions in an accurately defined plane by two specimen translation controls, generally mounted on either side of the column. Since it is not usually possible to manufacture the column parts to a sufficient degree of accuracy before assembly, means are generally provided to align some more critical lenses both radially and in tilt about the vertical microscope axis. These controls may be either mechanical or electrical.

We will now consider the functions of the various parts of the electron microscope column in greater detail.

6.6 The Illuminating System

This consists of an electron 'gun', which generates and accelerates the electrons used to illuminate the specimen, together with a condenser system to collect and direct the illuminating electrons on to the specimen. Means must be provided to vary the number of electrons per unit area falling on the specimen, because this determines the exposure time required to record the image on a plate or film. Very simple basic electron microscopes dispense with a condenser system entirely; this is only satisfactory for very low magnifications, since sufficient brightness for work at high magnifications (above about × 10,000) is difficult to obtain. Image brightness is regulated in this case by altering the emission from the electron gun. The provision of a condenser lens system enables the energy available at the specimen to be increased by at least a factor of 10 for an equivalent electron gun energy output. Image brightness may then be controlled in two ways: by controlling the electron gun, or by altering the focus of the condenser lens or lenses, which also alters the area illuminated.

Because of the very small lens apertures used in electron microscopes, the intensity of illumination from an electron source must be many orders of magnitude greater than that obtainable from the light source of a light microscope. If it were not possible to make electron sources at least 10^5 times as bright as light sources, the electron microscope would hardly be a practical possibility.

6.6.1 The Cathode

The supply of electrons for image formation is derived initially from a metal electrode called the 'cathode'. All metals are characterized by the fact that they contain positive ions and free electrons, which, although free to move within the metal, are prevented from leaving the surface by the attractive forces of the positive ions. To enable electrons to leave the metal surface more readily, their kinetic energy is generally increased by increasing the temperature of the metal; this process is known as 'thermionic emission'. To separate an electron completely from the surface, a certain amount of work must be done against the attractive forces. This is called the 'work function', and is characteristic of each metal. The work function, w, is defined by the relationship:

$$n = A \cdot T^2 \cdot e^{-w/kT}$$

where n is the number of electrons emitted per unit area of metal surface; T is the absolute temperature; and A and k are constants. The work function is measured in volts, and varies from about 1·8 V for a mixture of barium and strontium oxides to 4·6 V for molybdenum. The formula shows that the emission of electrons is very highly temperature-dependent. The lower the value of the work function, the more copious is the supply of electrons at any given cathode temperature. Fig. 6.5 is a curve showing electron emission plotted against temperature for a tungsten cathode. At the normal operating temperature of 2,600 °K, a temperature increase of 100 °K or 4% gives rise to a doubling of electron emission. It is therefore necessary to take steps to stabilize the emission from the cathode or the image brightness would fluctuate enormously for small temperature variations. The way in which this is done will be discussed in Chapter 8.

The form usually taken by the cathode is a hairpin filament of thin (0·1 mm dia.) tungsten wire, the free ends being spot welded to stout legs

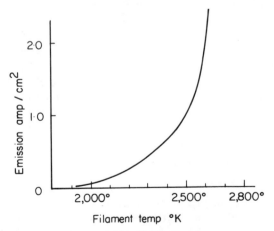

Figure 6.5   A curve showing the very rapid rise in electron emission from a hot tungsten filament as the temperature is raised

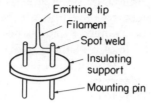

Figure 6.6   A typical hairpin filament assembly

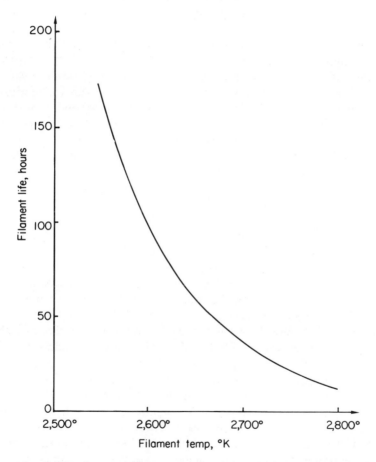

Figure 6.7   A curve showing the very rapid reduction in filament life consequent upon a relatively small increase in filament temperature. (Diagram redrawn from Haine, *The Electron Microscope*, Spon, London)

or pins held in a mica, ceramic or sintered glass support (Fig. 6.6). The whole assembly is arranged to plug into a socket in the electron gun insulator, so that it can be changed rapidly when the filament breaks or 'burns out'. The average life of a filament under normal operating conditions is 20–50 hours, but it is very readily burnt out by careless operation. Filament life depends very markedly on the temperature at which it is run. Increasing the filament temperature from 2,600 °K to 2,800 °K reduces the life by a factor of 10 (Fig. 6.7). The energy required to maintain the tungsten filament at its operating temperature of 2,600 °K is supplied by passing an electric current through it. The amount of power required is about 2 watts at 1·5 V, supplied as high or low frequency AC, or as DC.

Other materials with lower work functions have been used as cathodes, such as thoriated tungsten, lanthanum hexaboride and the mixture of barium and strontium oxides used in thermionic valves, but these cathode materials are very susceptible to damage by the damp air which must be let into the gun from time to time, and have under these conditions shorter lives than pure tungsten, even though they operate at a lower temperature.

For work involving the very highest resolutions in the TEM, and for the formation of very small spot sizes in the high-resolution STEM and SEM, the pointed filament (see p. 350) and the cold-cathode field emission (see p. 380) types of source are becoming increasingly used.

6.6.2 The Electron Gun

The heated filament emits electrons at random in all directions. The resulting electron cloud must be shaped or collimated into a beam, and must then be accelerated by falling through a potential drop of 20 to 100 kV. The high velocity electrons then pass into the system of condenser lenses, which focuses them on the specimen. The beam shaping and control of emission is effected by a second electrode called variously the 'Wehnelt cylinder', 'grid', 'shield' or 'bias shield', and the acceleration by a third electrode called the 'anode'. The function of the three electrodes—cathode, shield and anode—is analogous to the function of the three electrodes (cathode, grid and anode) of a triode valve (see Chapter 8). The tip of the filament hairpin is placed close to a circular aperture in the shield electrode (Fig. 6.8) which is maintained at a potential of between 100 to 500 V negative to the filament. This negative voltage serves to repel some emitted electrons back to the filament, and reduces the total emission and hence the brightness of the final image. The desk control which alters the shield potential is generally marked 'Emission' or 'Bias', and is one of the controls used to vary image brightness, the other being C_2 focus. For optimum gun brightness, the filament tip must be placed very accurately

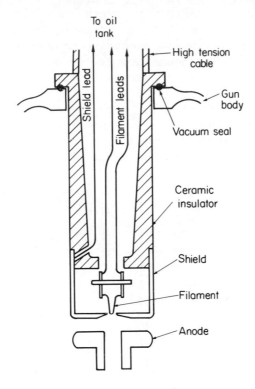

Figure 6.8 A diagram showing the main features of construction of a typical electron gun

at the centre of the shield aperture, and the distance between the filament tip and the inner surface of the shield must be set very critically (see p. 348).

The third electrode, the anode, is maintained at earth potential, since it is obviously impractical to have the whole microscope column at the high accelerating potential. The filament is therefore maintained at the full negative accelerating potential. The shield, which is negative with respect to the filament, is positioned close to the anode. A very strong electrostatic field therefore penetrates into the shield aperture. This is arranged to act as an electrostatic lens, focusing the true source (the space charge of electrons surrounding the filament tip) to form an image of the source below the anode, as shown in Fig. 6.9. All the electrons which emerge from the aperture in the shield pass through a plane of minimum cross-section where electron density per unit area is at its highest. This so-called 'gun crossover' is used as the actual source of electrons for the electron microscope, and forms the upper conjugate plane of the condenser-lens system. The shape of the crossover is not circular; it is elliptical. This is because the true source is a bent wire, and is a linear rather than a point source.

The design and performance of the electron gun is after the objective lens the most important single factor in the construction of a satisfactory electron microscope. On it depends the intensity of illumination of the image and hence its visibility. The stability of the image depends on the

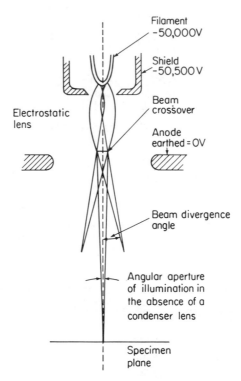

Figure 6.9   A diagram showing how the filament, shield and anode together form an electrostatic lens shaping the electrons emitted by the filament into a high-intensity 'crossover', which acts as the virtual source. (Diagram modified and redrawn from Siegel, *Modern Developments in Electron Microscopy*, A.P.)

stability of the gun, which must not be prone to electrical leakage or 'arc-over'. Gun design is still to a large extent empirical, and is constantly being improved.

6.6.3 Aperture of Illumination

In light microscopy, it is a fundamental principle that for optimum resolution and image contrast, the angular aperture α_i of the illuminating beam emerging from the condenser lens must be the same as the angular aperture of the objective lens α_0 (Fig. 6.10(a)). This is called 'Rayleigh's criterion'. If α_i exceeds α_0, the image will lose contrast due to glare; if α_i is less than α_0, the effective angular aperture of the objective lens will be reduced (Fig. 6.10(b)), which will lead to loss of resolution due to diffraction. Rayleigh's criterion does not apply to the electron microscope. As we have seen in Chapter 4, the smaller the angular aperture of the objective lens, the higher will be the resolving power, due to the reduction of spherical aberration. The condenser lens is used deliberately in the electron

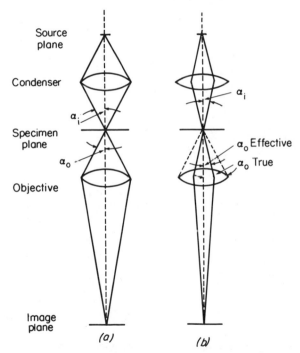

Figure 6.10 A diagram showing how the effective aperture of the objective lens is determined by the aperture of the condenser lens

microscope to reduce objective aperture, and at the same time to reduce the intensity at the specimen before a micrograph is taken.

The effect of varying the focus of the condenser lens on the angular aperture of illumination and hence on objective aperture is shown in Fig. 6.11. The physical sizes of the electron source (gun crossover) and the physical aperture in the condenser lens are fixed. When the source is focused on the specimen, a physical condenser aperture is selected which subtends an angle at the specimen rather greater than the angle subtended by the source. When the condenser is at focus (Fig. 6.11(c)), all the rays from the source will pass through the condenser aperture, and none will

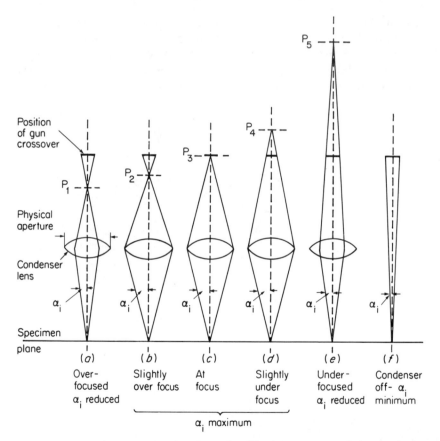

Figure 6.11 A diagram showing how the effective aperture of the condenser lens is dependent on the focal setting. Full aperture is only available over a short range on either side of focus; both above and below focus the illuminating aperture is reduced

strike it. At condenser focus, therefore, an elliptical image of the source itself (the gun crossover) will be projected through the imaging lenses on the screen. As the condenser lens is defocused to a plane below (over-focused; P_2 in Fig. 6.11(b)) or above the plane of the source (under-focused; P_4 in Fig. 6.11(d)), a focal point will be reached in each case where the angle subtended by the source at the specimen will be the same as the angle subtended by the physical aperture of the condenser at the specimen. Rays from the source will just strike the condenser physical aperture, and the diffuse elliptical spot will become bounded by a fairly sharp circle, which is the image of the condenser aperture. Between these two points of focus, the angular aperture of illumination at the specimen, α_i, is at a maximum. As the strength of the condenser is increased on either side of focus, the lens will image a plane P_1 below (Fig. 6.11(a)) or a plane P_2 above (Fig. 6.11(e)) the source. The angle subtended by the source at the specimen is then reduced and therefore the angular aperture of illumination, α_i, is reduced in each case. Defocusing the condenser

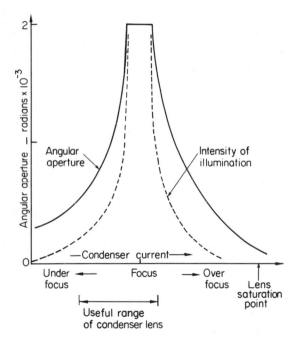

Figure 6.12   A diagram showing how both aperture and beam intensity fall off very rapidly on either side of condenser focus. (Redrawn from Cosslett, *Practical Electron Microscopy*, Butterworth)

above or below focus therefore reduces the angular aperture of illumination, and hence increases image resolution.

A curve showing the relationship between condenser lens current and angular aperture of illumination is shown in Fig. 6.12. The flat top corresponds to the conditions between Fig. 6.10(b) and (d) where angular aperture remains constant over a short range of lens strength. Angular aperture of illumination then drops off very rapidly on either side of the flat top. The limiting minimum values of angular aperture are obtained with the condenser lens switched off (Fig. 6.11(f)), or at lens saturation point (maximum lens current).

Intensity of illumination varies as the square of the angular aperture, since an illuminated area is involved, and therefore intensity falls off even more rapidly on either side of condenser lens focus, as shown by the broken curve in Fig. 6.11.

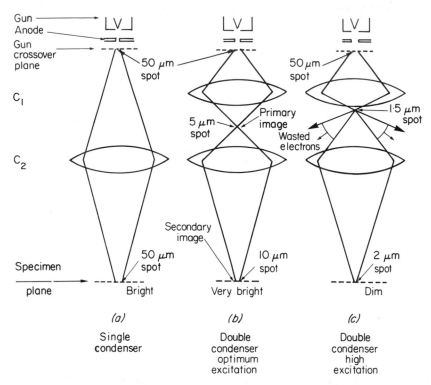

Figure 6.13   A diagram showing how the efficiency of the illuminating system and hence illuminating spot intensity change with the setting of C_1 lens

6.6.4 The Condenser Lens System

The purpose of the condenser lens system is to transfer the energy contained in the gun crossover to the specimen with the minimum of loss. It is perfectly possible, as we have just seen, to do this by using only one condenser lens. All early instruments had and most simpler modern ones have a single condenser. The main disadvantage of a single condenser is that the geometry of the system only allows a magnification of around unity, since the source–condenser distance is about the same as the specimen–condenser distance. The diameter of the gun crossover is dictated by the physical dimensions of the gun components, and for a reasonably bright source cannot be made much less than about 50 µm. This means that, with a single condenser, the minimum area which can be illuminated with the condenser at focus is also an approximate circle of 50 µm in diameter (Fig. 6.13(a)). If this is magnified $\times$ 1,000, it will cover a 5 cm diameter circle on the final screen. The area of a specimen illuminated will therefore be far larger than necessary when magnifications of $\times$ 10,000 or more are required. Unnecessary heating and contamination of the specimen will take place. The problem therefore is to reduce the size of the projected source image without losing intensity.

6.6.5 The Double Condenser

The most effective solution is to use a second lens between the source and the original condenser lens. The new condenser lens, which is designated 'C_1' because it is placed nearer the source, is used to project a diminished image of the gun crossover into the front conjugate plane of the original (now the second) condenser lens, which is now designated 'C_2'.

The action of the double condenser system is shown diagrammatically in Fig. 6.13. In (a), C_2 is mounted in its usual position, about midway between the source and the specimen. The gun crossover is a spot 50 µm in diameter. An image of this primary source is projected at unit magnification on to the specimen when C_2 is at focus. In (b), the new lens C_1 is placed about equidistant between the gun and C_2. Let us suppose that C_1 is now energized to give a diminution ratio between source and primary image of $1:10$. In order to focus this primary image on to the specimen, the strength of C_2 must be about doubled. It will therefore form at focus a secondary image magnified $\times$ 2 at the specimen. Notice, however, that because C_1 is closer to the source than C_2, it has a larger acceptance angle (aperture), and therefore collects more electrons from the source than C_2 alone. All these extra electrons are transferred to the specimen if the apertures of C_1 and C_2 are made equal, as is generally the case in practice. This means that the 10 µm final focused spot is far brighter than the 50 µm spot focused by C_2 alone. The efficiency of the illuminating system in mode (b) is therefore much greater than the efficiency in mode (a). This means that the brightness of the gun can be reduced with consequent increase in filament life. This increased illuminating efficiency is one advantage of the double condenser system over the single condenser.

The other, major advantage of the double condenser system is its ability to focus a spot on the specimen of precisely the diameter required at the magnification in use. The higher the magnification, the smaller the area under examination, and the smaller should the illuminating spot be to avoid unnecessary damage to other parts of the specimen. As the strength of C_1 is increased, so the primary diminution ratio is increased and the primary image becomes smaller. In diagram (c), C_1 strength has been increased to give a diminution ratio of 1:33. The primary image plane is now closer to C_1, so the strength of C_2 must be reduced to focus the final spot on the specimen. The final spot diameter is now about 2 µm, and the area illuminated between mode (b) and mode (c) has been reduced by a factor of 25.

This setting, used at very high magnifications, has one considerable disadvantage. The ray diagram (c) shows that a very large proportion of the electrons focused by C_1 cannot enter the aperture of C_2, and are therefore wasted. The efficiency of the illuminating system is therefore greatly reduced, and the gun brightness must be increased in order to compensate. This means a reduction in filament life; in practice this may be as little as one-tenth the life when used in mode (b).

6.6.6 Condenser Apertures

The physical aperture placed in the centre of C_2 is used to limit the angular aperture of illumination when the lens is defocused, as we have just seen. It is convenient to have several apertures of different diameters for different types of work. A holder or 'stick' of 3 apertures with centration mechanism is placed at the centre of C_2, the construction being similar to that adopted in the objective (Fig. 6.2). The aperture centration devices are generally made interchangeable. A typical set of aperture values would be 100, 200 and 500 µm. It is also necessary to place a physical aperture in C_1. This is generally fixed, although in some instruments a further centrable stick of 3 apertures is used. This refinement is not essential, since the size of the aperture in C_1 has little effect on the performance of the instrument.

The illuminating system as a whole is subjected to heavy electron bombardment, and must be constructed so that it can be dismantled readily for cleaning. The apertures, especially the C_1 aperture, act as very effective X-ray sources, which can be dangerous to the operator unless the whole illuminating system is effectively shielded. The whole of the upper part of an electron microscope is therefore surrounded by a lead or other heavy metal shield, and it is dangerous to run the instrument with the shield removed.

6.7 The Objective lens

6.7.1 Requirements

The objective lens is the most important single component in an electron microscope, just as in its light counterpart. Any defects in this lens will be further magnified by the rest of the optical system. In order to attain the highest resolving power, the focal length of the lens must be kept as short as possible. This requires considerable ingenuity on the part of the lens designer, because a number of auxiliary components must go into the objective lens. First and foremost, the specimen, mounted on a suitable carrier, must be held rigidly in a holder at the exact object plane. Next, it is necessary to provide two mutually perpendicular (orthogonal) lateral movements so that the specimen can be traversed from side to side while keeping it rigidly and exactly in the object plane. The device for traversing the specimen sideways is called the 'specimen stage'. It is then necessary to provide a contrast aperture in the back focal plane (see p. 97) to intercept the scattered electrons. The aperture must be held rigidly, but be capable of being centred from outside the column by means of two orthogonal lateral adjustments. For high resolution work, it is necessary to surround the specimen with a cooled surface, the 'anti-contaminator', to protect the specimen from the action of the residual gases in the column. It may also be necessary to provide a means for tilting the specimen up to $\pm 45°$ with respect to the column axis in all directions, while still keeping it in the object plane. Finally, it may be necessary, especially in metallurgical applications, to heat, cool, strain and rotate the specimen, either separately or together.

6.7.2 Specimen Supports

The primary support for the specimen is generally an extremely thin (200 Å or less) membrane of plastic or evaporated carbon. This in turn is supported on a copper disc punched with holes called a 'grid' (see p. 300). The preparation of support membranes is dealt with on p. 302. Grids are generally discs of electro-deposited copper 3·05 mm ($\frac{1}{8}$″) in diameter, about 50–100 μm thick. The central 2 mm of each disc is pierced with holes, either round or square, in a mesh of about 200 per linear inch. Some typical grid designs are shown in Fig. 13.1. The copper mesh supports the very thin support membrane, and also conducts heat rapidly away from it. This prevents thermal expansion and hence movement of the specimen under electron bombardment.

6.7.3 Specimen Holders

The grid must be clamped firmly at its edges to prevent distortion and to conduct heat away from the specimen. The grid must therefore be carried in a holder which can either drop down into the objective lens

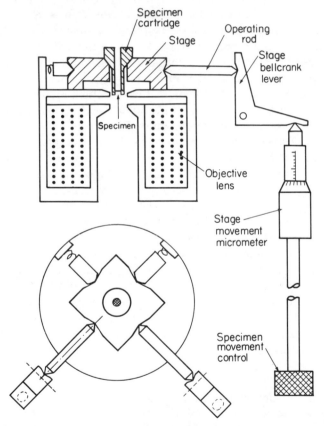

Figure 6.14   The top-specimen-entry, drop-in cartridge type of objective lens, stage and stage motion

from above (top insertion) or slide in between the polepieces sideways (side insertion). The design of a top insertion lens is shown in Fig. 6.14, and a side insertion lens in Fig. 6.15.

One disadvantage of a top insertion specimen holder is that only one specimen can be inserted into the lens at a time. A side insertion holder enables several (three to six) grids to be mounted side by side, so that the specimen can be changed without stopping operation and in a matter of seconds rather than minutes.

6.7.4 Specimen Stages

It is necessary to be able to move the specimen over the whole of its area (about ± 1 mm about the centre in any direction). The control of

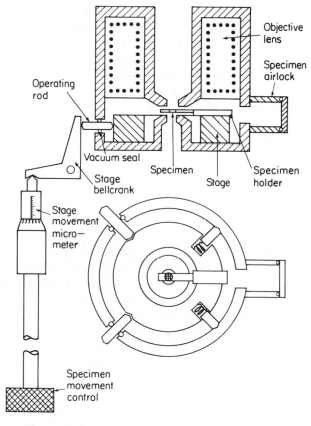

Figure 6.15 The side-specimen-entry, immersion type of objective lens, stage and stage motion

movement must be extremely precise; at very high magnifications the movements must be reproducible to within $\pm$ 100 Å (10 nm) or less. Lateral movement is by means of two controls which move the specimen stage. This is generally a massive block of non-magnetic material mounted in or above the objective lens, and which can be moved in two mutually perpendicular directions by two micrometer screws via bell-crank levers. The whole stage and its controls must remain stationary to within the resolving power of the instrument for the time needed to make a photographic exposure. The most logical position to place the stage is above the objective lens (Fig. 6.14). This position necessitates a top insertion specimen holder. The stage slides on the polished top surface of the lens. This design has the advantage of simplicity and rigidity, but it is difficult to achieve a focal length of less than 5 mm in this way. To reduce the focal length, the bore of the upper polepiece must be enlarged so that the specimen holder can drop down into the upper polepiece and still have the requisite 2–3 mm of lateral movement. This is the most commonly found pattern of objective lens construction in earlier types of instrument; a focal length of some 2 mm can be achieved.

A side insertion specimen holder gives the shortest possible focal length, as can be seen from Fig. 6.15. Values of 1 mm or less can be obtained. The stage is more difficult to design, and has to slide on a polished surface fixed below the lens polepieces. The energizing coil is now above the stage. The modern tendency is to use this type of lens construction.

6.7.5 Contrast Aperture Holders

The objective contrast aperture is a hole 20–50 μm in diameter drilled in a disc of platinum or molybdenum, which is usually about the same size as a grid. It is usual to have 3 apertures inside the lens, one in use plus two clean ones or two of different diameters. This entails the provision of a carrier and centration mechanism. The design is generally identical to and interchangeable with the aperture mechanism used in the condenser lens. Some instruments do not have individual apertures, but use a strip of molybdenum foil punched with a number of holes. This strip must be centrable from outside the column. The contrast aperture holder is placed after the specimen between the polepieces, close to the back focal plane of the lens.

6.7.6 Specimen Manipulators

These are devices for rotating the specimen through known angles (goniometer stages); tilting the specimen through known angles (tilt stages); and devices for heating the specimen electrically, cooling it with liquid nitrogen, stretching it and compressing it. The insertion of such devices into the already overloaded space between the objective polepieces generally forces the lens designer to widen the gap between the polepieces and hence to increase the focal length of the lens. Some instruments have specimen cartridges which are of almost incredible complexity and ingenuity in order to incorporate specimen manipulation facilities. Top entry stages have major advantages in this field, and are generally used when these special features are required. It is possible however to provide all these facilities in a side entry design.

6.7.7 Specimen Airlocks

The specimen carrier must be simply and rapidly exchangeable from outside the column. Only the minimum amount of air must be let into the column vacuum during specimen exchange, so that the operating vacuum can be restored within the minimum possible time. This requires the provision of a specimen airlock, otherwise the whole column would have to be brought up to atmospheric pressure and then pumped down again each time the specimen was changed.

Two types of specimen airlock are commonly used. One allows the entry into the column high vacuum of a very small amount (a few cc) of air at atmospheric pressure each time the specimen is changed. The other type seals off a larger volume enclosing the specimen chamber or part of it. This volume is then filled with air, the specimen in its carrier is exchanged, and the sealed-off volume is then pumped out independently of the main vacuum system. Each method has its own advantages and disadvantages, which will be discussed in the following chapter.

6.8 The Imaging System

This consists of the objective lens plus the remainder of the image-forming train of lenses. The total number of lenses is generally determined by the maximum electron optical magnification required, which is dependent solely on the ratio of the resolving power of the instrument to the resolving power of the eye. Since there is random 'noise' in very high magnification micrographs, it is usual to enlarge the finest detail up to about 1 mm. If the instrument can resolve 5 Å, the maximum overall magnification required is 2×10^6. It is unnecessary to provide such a high electron optical magnification, because the fine grain photographic emulsions used for recording the image will allow an optical magnification of at least $\times 10$ in an enlarger. The maximum magnification necessary on a high resolution instrument is about $\times 250,000$. The fluorescent screen, with a resolving power of about 0·1 mm, can then resolve 5 Å. To resolve

2 Å, a top magnification of × 500,000 is usually provided. For 10 Å, a top magnification of × 100,000 is ample.

6.8.1 High Magnification

This is generally provided by three lenses, although four are used in some designs. Two lenses (the objective and the projector) are of high power, while the central intermediate lens is a weak lens of variable power and is used to control overall magnification. The ray diagram of such a system is shown in Fig. 6.16. The values of magnification for each lens are generally around × 100 for the objective and projector, and 0 to × 20 for the intermediate. The current in the intermediate lens controls overall magnification. This can be indicated approximately by a meter marked in magnification units, a digital readout, or by the position of a stepwise

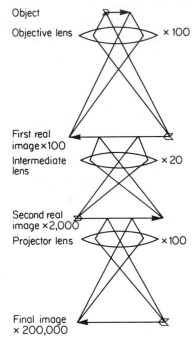

3 - real –image
High magnification system

Figure 6.16   A ray diagram of the 3-real-image medium and high magnification imaging system of a 3-lens microscope

'magnification control' on the operator's desk. As the power of the intermediate lens is changed, the power of the objective lens must also be changed in order to keep the plane of the first real image formed by the objective coincident with the object plane of the intermediate lens. This operation is called 'focusing', and must be performed by the operator each time magnification is changed.

This 3-real-image, 3-stage system works admirably at high magnifications, where all three lenses are working at small aperture. As the magnification of the intermediate lens approaches unity, its aperture increases and aberrations begin to appear in the final image. Distortion arises, and if a thick specimen is being used, chromatic change of magnification may give rise to an unacceptable out-of-focus effect at the periphery of the image (Fig. 4.11). We have the curious situation that it is easier to produce a high quality, high magnification image than it is to produce a distortion-free low magnification one.

6.8.2 Low Magnification

There are three commonly used solutions to the problem of obtaining aberration-free very low-power images. The simplest is to cut out one lens and reduce the power of the other two. Very low power images, used for scanning large areas of specimen when searching for suitable areas to examine at higher magnification, can easily be obtained by using the intermediate lens as a very long-focus objective. The ray diagram is shown in Fig. 6.17. A range of magnifications between $\times$ 50 and $\times$ 100 is obtainable, but resolving power is poor, due to the long focal length of the intermediate lens. Resolving power however is considerably better than that of the light microscope.

The second solution is to reduce the power of the projector lens. Three real images are still formed, and the ray diagram is basically the same as that shown in Fig. 6.16. This solution is not as simple as would appear at first sight, because the diameter of the final projected image is limited by the diameter of the bore of the projector polepiece. In order to have a powerful lens for high magnification, this bore must be small. If the power of the projector lens is reduced by reducing the current through it, a smaller and smaller circle is projected on the screen. Although final magnification is reduced, the area of specimen visible is not increased, and the reduced magnification is valueless. If this solution is to be used, the projector polepiece bore must be increased. In practice, this necessitates the provision of two or more interchangeable polepieces carried on a turret which can be operated without breaking the vacuum. Each pair of bores gives a different magnification range. Changing from one range to another necessitates an interruption of use, even though this may only be a few seconds. It is more convenient for the operator to have a continuous range of magnifications at the turn of a single knob.

The third solution to the low-power problem has been devised to give distortion-free single knob magnification control over the entire range

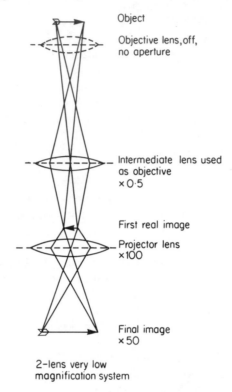

Object

Objective lens, off, no aperture

Intermediate lens used as objective ×0·5

First real image

Projector lens ×100

Final image ×50

2-lens very low magnification system

Figure 6.17   A ray diagram of the 2-real-image very low or 'scan' magnification range of a 3-lens microscope

from about × 500 to × 200,000. The basic idea is to weaken both objective and intermediate lenses so that the intermediate lens can be used as a diminishing and not as a magnifying lens. The ray diagram for this 2-real-image mode of operation is shown in Fig. 6.18. The power of the projector lens can be kept high and need not be reduced, although the projector pole-piece bores can be chosen so that the low-power range can be extended by reducing the projector current. This design has the advantage that the distortions arising in the intermediate and the projector lenses can be arranged to be in opposite senses, and thus to cancel out. Barrel distortion in the intermediate lens is compensated by pincushion distortion in the projector. Exact compensation can only be achieved at one particular magnification. The conditions for compensating distortion are not the

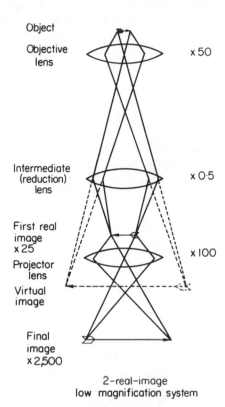

2-real-image
low magnification system

Figure 6.18   A ray diagram of the 2-real-image low magnification system of a 3-lens microscope

same as those for compensating chromatic change in magnification. A further slight disadvantage of the system is that the image suffers a reversal at the change-over point between the low magnification range (2 real images) and the high magnification range (3 real images). This sudden change to a mirror image is disconcerting at first, because the stage controls appear to operate in opposite directions. However, the 'rear-wheel-steering' effect becomes unnoticed by the operator surprisingly quickly.

A refinement of this system is to use a second intermediate lens placed closer to the objective lens, and to keep the power of the projector fixed over the entire magnification range. In this way, optimum conditions for the cancellation of aberrations are achieved. This system is necessarily more complicated and expensive than the three-lens method, since an additional lens must be provided. The extra low-power lens is placed in the optimum position for obtaining diffraction patterns, and is called the 'diffraction lens'. This system has a further advantage when obtaining diffraction patterns, because it enables the magnification of the pattern to be varied over a wide range (see p. 259). This arrangement also has the advantage that all four lenses may be used in series to give a 4-real-image magnification system, the top magnification of which can be up to $\times 10^6$.

6.8.3 *The Image Viewing and Recording Systems*

A real image of the specimen visible to the eye can be formed at any point below the projector lens on a screen covered with a material which fluoresces when bombarded with electrons. This final image will be in focus no matter where the screen is placed, due to the almost infinite depth of focus of the small-aperture imaging system (see p. 90). The material used to coat the screen usually fluoresces green, this being the colour to which the dark-adapted eye is most sensitive. This colour makes it easier to focus the image at the low brightnesses encountered at high magnifications. The grain of the screen material must be particularly fine, 100 μm or smaller, since it is generally viewed through a magnifier. The graininess of the screen, however, is always considerably greater than the graininess of a photographic emulsion, which is therefore capable of much higher resolution. This is why it is necessary to photograph the final image directly in order to obtain the maximum amount of information from it.

It is convenient to be able to prop the final screen at an angle, so that a magnifier may be used to examine the image. This obviously gives rise to a distorted main picture. This disadvantage can be overcome by keeping the main screen flat, and bringing a small auxiliary focusing screen into the beam, the plane of this screen being at the focus of the viewing telescope.

The viewing magnifier can be a simple lens, or a monocular or binocular telescope. The objective lenses of the telescope (which is more correctly a long working distance stereoscopic microscope) must be of very large aperture in order to avoid light loss. For minimum light loss, the erecting lenses or prisms are sometimes omitted; this is somewhat inconvenient, as the image moves in the opposite direction when seen through the telescope. This again is compensated for surprisingly quickly by the operator. The magnifying power of the telescope is generally between × 5 and × 10. The main advantage of the magnifier is that little image brightness is lost, whereas if the extra magnification were obtained directly by increasing it electron optically, image brightness would fall off drastically. This is a simple consequence of the increased optical aperture of the telescope over that of the eye.

The image is recorded on a fine-grain blue-sensitive bromide emulsion, which may be coated on glass plates, roll film or sheet film. Each material has its own advantages and disadvantages. Photographic technique will be dealt with in Chapter 14. Plates and cut sheet film are used more than roll film because of their greater dimensional stability and ease of processing. Plates or film are loaded into cassettes, and are placed in a camera, which can be placed either above or below the viewing screen. If 35 mm film is used, the camera is generally placed close to the projector lens, so that the full size of the image may be recorded. Provision must then be made for moving the camera in and out of the beam. Plates and cut film are generally carried in a camera placed below the viewing screen. This position has the advantage that the screen shields the camera from stray radiation.

The exposure time for a plate of correct density varies from 0·1 to 10 seconds, depending on the magnification and gun brightness. For exposures greater than 3 seconds, the tilting fluorescent screen can be used as a shutter. This simple method was used on all early instruments. It has the disadvantage that the plate is unevenly exposed unless the exposure is made unduly long, when there is the possibility of image blurring due to specimen and circuit drift and vibration. It is best to keep exposure times as short as possible; some form of mechanical blade-type shutter is therefore essential. This is generally mounted below the projector lens, and is operated electromagnetically. The designer must take great care that the magnetic field for the shutter operation does not interfere with the electron lenses.

It is an advantage to have an airlock valve between the camera chamber and the viewing chamber, for it is then unnecessary to admit air to the whole column in order to change the plates in the camera. The valve can be closed and operation continued, although micrographs cannot then be taken. Most instruments now have this provision.

6.9 Alignment and Astigmatism Correction

6.9.1 Alignment

The electron beam, as it passes down the column, is a very fine pencil of electrons which does not exceed a few micrometres in width as it passes through each of the lenses. The polepiece bores of the powerful lenses—C_1, objective and projector—are only a millimetre or two in diameter. It is therefore necessary to have each lens aligned radially about the axis of the electron beam to an accuracy of a few micrometres; a few tenths of a micrometre in the case of the objective lens. It is also necessary for the lenses to be aligned axially so that the axis of each lens is not tilted away from the axis of the electron beam. Ideally, therefore, each lens should be provided with very accurate and reproducible means for moving it radially in any direction, and also for tilting it about the axis of the beam. This would necessitate a total of 12 separate pairs of adjustments on a 5-lens instrument; 2 for the gun and 10 for the lenses. In practice, it has been found that some adjustments are far more critical than others. Also, the accuracy with which the alignments need to be made depends to a large degree on the ultimate performance of the instrument. A medium-performance microscope is more tolerant of slight misalignment than a high-performance one, if both are being operated at their limits.

The most important alignment is to ensure that the axis of the illuminating beam coincides with the axis of the imaging system. This can be accomplished either by tilting the illuminating system to the objective or vice-versa. The illuminating system can be tilted either by tilting the whole gun-lens system mechanically, or else by tilting the beam emergent from it either by electrostatic or magnetic fields which can be suitably adjusted by the operator. The objective lens must be tilted mechanically, generally by moving one polepiece radially with respect to the other. The tendency in modern designs is to reduce the number of mechanical adjustments to the minimum, since these generally entail moving vacuum seals, and to prealign the system in the factory as far as possible. Beam-tilting electrodes or coils are then provided for the final adjustment to be made by the operator. This makes the column more trouble-free, and the tilting adjustments, made simply by rotating two electrical controls, are much more easily learnt and performed correctly by the operator. For aligning the imaging system, either the projector lens can be fixed and the intermediate lens made radially centrable, or vice versa. The former system is more common except on instruments having a turret of projector pole-pieces, in which case the turret is made mechanically centrable.

A multiplicity of centration controls is of considerable advantage to a highly skilled operator, but they generally ensure a permanently misaligned column in the hands of operators who are not fully conversant with their use. The problems of alignment are discussed at greater length in Chapter 11, where a scheme for alignment which is universally applicable to all columns is described.

6.9.2 Astigmatism

Astigmatism, as we have seen on p. 86, is caused by radial asymmetry in a lens, giving it a focal length in one axial plane which differs from the focal length in a perpendicular axial plane. However carefully the lens is assembled during manufacture, a small degree of basic astigmatism is bound to remain, and some means must be provided to cancel it out. Astigmatism also arises in operation due to particles of dirt and other forms of so-called 'contamination' (see p. 159) which form on the apertures and lenses; this additional astigmatism must also be correctable during operation.

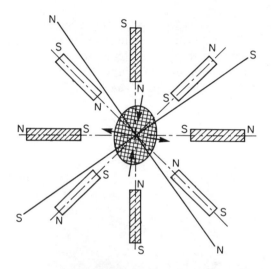

Figure 6.19 The principle of the octupole magnetic stigmator. The 8 separate electromagnets N–S are wired in 2 opposing sets of 4. When energized equally, they form a symmetrical magnetic field around the lens. The astigmatism correction is then zero. When energized unequally, they form an elliptical field shown cross-hatched, the strength and axial direction of which can be controlled by the magnitude and direction of the current passing through the 2 sets of magnets. This elliptical field can be made equal and opposite to the astigmatism of the lens, thus cancelling it out. (Diagram courtesy Philips Eindhoven)

The devices which correct astigmatism are called 'stigmators'. As has been explained in Chapter 4, a stigmator is essentially a weak cylindrical lens of variable azimuth (direction) and amplitude (strength) mounted so that an equal and opposite asymmetry can be applied to the electron beam emerging from an astigmatic lens. Stigmators are invariably provided beneath the objective lens, since the ultimate resolution of the instrument depends on this component. A stigmator is generally provided below the illuminating system, so that the illuminating electrons can be formed into a circular spot of maximum brightness and not be dissipated in an ellipse. A stigmator is occasionally found below the intermediate lens so that the diffraction spot can be made truly circular for high-resolution diffraction work. The projector lens can be manufactured sufficiently accurately not to need a stigmator.

Stigmators are of three types: electromagnetic, electrostatic and mechanical magnetic. The last type is no longer found on modern instruments. The principle of an electromagnetic stigmator is shown in Fig. 6.19. Eight small electromagnets are mounted radially about the electron beam, which is shown as a cross-hatched ellipse. The windings of the electromagnets

Figure 6.20   The electromagnetic octupole stigmator of the Philips EM 300 objective lens. The polepieces of the eight small electromagnets can be seen in the centre. The whole assembly together with its associated wiring is cast into a solid epoxy resin block for maximum mechanical stability and electrical insulation. (Photo courtesy Philips Eindhoven)

are wired so that when they are energized, pairs of north and pairs of south poles are facing one another. The electron beam is attracted by one pair and repelled by the opposing pair. Four of the magnets cause the beam to become elliptical in one direction, and the other four make it elliptical in a direction perpendicular to the first four. It can be seen how by varying the strength and the polarity of the pairs of electromagnets, a compensating field can be formed in any direction or azimuth, the strength or amplitude of which is dependent on the currents in the electromagnets. The currents in the two sets of four electromagnets are controlled by two potentiometer knobs, by operating which the user can virtually eliminate first-order astigmatism, and thus ensure that the full resolving power of

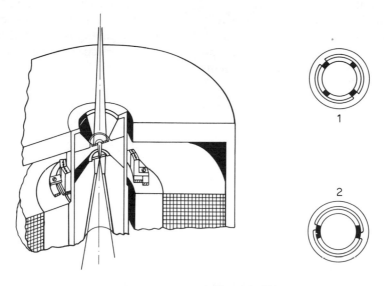

1 Zero position
2 Position of maximum intensity

Figure 6.21   The principle of the quadrupole mechanical magnetic stigmator. 4 soft iron slugs are mounted on a circular carriage which can be rotated about the lens. Two of the slugs can be rotated through 90° with respect to the other two. When the slugs are equidistant (1), there is zero correcting field. When the slugs are paired together (2), the elliptical correcting field is at a maximum. The distance between the pairs of slugs (strength) and the axial direction (azimuth) can be adjusted independently by means of two control knobs mounted on the outside of the column. The elliptical correcting field can thus be made equal and opposite to the astigmatic field. (Diagram courtesy Siemens Berlin)

the microscope is available. Stigmators of this type are called 'octupole electromagnetic' stigmators. Octupole electrostatic stigmators operate on the same principle, although in this case an electrostatic field is created with eight metal pins placed beneath the lens. A similar effect can be obtained with four electromagnets or pins, although the accuracy of compensation is not so great. This type is called a 'quadrupole' stigmator.

Mechanical magnetic stigmators consist of two pairs of soft iron slugs (Fig. 6.21) which can be moved together around the lens by means of one control (azimuth); also one pair can be moved relative to the other pair by means of another control (amplitude). When the two pairs are touching one another, the amplitude is at a maximum; when they are 90° apart the amplitude is zero. Having found the correct strength to apply, the azimuth control is then used to rotate the whole unit around the lens, until the correct direction is found. This type of stigmator is more difficult to use and more expensive to construct than the electrical types and so is only found on older instruments.

CHAPTER 7

The Vacuum System

7.1 Introduction

The vacuum system of an electron microscope is designed to remove air and other gases from the column. As much gas as possible must be removed, for the following reasons. Firstly, gas molecules interact with high velocity electrons and scatter them randomly, thus giving rise to 'glare' or reduced contrast in the image. Secondly, the presence of gas in the gun chamber gives rise to ionization and random electrical discharges, causing instability or 'flicker' in the electron beam. Thirdly, residual gases combine with the white-hot filament, eroding it away. Fourthly, residual gases condense on the specimen and contaminate it.

7.2 Units and Terminology

Vacuum terminology in common use is very confusing to the beginner. Vacuum can be 'high', 'roughing', 'fine' and many other imprecise descriptions. Vacuum is, of course, the inverse of pressure. A 'high vacuum' is therefore a 'low pressure'. It would be more logical always to speak of degrees of pressure and drop completely the term 'vacuum'. Unfortunately, the term is now so hallowed by usage that this is not possible.

Vacuum is measured simply in terms of pressure. The unit is generally the millimetre of mercury, nowadays called the 'torr', to commemorate Evangelista Torricelli, the seventeenth-century Italian discoverer of the mercury barometer. 'Normal' pressure is one standard atmosphere at sea level, which will support a column of mercury 760 mm high. 1 mm of mercury (1 torr) is therefore $1 \cdot 32 \times 10^{-3}$ atmospheres, or roughly one-thousandth of an atmosphere pressure. A 'high' vacuum is about 10^{-3} to 10^{-6} torr. A 'low' or 'backing' or 'rough' vacuum is between about 10 torr and 10^{-2} torr. A 'very high' vacuum is between 10^{-6} and

10^{-8} torr; an 'ultra-high' vacuum is a pressure of less than 10^{-8} torr. The highly confusing term 'micron' (μ) is frequently used for 10^{-3} torr. Since to microscopists the micron is a unit of length, it will not be used in this book. It is of interest to note that one cubic centimetre of air at a pressure of 10^{-6} torr—the best vacuum that can be achieved in a conventional electron microscope column—contains 3×10^{10} molecules. At this pressure, the average distance a molecule will travel through this cloud of its fellows before hitting another molecule (the 'mean free path') is no less than 50 metres at room temperature. A 'perfect vacuum' is apparently impossible; even outer space contains several molecules per cubic metre.

7.3 Vacuum Requirements

High-velocity electrons such as are used in electron microscopy interact with gas molecules just as gas molecules interact with one another. The presence of gas in the column causes them to be scattered away from a straight path.

The degree of scattering is proportional to the 'mass-thickness' (see p. 96), which is the product of the density of the substance through which the beam passes and the path length traversed. For a given degree of scatter, as density is reduced, so path length is increased. The primary purpose of the column high vacuum is to reduce the mass-thickness of the residual air in the column so that electron beam scatter by residual gas molecules is reduced to negligible proportions.

The statistical mean distance an electron can travel before it is deflected by a gas molecule is called the 'mean free path' of the electron. Since the electron path length in an electron microscope column is about 1 metre, the highest pressure that can be tolerated is that for which the mean free path of an electron is also 1 metre. This pressure can be calculated from Avogadro's number and the gas laws, and is about 10^{-3} torr. However, other factors have to be taken into consideration when working at high voltages, notably high tension stability and freedom from specimen contamination. The vacuum requirement for an electron microscope can in fact be summed up by saying that the better the vacuum, the better the results. The specimen chamber vacuum is generally 10^{-4}–10^{-5} torr, which is 10^{-7}–10^{-8} of atmospheric pressure.

7.4 High Vacuum Production

The ideal way of operating an electron microscope would be to pump out all the air from the column, seal it off, and switch off the pumps.

Unfortunately, this is not possible. Gases are continuously finding their way into the system through minute leaks and by the evolution of gas and vapour from the surfaces inside the column, especially from the photographic materials which have to be placed in it. These gases, mainly air and water vapour, must be removed continuously.

In the present state of vacuum technology it is not possible to create the necessary high vacuum in one stage. Two stages have to be used in series, each stage being produced by a quite different type of pump. The first stage is the creation of a low vacuum (about 10^{-1} torr) from atmospheric pressure, involving a pressure reduction of about 10^4. The second stage is the creation of the high vacuum (about 10^{-5} torr) from the low vacuum, involving a further pressure reduction of 10^4. The first stage is produced by a basically simple, though somewhat sophisticated, rotating mechanical pump; the second stage is produced by diffusion pumps, which use the kinetic energy of fast-moving, heavy molecules of oil or mercury to drive the relatively light molecules of air and water vapour in one direction, thus removing them from one part of the vacuum system to another. These high-speed molecules act in the same way as the piston in a piston pump. The principle and construction of each of these pumps will be considered in turn. The high-vacuum (diffusion) pump, being simpler, will be considered first.

7.5 The Diffusion Pump

This elegant and ingenious apparatus (Fig. 7.1) was invented by Gaede in 1915, and developed during the early twenties by Irving Langmuir. It contains no mechanical moving parts. The gas in the electron microscope column mounted above the pump is entrapped by a fast-moving stream of heavy oil or mercury molecules moving in a downward direction. These high-speed vapour molecules are generated in a boiler at the base of the pump and rise upwards through a central tube. As they emerge from the top of the tube, they are deflected through almost 180° by the 'umbrella', and enter the vacuum space with a high velocity in a direction away from the column. Any gas molecule which diffuses down from the column will be struck by an oil molecule, and will receive an impetus away from the column. Gas is thus free to diffuse from the column into the pump, but is prevented from diffusing back again. A pressure difference is therefore established across the pump umbrella, which gives rise to the pumping action. The actual pressure difference depends on a number of factors, mainly on the clearance between the umbrella and the pump casing. The larger this annular area is made, the more rapidly will the pump be able to

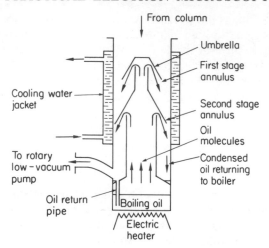

Figure 7.1(*a*)   A cross-section of a typical two-stage oil diffusion high vacuum pump

Figure 7.1(*b*)   A three-stage oil diffusion pump as used on an electron microscope. (Courtesy GEC/AEI Ltd.)

reach its ultimate vacuum, and the higher will the ultimate vacuum be. But, if the gas pressure in the space below the umbrella rises above a critical value, then the gas which has already been removed will diffuse back into the column across the barrier of moving oil molecules, and the pump will cease to operate. It is therefore necessary to keep the space below the annulus continually pumped. This is done by having a second-stage umbrella mounted beneath the first. The annular space is made smaller, so that the concentration of oil vapour molecules is increased, and this stage of the pump is able to transfer gas from the intermediate pressure region to the space below. This process may be repeated a number of times, the annular space being reduced in area each time. The limit is reached when the distance between the last umbrella and the casing has reached the reasonable limit of machining tolerance.

When the oil vapour has crossed the annular space of each stage, it strikes the pump casing, which is maintained at as low a temperature as possible by circulating cooling water either through copper piping wrapped around the outside of the casing, or through a jacket brazed to the pump. It then condenses to a liquid and returns by gravity to the boiler, where it is vaporized and recycled. The gas which accumulates below the final stage must be continuously removed; this is done by the mechanical pump of the two-stage system.

In a well-designed diffusion pump using a mineral oil as the pumping fluid, each stage can deal with a pressure difference of about a factor of 10 across it. Four stages will therefore give a total pressure difference of 10^4, or a high vacuum of 10^{-5} torr if the low vacuum or 'backing vacuum' into which the diffusion pump works is 10^{-1} torr. The ultimate vacuum of the pump is determined by the vapour pressure of the pumping fluid at the temperature of the walls of the condenser. With the oils in common use, this is about 10^{-6} torr, and is sometimes achieved in a really clean, leak-free column from which all photographic material has been removed. It is possible to improve the ultimate vacuum by placing a secondary con-denser or 'vapour trap', which is simply a surface cooled by liquid nitrogen, between the diffusion pump and the column. This type of cooled baffle reduces pumping speed by its choking action, and is seldom justified in normal transmission biological electron microscopy.

7.6 The Mechanical Pump

This is generally an oil-immersed, eccentric-vane rotary pump (Fig. 7.2). A cylindrical rotor, bearing along its length a pair of spring-loaded vanes, is mounted inside a cylindrical casing of somewhat larger diameter. The rotor is eccentrically mounted so that it just touches the casing at one point, giving rise to a line contact. The spring-loaded vanes also bear on the casing in a pair of line contacts. The space between the rotor and the casing is thus divided into three compartments by the two line contacts between the vanes and the casing, and the one line contact between the rotor and the casing. These three spaces are shown 1,2 and 3 in Fig. 7.2(*a*).

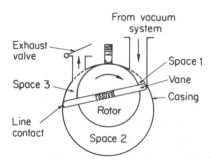

Figure 7.2(*a*)   A diagram of a cross-section of a rotary vane low vacuum pump showing the operating principle

Figure 7.2(*b*)   A single-stage rotary pump as used on an electron microscope. (Courtesy GEC/AEI Ltd.)

Consider now what happens when the rotor revolves. Space 1 becomes larger; therefore gas will be drawn in through the inlet pipe. Space 2 remains approximately the same size, but space 3 becomes smaller. Any gas entrapped in space 3 will therefore become compressed, and will be forced out of the exhaust pipe, lifting the exhaust valve as it goes. However, as soon as the tip of the vane passes the orifice of the exhaust pipe, space 2 becomes space 3, and the gas entrapped in space 2 at the beginning of the cycle is now in its turn forced out of the exhaust pipe. But, at the same time, as soon as the tip of the uppermost rotor clears the orifice of the inlet pipe, the original space 3 becomes space 1, and begins to refill with gas. What was originally space 1 has now become space 2, awaiting its turn to become space 3 and be exhausted to the atmosphere. There is thus a continuous cycle of suction, idle and exhaust as the rotor is turned by the electric motor.

A rotary pump can be used either as a pressure pump or as a vacuum pump. The ultimate vacuum which it can achieve on the suction side is dependent on the rate at which gas can leak across the vane seals and the rotor-to-casing seal. The gas tightness is greatly improved if the whole apparatus is immersed in oil; in this case, if the seals are perfect, the theoretical ultimate vacuum is the vapour pressure of the pump oil. These pumps are therefore very susceptible to mechanical wear and to contamination of the oil with substances of high vapour pressure. They are also susceptible to overheating, since this increases mechanical clearances by expansion. The ultimate vacuum to be expected from a single-stage rotary valve pump in good condition and filled with clean oil is between 10^{-2} and 10^{-1} torr.

A lower ultimate vacuum can be obtained if two rotary pumps are connected in series, so that the exhaust of the primary pump is maintained at a pressure of 1 torr or less by the action of the secondary pump, which exhausts into the atmosphere. Such two-stage pumps have the two units mounted on the same spindle and contained in the same oil-filled casing. The use of two-stage pumps is not generally justifiable in electron microscope vacuum systems.

An oil mist trap should be provided at the exhaust of a rotary pump to arrest the oil spray which is produced while the pump is actually pumping air from the system. It may be possible to arrange for the pump to exhaust into a large diameter polythene pipe which is led out of the microscope room into the open air. This is the most satisfactory way of preventing the accumulation of pump oil on surfaces adjacent to the pump, which has the undesirable effect of attracting dust. Oil vapour is also said to be carcinogenic.

7.7 Condensable Vapours

It is a characteristic of the mechanical type of pump that it is unable to pump the vapours of readily condensable substances, e.g. water and organic solvents. When such vapours are compressed in space 3, they tend to liquefy, and the liquid is dragged by the vanes back into the suction space 1. If condensable vapours are present in the vacuum system to any significant extent, they must either be absorbed or condensed before reaching the mechanical pump; or they must be swept out of the pump by a method known as 'gas ballasting'. In the latter case, a port open to atmosphere via a valve is arranged in some designs of pump to open into the exhaust space. Since this space remains at a negative pressure until just before the exhaust valve opens, air will then be drawn into the exhaust space and will thus 'wash out' contaminating vapours. However, this method increases greatly the pressure on the exhaust side of the pump, and thus increases the leakage rate across the rotor, which impairs the vacuum at the suction side of the pump. Gas ballasting increases the pressure at the vacuum side of the pump from 10^{-2} torr to 1 torr; this clearly cannot be tolerated. The gas ballast valve should only be opened when the high vacuum valve is shut, and should be used as an occasional method for getting rid of any contamination of the rotary pump oil with water or organic solvents, accidental or otherwise.

Although organic solvents can be kept out of the column by the exercise of reasonable care, water vapour always enters from the laboratory atmosphere during specimen and plate changes, however carefully the operator endeavours to prevent this from happening. Water vapour can be removed from the system by an absorption trap containing phosphorus pentoxide (P_2O_5) placed in the column or between the diffusion pump and the mechanical pump.

7.8 The Low Vacuum Reservoir

The mechanical pump always gives rise to a certain amount of vibration while in operation. If this vibration is transmitted to the column, micrographs taken while the mechanical pump is operating will be blurred; the degradation of the image is especially noticeable at high magnifications. The mechanical pump can either be mounted well away from the column console, or else arrangements must be made for stopping the pump while a micrograph is being exposed. The oil diffusion pump can be designed to exhaust into a pressure at least ten times as high as the ultimate vacuum of the mechanical pump, so if a large volume is interposed between the two pumps, the diffusion pump can discharge into this while the mechanical pump is turned off. This reservoir of low vacuum, often termed the 'backing tank', is generally a stout steel cylinder of 10 l or more capacity. If the column is free from leaks and excessive outgassing, the mechanical pump can remain off for 20 minutes or more. Some means of monitoring the pressure in the reservoir must be provided, either by means of a gauge or an alarm bell, so that the operator can turn on the mechanical pump before the reservoir pressure becomes so high that 'backstreaming' occurs through the diffusion pump. The switching on and off of the mechanical pump is better done automatically by means of a pressure-sensitive servo system attached to the low vacuum reservoir. This has the disadvantage that if the mechanical pump is turned on automatically during an exposure, the current surge taken by the pump motor on starting upsets the lens and high tension stabilizers sufficiently to blur the image. In practice, this need not happen, since the operator can generally observe the pressure in the reservoir with a gauge, and can switch the mechanical pump on manually if necessary.

7.9 The Desiccator

Since the mechanical pump does not remove water vapour from the system, it is highly desirable to avoid its introduction. The gelatin base of photographic emulsion contains a high percentage of water, which is normally removed before introducing photographic materials into the column. This is done by allowing the plates or film to stand for some time in the presence of a water absorbent such as P_2O_5. The process of water transfer is very much more rapid if the air is removed; therefore many electron microscopes have a vacuum desiccator attached to the vacuum system. Only low vacuum is required, so the mechanical pump is put to further use by evacuating the desiccator. This requires further piping and valves, which must be placed so that they can be operated independently of the low vacuum valves serving the column and diffusion pump. Plates should be allowed to stand under vacuum in the presence of fresh desiccant for at least 2 hours; the capacity of the desiccator must be at least 2 loads of fresh plates. If plates are used at a more rapid rate than the microscope desiccator can cope with, it is advisable to dry out further supplies in a separate vacuum desiccator, or in an evaporator (see Chapter 19).

7.10 The Three-Stage System

The two-stage pump system has the disadvantage that the vacuum reservoir must be pumped at frequent intervals with the mechanical pump.

This can be overcome if the diffusion pump is designed to operate into a higher backing pressure. This can be done if the mass of the pumping fluid molecules is increased by using mercury. Mercury cannot be used as the fluid for the pump serving the column directly, since its vapour pressure at the temperature of the cooling water is only 10^{-3} torr. A mercury diffusion pump can however be interposed between the oil pump and the mechanical pump in such a way that the oil pump acts as a shield to prevent mercury molecules from diffusing or 'backstreaming' into the column. The mercury pump is designed to operate between 10^{-1} torr, the pressure at the exhaust side of the oil pump, and 5–10 torr in the reservoir. In this way, the reservoir can receive the throughput of air resulting from 20 or more specimen changes before the mechanical pump needs to be switched on. In practice, this means that the mechanical pump is only switched on when the plates are being changed. It is not normally necessary to use it while observing or photographing a specimen, and so it can be built into the desk unit.

Great care must be taken to ensure that no mercury reaches the oil diffusion pump, otherwise it will condense into the oil and contaminate the column. A complex baffle system is interposed between the two, which tends to reduce the pumping speed of the combined system. The disadvantages of additional expense and complexity and possible contamination of the column with mercury must be weighed against the advantages of having the mechanical pump concealed within the console, and reduced wear and servicing.

The main advantage of the three-stage system is that it requires only one mechanical pump, since the latter can be used for ancillary 'roughing' duties when it is not required for backing the diffusion pump. The term 'roughing' is applied to the removal of air from the column after a filament or aperture change or from the camera after a plate change or from the vacuum desiccator.

The best solution to the problem of pump vibration is to use a mechanical pump designed for minimum vibration, mounted on antivibration mountings well away from the column and operating continuously. In this case, a second mechanical pump must be provided for roughing the column, camera and desiccator. This system has the advantage that the roughing pump is used solely for pumping air and water vapour out of the system, thus protecting the backing pump from possible contamination.

7.11 Pumping Speed

The laws governing the flow of gases in enclosed systems change when the pressure in the system is such that the mean free path of the gas molecules is of the same order as the diameter of the pipes employed. The normal 'viscous' or 'laminar' flow gradually becomes 'molecular' flow, and intermolecular collisions become less important than collisions of gas molecules with the walls of the system. Under these conditions, the conductance of a pipe of circular cross-section is directly proportional to the cube of the diameter and not to the fourth power as is the case for laminar flow, and the constricting effect of baffles and bends on gas flow is correspondingly less. Under conditions of molecular flow, the conductance of a pipe is independent of pressure.

Diffusion pumps operate under conditions of molecular flow, and the pumping speeds are at first sight extremely high. The speed of the main oil diffusion pump of an electron microscope is generally of the order of 100 l/s. However, it must be remembered that 100 l of gas at 10^{-5} torr represents only 0·001 ml approximately of gas at atmospheric pressure. From this it will be seen that a gas ingress rate of 0·001 ml/s is as much as the vacuum system will cope with if it is to maintain a vacuum of 10^{-5} torr; this does not represent a very large leak. In fact, it is about the rate at which adsorbed gas is evolved from a normally clean column.

Mechanical pumps are rated in terms of total displacement at atmospheric pressure in l/min; the usual size fitted to an electron microscope vacuum system displaces 50 l/min. Since the total output of gas from the diffusion pump should not exceed 0·1 ml/min, such a pump has a capacity greatly in excess of requirements. The reason for such a large capacity is for it to be able to evacuate the comparatively large volume of the column (several litres) down to the pressure at which the diffusion pumps can take over in a reasonable time—less than 5 minutes. If the capacity of the mechanical pump is increased, this merely reduces the roughing-out time.

7.12 Vacuum Indicators

It is helpful for the operator to know the state of vacuum in both parts of the system, and preferable for him to be able to monitor both continuously. All that is strictly necessary is for a high-voltage discharge tube ('Geissler tube') to be placed in the system close to the column to indicate the high vacuum, and to connect a simple mercury manometer with rubber tubing to the low vacuum reservoir so that the operator can see it easily. This was all that was provided on early instruments, and is perfectly satisfactory provided the operator learns and understands the somewhat ambiguous indications given by the discharge tube. However, such simple equipment makes it a very difficult operation to localize leaks. It is preferable to use more sophisticated indicators, which amplify the almost imperceptible movement of the mercury manometer and the subjective indication of the discharge tube and display them as electrical meter readings.

7.13 The Pirani Gauge

This is used to measure low vacuum; its range is between 10 and 10^{-3} torr, and parallels that of the mechanical pump. The principle is as follows. If a heated wire is suspended in a gas, the gas molecules will conduct heat away from it and thus reduce its temperature. If the energy supply to the wire is constant, then its final temperature will be a function of the number of gas molecules, and hence the pressure of the gas. Since its electrical resistance is a function of temperature, then the resistance can be monitored continuously on a meter scaled in terms of pressure or vacuum.

Fig. 7.3(a) shows the gauge head, which is simply a length (about 10 cm) of fine, coiled tungsten wire mounted on two electrically conducting supports in an open-ended glass or metal tube, the open end of which is sealed into the low vacuum reservoir. The wire is made one arm of a Wheatstone bridge (Fig. 7.3(b)), which is fed from a low-voltage

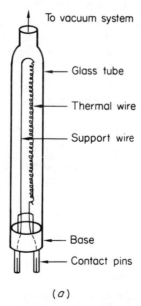

(a)

Figure 7.3(a)   A diagram of a Pirani gauge used to measure low vacuum source (generally a 6 V transformer). The values of the fixed resistors are chosen so that the current through the wire is such as to maintain it at about 100 °C when the vacuum in the tube is better than 10^{-2} torr, and the sensitivity of the indicating meter is chosen so that full scale deflection is given for a change in wire resistance corresponding to a change from

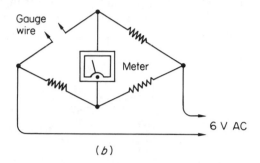

(b)

Figure 7.3(b) The circuit used to operate a
Pirani gauge

room temperature (atmospheric pressure) to 100 °C. There is a curious convention followed by some manufacturers of these meters in that full scale deflection is high vacuum, while zero deflection is atmospheric pressure; this is the inverse of what would be expected, and is the inverse of the indication given by the Philips gauge, described in the next section.

The Pirani gauge is used almost exclusively as a low-vacuum meter in Europe; in the U.S.A. a related type of gauge, the 'thermocouple gauge', is preferred. This gauge is identical in principle to the Pirani gauge, but the actual wire temperature is measured by a miniature thermocouple connected to an indicating meter scaled in terms of vacuum. Each type has minimal advantages and disadvantages.

7.14 The Penning or Philips Gauge

This is used to measure high vacuum; its range lies between 10^{-6} and 10^{-3} torr. The principle is based on the measurement of the electrical conductivity of a gas at low pressure. The device resembles the bulb of a thermionic vacuum valve with a tube sealed into the vacuum system at one end, and a multi-pin plastic base at the other. The bulb contains a pair of electrodes (Fig. 7.4) which are placed in the field of a powerful externally-mounted permanent magnet. An external circuit containing a current-measuring meter supplies the electrodes with a high voltage (1 to 2 kV).

Cold emission of electrons from the cathode initiates ionization in the gas between the electrodes. The positive ions thus formed are attracted towards the cathode, but are constrained to travel in spiral paths by the applied magnetic field. Due to the increased path distance traversed by the gas ions, multiple ion production by collision is greatly increased, and hence the total current at any given gas pressure through the gauge is increased. The magnet therefore increases the sensitivity of the device by a large factor. The total current through the gauge is a function of the

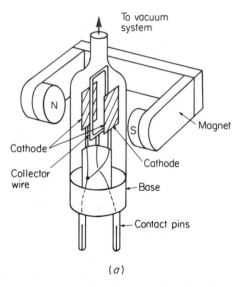

(*a*)

Figure 7.4(*a*)   A diagram of a Penning (Philips) gauge used to measure high vacuum

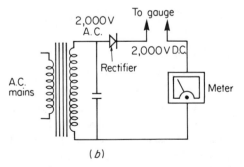

(*b*)

Figure 7.4(*b*)   The circuit used to operate a Penning gauge gas pressure in it; this is displayed on a meter which is calibrated in terms of pressure.

Neither of the gauges is particularly accurate in the simple form used on electron microscopes; all that is required is a reliable and reproducible method of indicating orders of magnitude of vacuum. All that is strictly necessary is that the indicating gauge should bear a mark on the scale, so that the operator knows that the instrument may not be used if the vacuum does not come up to this mark.

7.15 Design of the Pumping System

Having described the individual components of the pumping system, let us now consider the requirements of the system as a whole.

The main problem in the design of a complete vacuum system is the interconnection of various parts by means of valves. It is necessary to make provision for all parts of the system to be filled with air in order that it may be dismantled for maintenance purposes. All parts of the system must first be pumped to a low vacuum by the mechanical pump at some stage of operation, and all those parts which are normally at high vacuum must be pumped to low vacuum or 'roughed' before the high vacuum pump can be used. This demands a somewhat intricate system of valves which can be opened and closed in the correct sequence either by the operator or automatically.

During the normal course of operation it is necessary to change the specimen, the photographic plates or film, the apertures and the filament. To do these things, part or the whole of the column must be isolated from the rest of the vacuum system and brought up to atmospheric pressure. It is best if the camera, electron gun or specimen chamber can be completely isolated and pumped individually first to low and then to high vacuum after the completion of the manipulation, but few instruments offer such complex facilities. It is generally sufficient if the whole column is brought up to atmospheric pressure when the plates or filament are changed, and to design the specimen changing apparatus so that only a small quantity of air needs to be pumped away each time the specimen is changed.

If a valve is interposed between the camera and the rest of the column, the camera can be isolated from the high vacuum system and the plates can be changed without interfering with specimen viewing or contaminating the rest of the column with air. A camera airlock valve is therefore incorporated in most modern instruments.

In general, the specimen is changed more frequently than either plates or filament. It is therefore best if this procedure can be made simple and rapid. Properly designed diffusion pumps are able to deal with a few ml of air if this is suddenly let into the high vacuum space, provided the column pressure does not rise above the backing pressure. The specimen is best introduced through a small airlock, so that it is not necessary first to isolate and then to prepump the specimen chamber when changing a specimen. The latter procedure complicates column construction because further internal valves have to be provided, and also makes the specimen changing time longer, because the specimen chamber has to be prepumped.

The filament and apertures seldom need to be changed if proper operating procedures are followed. The provision of a gun airlock is therefore a minimal advantage, especially as apertures are best changed at the same time as the filament, and the whole column must in this case be brought up to atmospheric pressure.

7.16 Operation of the Pumping System

Let us now consider the operation of the complete pumping system shown in Fig. 7.5. This shows the bare, basic essentials necessary to operate an electron microscope. When the system is first assembled, it starts off full of air, which must be removed from the diffusion pump before its heater its turned on. First, all the valves are closed. Then the mechanical pump is started and the backing valve C is opened. After 2–5 minutes, the low vacuum indicator will show that the pressure in the low vacuum reservoir is low enough to turn on the diffusion pump heater. The pump cooling water is turned on at the same time. Next, the air must be removed from the column. Accordingly, the backing valve C is closed, and the column roughing valve D is opened. After a further 2–5 minutes, the low vacuum indicator will show that the pressure in the column is as low as the mechanical pump can reach, and the high vacuum gauge will show that the diffusion pump is beginning to operate. The low vacuum reservoir will now need pumping again, so the column roughing valve D is closed and the backing valve C is opened. After a further 5 minutes or so, the diffusion pump will have reached maximum speed, and the high vacuum gauge will show that the ultimate vacuum is being reached in the space above the diffusion pump. The high vacuum valve B is now opened, and the diffusion pump transfers the greater part of the air molecules remaining in the column into the low vacuum reservoir. The readings of both gauges will rise for a minute or so, and will then begin to drop. The system is left in this condition until both gauges show that both high and low vacuum sides of the system have reached their ultimate vacuum. The electron microscope can now be switched on and used with the mechanical pump still running. If it is necessary to reduce vibration when taking a micrograph, the mechanical pump must be stopped. Before switching off the pump motor, the backing valve C must be closed to prevent air from entering the low vacuum reservoir through the pump. After the pump has stopped, the air inlet valve E must be opened in order to prevent the pump oil from being forced into the low vacuum line

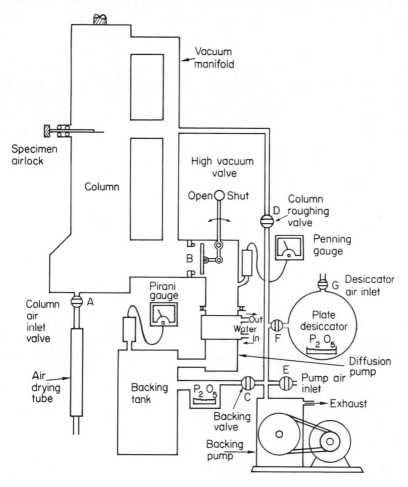

Figure 7.5 A schematic diagram of a typical fully manually operated electron microscope vacuum system with a plate desiccator, using a single oil diffusion pump and one single stage rotary pump for both backing and roughing services

by air pressure across the pump. The vacuum system will now continue to function until the pressure in the low vacuum reservoir rises enough to impede the working of the diffusion pump. This will be shown by a rise in the high-vacuum indication. The mechanical pump must now be allowed to pump the reservoir, but first the air in the low vacuum line must be pumped away. Accordingly, E is closed, the pump is started, and the operator waits until the low vacuum gauge shows that the backing valve C can be opened. After a few minutes pumping, the mechanical pump can be shut down again and the pumping cycle repeated. If only one diffusion pump is used, the cycle will have to take place several times an hour, but if the oil pump is backed by a mercury pump, the cycle will only need to take place perhaps twice a day, provided the specimen is not changed. Each time the specimen is changed, a small amount of air will enter the column through the specimen airlock and the pressure rise in the low vacuum reservoir will stop the single diffusion pump. It is therefore necessary in this simple system to pump the reservoir each time the specimen is changed, which means that if the instrument is in continuous use, the mechanical pump will rarely be stopped, and it will have to be mounted so that its vibration does not interfere with the column.

When it is necessary to reload the camera, or to change the filament or apertures, the column must be filled with air. Accordingly, the high vacuum valve B is closed and the column air inlet valve A is opened. After the change, A must be closed followed by C before D is opened to rough out the column. When the column is down to low vacuum, D is closed, C is opened followed by the high vacuum valve B.

When shutting the instrument down for the night, the column must be isolated to protect it from contamination by backstreaming of oil vapour from the cooling diffusion pump, and the diffusion pump must be isolated from the mechanical pump, also to prevent contamination. Accordingly, B is closed followed by C; a check is made that D is shut; the diffusion pump heater is shut off; the mechanical pump is stopped and E is opened. It is necessary to allow the cooling water to continue to flow around the diffusion pump condenser until the boiler has cooled to room temperature, otherwise oil vapour will rise and contaminate the high vacuum valve and gauge.

7.17 Valves

The valves used on the low vacuum side of the system are generally poppet valves operated by a cam or solenoid. Rubber diaphragm valves (Saunders valves) are sometimes used. The high vacuum valve must offer as little resistance as possible to the air molecules diffusing down from the column, and is in the form either of a plate lifted up from a rubber ring seating, or better a plate which can be slid aside so that it is removed completely from the throat of the diffusion pump. Cooled baffles and traps are sometimes interposed between the diffusion pump and the column. In automatic vacuum systems, the valves can be motor-driven or solenoid-operated.

7.18 Vacuum Automation

It will be seen from the foregoing that a simple vacuum system such as has been described is very liable to damage by operator error, since precise sequential valve operation is essential. Most manufacturers simplify the operator's task by linking the valves together mechanically so that they can be operated by one or at most two main controls, and by adding safety measures to reduce damage caused by mishandling. A simple mechanical linkage between the valves cannot guarantee freedom from damage to the vacuum system, because it is necessary to wait a certain length of time between successive operations to allow air to be pumped from one part of the system to another.

The ideal system is one in which the complete sequence of operations is carried out automatically after a single button is pressed. This system requires motorized valves and pressure sensors in various parts of the pumping lines. Fully automatic systems are now offered by many manufacturers, and will doubtless be a standard feature of electron microscopes of the future.

Care must be taken to distinguish between 'automatic' and 'automated' vacuum systems. The former type will perform a complete sequence of operations without the intervention of the operator; the latter type requires the operator to initiate each successive step by pressing a button when instructed to do so by a signal light. Pressing a single button on a fully automatic system will, for example, set in train the complete sequence of operations necessary to start the instrument from cold and bring the column to high vacuum ready for the immediate insertion and examination of a specimen. The operator is completely free to carry out other tasks while waiting the necessary 20 minutes or so. An automated system, on the other hand, breaks the full sequence down into a series of simpler operations and requires the operator to initiate each in turn, generally by pressing the next button of a set.

An automated system is naturally not as foolproof or as convenient as an automatic system, but is generally a lot cheaper to manufacture, and has the advantage of being inherently more flexible provided always that the operator is fully conversant with the operation of the system.

7.19 Safety Systems and Alarms

Even fully automatic vacuum systems may not be proof against damage arising from mains electricity or water pressure failure. These are not under the operator's control, and failure of either in a simple vacuum system such as has been described would be disastrous if the instrument were unattended.

The main possibility of damage to a vacuum system arises from the presence of various pumping fluids in it, combined with the extreme susceptibility of the column to contamination by insulating materials such as pump oils. The system is designed to pump air from the column to the atmosphere; if air enters the system in the opposite direction the result can be disastrous.

If the main electricity supply fails in the basic system, the rotary pump will stop with all the valves between it and the column in the open position. The trouble now arises from the fact that the seals in the rotary pump cannot be made perfect. Air will leak back through the pump, giving rise to a foam of air and oil, which slowly fills the vacuum reservoir, the diffusion pump and then the column. The operator then returns to the unattended instrument to find the column full of oil foam. If the system also contains a mercury pump, liquid mercury may also have been forced into the column. The column must then be completely dismantled together with the whole vacuum system in order to clean it out. If mercury has entered the column, it might amalgamate with the soft iron polepieces of the lenses, in which case they will be useless for high resolution work. It can be seen, therefore, that a mains failure in an unattended simple vacuum system can lead to a disaster of the first magnitude.

Several things can be done to prevent this. In the first place, the instrument should never be left unattended with the rotary pump running and the backing valve open. Secondly, an automatic, float-cum-gravity valve can be fitted between the rotary pump and the backing valve, so that it will rise and cut off the rotary pump from the rest of the system if air or oil leaks past. Thirdly, switches can be fitted to the valves in order to operate a warning bell through a relay if the mains fails with any of the valves in a potentially dangerous setting. The bell must, of course, be energized from a separate battery and not the mains. Automatic and automated vacuum systems, which can be designed to 'fail safe', should be the answer to this problem.

Another possible cause of heavy column contamination with the simple, unprotected system is failure of the cooling water supply to the diffusion pump. In this case, the diffusion pump condenser will no longer condense the pumping oil, which will diffuse up into the column and contaminate it heavily. This can be guarded against by arranging switches in the diffusion pump heater circuit which cut off the heater current if the water supply fails. These switches can be operated by water pressure, water flow or pump temperature and can also include the warning bell circuit.

These safety devices cannot protect the instrument from mishandling by the operator. The column can very easily be contaminated if, for instance, the backing valve is opened when the low vacuum line is full of air, the diffusion pump is running and the high vacuum valve is open. The rush of air back through the diffusion pump will also cause a cloud of pump oil vapour to ascend into the column. Constant vigilance is therefore essential on the part of an electron microscope operator if expensive shut-downs for column cleaning are to be avoided.

Other safety devices are frequently fitted to prevent the operator from injuring himself by coming into contact with the high voltages employed in electron microscopes. Switches are fitted to electronic cabinet doors so that the unwary are protected from the consequences of prodding in circuits in the hope of locating a fault, and mechanical and pressure switches are fitted to the gun together with earthing devices so that the high voltage cannot be turned on while the gun is open.

The operator must familiarize himself with the position and action of all these safety devices, otherwise he may spend hours fruitlessly seeking a fault when the fault is simply due to a door having been inadvertently left open. The operation of the water switch and alarm bell must be checked every morning when switching on, otherwise it is possible for a switch to become inoperative or the bell battery to become useless without being noticed. Expensive damage might then arise.

7.20 Contamination

However leak-free the vacuum system may be, it can only pump down to a certain limiting pressure. This is determined mainly by the vapour pressure of the fluid used in the diffusion pump, and is between 10^{-5} and 10^{-6} torr. The pump fluid is generally a mineral oil, and so the vacuum in the column contains a large number of hydrocarbon molecules, which are also derived from the grease used to lubricate movable rubber vacuum seals. The vacuum also contains residual gases, which are mainly water vapour together with nitrogen and carbon monoxide. Small amounts of hydrogen, methane, oxygen and carbon dioxide are also found in the column vacuum.

The presence of these residual gases in the electron microscope column gives rise to two undesirable effects. Firstly, the hydrocarbon molecules land on the surface of the specimen, forming an amorphous covering layer and thus reducing specimen contrast. When the imaging electrons strike the hydrocarbon molecules adsorbed to the specimen surface, they decompose them into carbon and hydrogen. The hydrogen reenters the vacuum, but the carbon remains on the surface of the specimen. This carbon 'contamination' as it is commonly termed can build up at a rapid rate; an increase in specimen thickness of 5 Å/s is not uncommon in a dirty column. A method for measuring contamination rate is described in Chapter 13. Contamination simply reduces image contrast at medium and low magnifications, but at very high magnifications the structure of the contaminating layer is imaged with the specimen, and effectively destroys resolution. Fig. 7.6 is an electron micrograph showing the deposition of

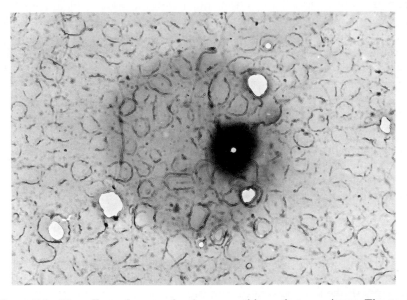

Figure 7.6   The effect of contamination on a thin carbon specimen. The two contaminated areas were each irradiated for 3 min, the larger with C_2 defocused and the smaller with C_2 at focus. The heavier deposit at focus shows how contamination rate is dependent on beam intensity at the specimen. Taken in a dirty column with the anticontaminator at room temperature. × 1,000

carbon on the surface of a specimen after irradiation, at both low and high current densities.

The second undesirable effect is the ionization of the residual water vapour molecules in the vacuum by the electron beam. The resulting hydroxyl ions are very reactive and attack the carbon in the specimen, vaporizing it in the form of carbon monoxide. They are beneficial in one way, in that they remove the contaminating layer of carbon, but they are detrimental in that they also 'burn away' the specimen. This phenomenon is called 'beam damage' or 'stripping'. Fig. 7.7 shows the effect of stripping on a specimen. A high-intensity beam of small cross-sectional area was focused on to a hole in a carbon film. The beam has thinned down the carbon film at the centre, where it was most intense, and has enlarged the hole on which it was focused, giving it a ragged edge. The effect is the opposite of contamination, where the film is thickened and the hole filled up. Around the edge of the beam, where it was less intense, a ring of contamination has built up, surrounding the central burnt-away region.

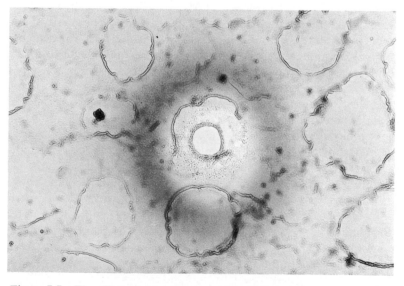

Figure 7.7 The effect of stripping on a thin carbon specimen. The focused beam has etched much of the carbon film away over the central high intensity area, but contamination has been deposited at the beam edge, where intensity is less. Taken in a dirty column with anticontaminator at liquid nitrogen temperature. × 20,000. (From Meek, *J. R. Microsc. Soc.,* **88**, 419 (1968). Reproduced by permission.)

Contamination affects all parts of the system which are struck by electrons. The most susceptible components to the build-up of contamination are the movable apertures in the condenser and objective lenses, and to a lesser extent the bores of the lens polepieces. Contamination on dirty apertures can easily be seen with the naked eye as a black area, generally about 1 mm in diameter, surrounding the hole. Contamination is also deposited inside the aperture, and gives rise to astigmatism. The oily deposit, being a bad conductor of electricity, charges up and repels the electron beam passing through. Since the layer of contamination builds up unevenly, the repulsion is asymmetrical and the lens behaves as though it were astigmatic. When the effect is no longer correctable by the stigmator, the aperture must be changed. This is the reason why a stick of three apertures is generally provided. Aperture contamination can be minimized by using a thin molybdenum foil pierced with a hole as an aperture. The foil becomes heated by the bombarding electron beam, and the contamination tends to be driven off. This type of aperture, available either as single units or multiple aperture foil strips, is becoming known as a 'self-cleaning' aperture, and is fitted as standard to high-performance TEMs.

Specimen contamination and beam damage can both be greatly reduced by reducing the intensity and cross-sectional area of the electron beam. This has the disadvantage that the image is very dim and difficult to focus. It also increases the exposure time needed to record it, and increases the 'graininess' of the resulting micrograph, due to the greater randomness of the electron beam. A more practical way of reducing the effects of contamination and beam damage is to prevent hydrocarbon and water vapour molecules from striking the surface of the specimen. This can be done in part by a trap or baffle, cooled down to liquid nitrogen temperature ($-180\,^\circ$C), between the diffusion pump and the column. This unfortunately reduces pumping speed drastically, and does not deal with hydrocarbons derived from the lubricated column seals. A better method is to place a metal surface cooled to liquid nitrogen temperature as close to the surface of the specimen as possible. This device, called an 'anticontaminator', is very effective when properly designed, and reduces contamination rate to a very low value.

7.21 Anticontaminators

The anticontaminator consists simply of a cooled surface placed as close to the specimen as possible, the idea being that the hydrocarbon molecules will condense preferentially on it. The usual design for a side-entry specimen holder is shown in Fig. 7.8 and consists of two copper blades placed above and below the grid carrier and as close to it as possible. Holes are drilled in the upper and lower blades to allow the electron beam to pass through the specimen. The blades are part of a copper bar which passes through a vacuum seal out of the column. Fastened to this horizontal copper bar is a vertical copper bar or a brush of copper wires which dips

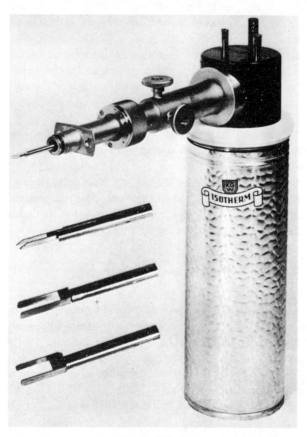

Figure 7.8   A typical anticontaminator with 3 pairs of blades (shown at higher magnification). The blades sandwich the specimen as closely as possible, and are provided with centration and axial adjustments. The wider blade spacings allow for specimen manipulation. This type of anticontaminator is used with side-entry stages. The electrical connections at the top are for a thermocouple temperature gauge and an electrical heater.
(Philips EM 200. Courtesy Philips, Eindhoven)

into a Dewar vessel kept filled with liquid nitrogen. Heat is thus conducted rapidly from the anticontaminator blades, which are maintained at about $-150\,°C$.

A well-designed anticontaminator will reduce contamination rate to a negligible and barely measurable value. However, anticontaminators have their disadvantages. They consume liquid nitrogen at the rate of a litre or more per day, although if a somewhat higher contamination rate can be tolerated, the Dewar flask can be filled with a mixture of crushed solid carbon dioxide and alcohol. This gives a blade temperature of about $-50\,°C$, and tests have shown that this reduces the contamination rate by at least a factor of 10. With liquid nitrogen under the same conditions, the reduction factor was nearer 100. The advantage of the solid CO_2 trap is that it only needs filling once per working day, and it only consumes about a pound of coolant.

Great care must be taken when working with a cooled surface inside the microscope to make sure that damp air does not enter the column, otherwise ice crystals may form on the anticontaminator blades. Since these are close to the electron beam, the ice can give rise to gross astigmatism as well as poor vacuum. Air entering the specimen chamber must therefore be dried. If the column is to be filled with air for any length of time, e.g. in order to change the filament, the Dewar flask should be removed from the anticontaminator and the copper rod allowed to warm up.

7.22 Cryopumps

The success of the simple blade-type anticontaminator in keeping down contamination and stripping, and the realization by designers of the importance of maintaining the best possible vacuum in the specimen chamber have led to the introduction of much larger surfaces cooled to liquid nitrogen temperature into the specimen chamber and its adjacent pumping lead. These large cooled surfaces (Fig. 7.9) are termed 'cryopumps'. The present tendency is to design them so that a continuous flow of coolant can be passed through a hollow box presenting as large an area as possible to the specimen chamber. A well-designed cryopump has been shown to reduce the pressure of undesirable vapours in the specimen chamber by several orders of magnitude, thus tackling the problem of contamination at its source.

7.23 Vacuum Seals

The various components of an electron microscope column have to be constructed in such a way as to be readily demountable for servicing. The major parts are made of metal, and it is not practicable to machine metal-to-metal joints which are vacuum tight over relatively large areas. Consequently, rubber gaskets have to be interposed between the surfaces of the metal components. These gaskets, known as 'O-rings', are of circular cross-section and are generally made of a soft synthetic rubber. In order to make a vacuum-tight seal between, for example, two lenses, a circular groove of square cross-section is machined in the top mating surface of the lower lens, and is polished to a fine finish. The lower mating surface of the

Figure 7.9 A 'cryopump' type anticontaminator, which completely surrounds the specimen in the specimen chamber. It is interchangeable with the 'cold-finger' blades shown in the previous figure. (Philips EM 300. Courtesy Philips, Eindohoven)

upper lens is left flat, and is also ground to a fine finish. The O-ring is placed in the groove, and stands proud when the joint is uncompressed. The upper lens is then lowered on to the ring, which is compressed so as to fill out the groove. The precise location of the joint is by metal-to-metal contact, but the rubber ring prevents air from leaking across. It will be seen that both the weight of the lens and the pressure of the atmosphere serve to pull the joint up more tightly. O-ring joints if correctly made in the first place require no grease or other sealing compound. Since grease gives rise to column contamination, as little use as possible is made of it when assembling the column. This type of O-ring seal remains vacuum-tight unless disturbed, or unless the rubber loses its elasticity due to age or bombardment by X-rays.

Other seals have to be provided which allow movement of one part relative to another. Small amounts of movement such as the few degrees required to align the gun, can be accommodated by machining spherical mating surfaces. A large amount of movement, such as is required in seals on valve shafts, specimen insertion mechanisms, camera drives etc., require a somewhat different design. An O-ring is still used as a gasket, but the metal surfaces do not come into contact with one another. Movable seals are by their nature much less satisfactory than plain seals, and require a certain amount of lubrication in order to operate with the minimum of friction. This entails the use of a special vacuum grease of very low pressure, which must be applied as sparingly as possible. The grease must never be allowed to come into contact with the electron beam. Movable seals require a certain amount of maintenance, but, like plain seals, should be disturbed as little as possible.

CHAPTER 8

The Electronic System

8.1 Introduction

This chapter is intended to give an account of the general arrangement and function of the electronic circuits used on electron microscopes, followed by an account of the basic principles underlying the operation of the circuits. The latter section is intended for readers with some knowledge of electronics. It must be stressed that electronics is now a highly complex technology in its own right, and that the circuits used on modern electron microscopes, while they may be basically simple in principle, are in practice highly sophisticated and require expert attention for their repair. The information given in this chapter is not sufficiently detailed to enable a reader with no knowledge of electronics to repair major faults on circuits. It should however assist in tracking down and repairing minor faults involving the changing of faulty valves or fuses. Further information on maintenance and fault finding is given in Chapter 15.

The electron microscope requires two separate sources of electrical power: a high voltage, low current supply to accelerate the image-forming electrons; and a low-voltage, high current supply to energize the magnetic lenses which focus the beam and form the image. Once the instrument has been set to give a focused image, any fluctuation in either the high voltage or the lens supply will give rise to both image movement and focal change at the image plane, and hence resolution will be lost. The power supplies must be designed so that the maximum permissible lens current and H.T. voltage fluctuation during the interval of time needed to expose a plate will cause a loss in resolution which is less than the ultimate resolving power of the objective lens.

8.2 Stability Requirements

Slow, low-voltage electrons are imaged closer to the lens than fast, high-voltage electrons, as we have seen on p. 87. This effect is the electron

167

analogue of chromatic aberration in optical systems. It is essential, therefore, to use a 'monochromatic' beam of electrons to illuminate the specimen. The high voltage used to accelerate the electron beam must be prevented from fluctuating by more than a very small amount. The maximum permissible fluctuation is determined by the ultimate resolution required from the instrument. In order to obtain 5 Å, a stability of the order of 1 part in 50,000 in the high voltage is necessary over a period of, say, 5 seconds.

We have also seen in Chapter 4 that any change in lens current will give rise to a change in focal length of the lens, and therefore the position of the focal plane will move out of the plane of the photographic plate.

A more serious defect, however, arises from the fact that the electrons passing through a magnetic lens suffer rotation. This means that if the current through any of the imaging lenses varies, the image rotates about a point called the 'current rotation centre', which should be coincident with the centre of the screen. A micrograph taken under these circumstances will be blurred at the edges and sharp at the centre. If the accelerating voltage changes, this also gives rise to an alteration in magnification; the image grows or contracts radially about a point called the 'voltage centre', which should also be coincident with the centre of the screen. A micrograph taken with a fluctuating H.T. voltage will also be blurred at the edges and sharp at the centre.

The focal length of a magnetic lens is a function of the square root of the current, so the lens current stability requirement is twice that of the accelerating voltage. If the accelerating voltage has to remain stable to within 1 part in 50,000, then the objective lens current must be stabilized to within 1 part in 100,000 over a period of 5 seconds. The stability requirement of the other imaging lenses is less, because their aberrations are less highly magnified. In practice, the stabilities of the other imaging lenses need to be of the order of 1 part in 10–20,000, and that of the condenser lenses can be even less.

Early electron microscopes were limited in resolution by lens design to 100 Å or so, and did not require particularly high stabilities in the power supplies. The currents for the lens coils were supplied from lead accumulators, which can give high currents at low voltages with a very high degree of stability over a short period of time. They have the disadvantage of considerable voltage drift over longer periods of time, and require constant maintenance. The accelerating voltages in early instruments were generated from the A.C. mains after partial stabilization by a constant voltage transformer (see p. 173). This simple device is still used for initial mains voltage stabilization in many instruments.

As basic lens design progressed and microscopes became capable of higher resolution, so demands for stability became more stringent. The idea of using the domestic electric mains supply as the sole power source and stabilizing it electronically is due to Zworykin, who first applied this now almost universally used method in the R.C.A. EM–B instrument designed in 1941.

The commercial electric mains have the great advantages of continuous availability, convenience and cheapness, but suffer from considerable voltage fluctuations, which can be as much as $\pm 20\%$ over a period of 24 hours, due to the effect of industrial and domestic load variations on the main generating plant. The short-term stability can also be affected by sudden load variations in the laboratory building causing a voltage drop in the internal distribution wiring. This is why it is worth while making great efforts to install a separate power supply for a high resolution electron microscope direct from the supply to the building, and not simply to plug it into the nearest wall socket.

8.3 Power Supply Requirements

Electromagnetic objective lenses require a very powerful magnetic field across the gap in order to focus the beam of high-energy electrons. This necessitates a coil of several thousand ampere turns. Since it is easier when using conventional valve circuits to stabilize a small current than a large one, the lens coils are wound with as many turns as practicable so that the lens current may be reduced to a readily stabilized value. In valve circuits, coils of 10,000 or more turns are used with a current of about 0·2 A flowing through them with the lens at focus, the objective lens generally requiring a higher current than the other lenses. The resistance of such large, multiturn coils is high; hence a potential of several hundred volts must be imposed across the lens coil to obtain the required current. The problem, therefore, is to supply 5 lenses with a total of 1 A or more at a voltage of around 500 and a stability of 1 part in 100,000, this direct current supply to be derived from the alternating current mains, which is known to fluctuate by $\pm 20\%$ or more.

Transistorized power supplies are now becoming more common. Their load requirements are opposite to those of electronic valves; large currents at low voltages are easier to stabilize with transistors. 24 volt transistorized lens supplies generating tens of amperes are now becoming common.

The high tension supply for the electron gun presents the inverse problem to the lens supply; in this case, a high voltage (100,000 V) at a low current (500 μA) has to be generated, with a stability of the order of 1 part in 50,000.

It is also necessary to supply electrical power for various other circuits. These are the stigmators and beam deflectors; the circuits operating the shutter, exposure meter, camera, focus wobbler and modulation oscillators if provided; the energizing currents for the electromagnetic valves and motors if an automated or automatic vacuum system is used; the power for the various safety devices and relay switches; and the power for the heaters of the oil diffusion pumps, the rotary pump motor and the

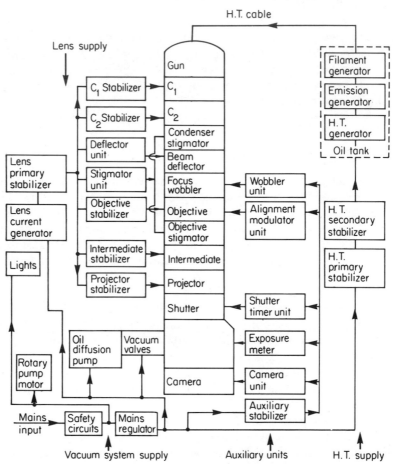

Figure 8.1 A block diagram showing the electronic circuit units used to supply the various power and voltage requirements of a modern high-performance electron microscope

desk and room lighting. Some of these auxiliaries such as the stigmators require stabilities as high as the lenses and high tension; others such as the rotary pump motor and the room lighting will work perfectly satisfactorily over the whole range of mains fluctuation and hence require no stabilization. Others, such as the oil diffusion pump heaters, require a moderate degree of stabilization.

A block schematic diagram of the various circuits supplying an electron microscope column is shown in Fig. 8.1. This is typical of the arrangement found on a complex and sophisticated modern high-performance instrument. It is an example of modern practice; the arrangement differs in detail from instrument to instrument.

The complete electrical system can be divided into 6 parts:

1. The safety system;
2. The mains regulator;
3. The vacuum supply system;
4. The lens supply system;
5. The H.T. supply system;
6. The auxiliary supply system.

The mains electricity first passes to a safety system consisting of a set of switches or breakers actuated by signals from sensors in various parts of the system, such as the cooling water supply. The safety circuit is arranged so as to isolate the whole instrument from the mains in the case of an emergency such as failure of cooling water pressure or mains supply. In the case of automatic or automated vacuum systems, all the spring-loaded vacuum valves will return to the 'fail-safe' position (all valves closed and rotary pump air admittance opened). If the microscope is not fitted with servo-operated vacuum valves, an alarm bell will then ring, and the operator must return the vacuum valves manually to the 'safe' position without delay. The safety circuit will also isolate parts of the system, e.g. the H.T. cannot be switched on if the gun is opened, or if the vacuum is not good enough, or if the door of the H.T. generator cabinet is left open.

It is usual to take the mains supply for the rotary pump motor and general lighting after the safety breaker and before any stabilization is applied. The rotary pump motor takes a very heavy current on starting, which might overload the mains stabilizer and cause a fluctuation in H.T. and lens currents.

The mains electricity is next roughly stabilized by means of a mains regulator, which reduces the possible $\pm 20\%$ fluctuation to about $\pm 2\%$, i.e. a stabilization factor of 10 is introduced. This degree of stabilization is ample for the diffusion pump heaters and the motors and solenoids operating the vacuum valves in automatic systems.

The remaining three systems to be stabilized are first the H.T., filament and gun emission; secondly the lens, deflector and stigmator currents; and thirdly the power for the auxiliary systems. The first two require the highest stabilities which can be achieved; the third requires only moderate stability.

The gun potentials have to be generated in a tank containing transformer oil, since air is not a sufficiently good insulator to prevent arcing-over at the voltages employed. Because of the very high voltages employed and the consequent necessity to isolate the stabilizer circuits (which have to run at a low potential for safety reasons), it is usual practice to isolate the stabilizer circuits from the actual H.T. generator. This is done by stabilizing the input to the generator to the required degree, and then generating the various high potentials from the prestabilized input. The process takes place in 4 stages; mains regulation; primary stabilization; secondary stabilization; H.T. generation. In practice, the mains regulator will reduce a 20 % input fluctuation to 2 % or one power of 10; the primary stabilizer by a further factor of 100 and the secondary stabilizer by another factor of 100. The total factor of stabilization is the product of the factors introduced by all 3 stabilizers: $10 \times 100 \times 100$ or 100,000. Therefore, if the mains input fluctuates by 20 %, the H.T. will fluctuate by only 1 part in 500,000.

Stabilization of the lens currents takes place in a broadly similar fashion, but here it is usual to generate the lens current first and to stabilize it afterwards. A high current primary generator is supplied from the mains regulator, and provides sufficient current for all the lenses. This high current is then stabilized by a primary stabilizer with a factor of about 100. Each individual lens then has a separate secondary stabilizer. For the lenses requiring a lower stability than the objective, a further factor of about 20 is introduced, giving a total of $10 \times 100 \times 20 = 20,000$. A more complex stabilizer is generally used for the objective, giving a factor of 100 or more, and a total stabilization factor for the objective of $10 \times 100 \times 100 = 100,000$. The stigmators and beam deflectors, which require stabilities of the same order as those demanded by the lenses, are generally fed from the lens circuits.

The auxiliary circuits, which power the shutter, camera, exposure meter, alignment modulators and focus wobbler, do not require such a high degree of stabilization as the lenses. They are usually operated from a small auxiliary stabilizer with a factor of 20 or so fed from the mains regulator, giving an overall factor of some 200.

8.4 Stabilizers

The foregoing account gives a summary of the power requirements of a modern electron microscope. Let us now see how these are provided in practice.

8.4.1 Mains Regulation

Fluctuation in mains supply voltage can be reduced by introducing a constant-voltage transformer or a motor driven variable transformer between the mains and the instrument.

(A) Constant Voltage Transformers These are sometimes known as 'saturated reactors', and consist basically of a transformer designed so that the iron core is magnetically saturated over the expected range of mains voltage variation. The magnetic field induced in the core by the current in the primary winding is then almost independent of the mains voltage across the primary winding. Since the voltage appearing across the secondary winding is a direct function of the magnetic field in the core, the secondary voltage is almost independent of the primary voltage, provided that the primary voltage is high enough to saturate the iron core. Saturated reactors have the great advantage of a very short time constant of operation, but have the disadvantage that they distort the alternating waveform from a true sine wave, and thus introduce unwanted harmonics which may affect subsequent stabilization stages. Also, the output voltage changes with frequency. They are not true automatic voltage stabilizers, being rather non-linear devices which do not obey Ohm's Law.

(B) Motor-Driven Regulators The simplest automatic voltage regulator is the motor driven rotary variable transformer or 'Variac'. Here, the incoming mains voltage is compared with a constant reference voltage. If the mains voltage varies appreciably from this reference, a signal is generated, which is first amplified, and is then applied to a small electric motor which is arranged to rotate the wheel controlling the Variac in the correct direction to reduce the variation. This type of automatic voltage stabilizer is very slow, requiring several seconds to operate. It is also very prone to 'hunting', a form of very low frequency oscillation (see below), and cannot be made very sensitive. It is, however, capable of ironing-out large, slow mains fluctuations such as occur during load-shedding at the mains generating station. The reference potential is a hot-wire device known as a 'baretter' consisting of a thin iron filament in a bulb containing hydrogen at reduced pressure. When operating, it resembles a

carbon filament lamp bulb. It acts as a non-ohmic resistor or constant current device, and is used in a bridge circuit in such a way that voltage fluctuations caused by mains variation are made to drive the Variac motor in the correct sense.

8.4.2 Electronic Regulators

These are true automatic voltage regulators, and comprise two separate units: a measuring unit and a control unit. The measuring unit detects a change in the input or output voltage of the regulator by comparing it with a voltage standard, and produces a signal. The signal then operates the control unit in such a way that the detected change is minimized. The simplest way in which such a device could be operated is manually, and the action of an automatic voltage stabilizer can most readily be visualized by imagining the way this would be done. Let us suppose (Fig. 8.2) that the lens coil of an electron microscope has an ammeter in series with it, and that the supply circuit has a variable resistor, also in series. A man

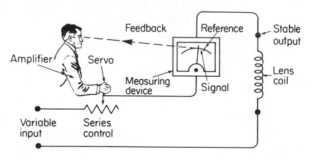

Figure 8.2   A diagram showing the mode of operation of a negative feedback stabilizer and explaining the terms used

sits with his hand on the control knob of the variable resistor, and keeps his eye fixed on the ammeter. There is a mark on the ammeter dial, and as soon as he sees the needle move away from the mark (this is the signal), he moves the variable resistor in the correct direction to bring the ammeter needle back to its original position. The ammeter thus acts as the measuring and reference unit; the man's eye is the detector; his arm is the signal amplifier; and the variable resistor is the control unit. It will be appreciated that such a system cannot maintain perfect stability of current through the lens coil. The man cannot act until his eye detects a signal; therefore the current must change before it is brought back again, so there is bound to be a fluctuation. There will be a time lag before a correction is applied, and the correction cannot be perfect, because there is a limit to the accuracy with which the man can read the ammeter. Also, his applied correction may cause the current to overshoot in the opposite direction, and he will have to apply a reverse correction. This is the phenomenon of 'hunting'.

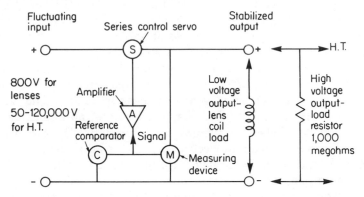

Figure 8.3   A block diagram showing the circuit elements of an electronic stabilizer, using the same terminology as in the preceding figure

Since men are expensive, inaccurate and subject to fatigue, automatic measuring devices have to be found to take their place, together with automatic servo controls to vary the current. The basic circuit of an electronic stabilizer is shown in Fig. 8.3. It is known as a 'degenerative' or 'negative feedback' stabilizer. A measuring device M capable of detecting small changes in output voltage from a predetermined value is connected across the output of the stabilizer, which is fed by a fluctuating input voltage. The voltage detected by M is compared with the voltage of a comparator C, which must remain as stable as possible, since the performance of the circuit depends on its stability. Any fluctuation, either positive or negative, is then amplified by an amplifier A and is then passed to the series or servo control S, which acts in such a way ('negative feedback') as to nullify the effect of the fluctuation detected by M. It will be seen that the circuit cannot operate in the absence of a signal, and that its sensitivity will depend on the amplification provided by A.

To make this basic circuit work, we must provide devices which will act as signal detectors, amplifiers and controllers. Fortunately, all these functions can be performed by one basic unit, the 3-electrode 'valve' (or 'vacuum tube' in American parlance). Modern technology has lately developed the 'transistor', which is a more robust and reliable solid state device capable of performing the function of a valve. The way in which transistors work will be discussed on p. 187.

8.5 The Electronic Valve

A valve is a device which controls the flow of a stream of electrons across an evacuated space by means of the application of a voltage. In this way, a pure voltage signal, devoid of power, can be made to control large amounts of power in the form of an electric current in an external circuit. This property also enables a valve to amplify voltages and thus to increase the intensity of a signal in an external circuit.

A diagram of the construction of a three-electrode or 'triode' valve is shown diagrammatically in Fig. 8.4. It consists of an evacuated bulb, generally made of glass, containing the three electrodes. The first, the 'cathode', emits a stream of electrons and is placed at the centre of the bulb. Its function is similar to that of the filament in the electron gun. The second electrode, the cylindrical 'anode', surrounds the cathode symmetrically and at some distance from it. It is identical in function to the anode of the electron gun. The third is the control electrode or 'grid', an openwork wire mesh cylinder closely surrounding the cathode. The grid corresponds to the bias shield in the electron gun. Wires leading from the electrodes are brought out from the vacuum through a glass 'pinch' to pins held in an insulating base cemented to the bottom of the glass bulb. The connecting pins are arranged so that the valve can be plugged into a socket from which further wires connect the electrodes to the elements of the external circuit.

The cathode is a thin metal tube coated with a mixture of rare earth oxides, which have a very low 'work function' (see p. 112). It emits electrons very copiously at a dull red heat. The power necessary to keep the cathode hot is provided by the electrical resistance heating of a thin insulated wire, the 'heater', which forms a hairpin inside the tubular cathode. The heater is generally designed to operate at 6.3 V and to dissipate about 2 watts of power.

When the valve is in operation, electrons 'boil off' the red-hot cathode and their mutual repulsion causes them to form a 'space charge' close to the surface of the cathode. If a potential positive with respect to the cathode is now applied to the anode, electrons will be drawn from the space charge to the anode, and a current will flow in an external circuit.

Let us now consider the function of the third electrode, the grid. It is generally a closely wound spiral of fine wire placed very much closer to the cathode than to the anode. Because it is so close, it exerts far more

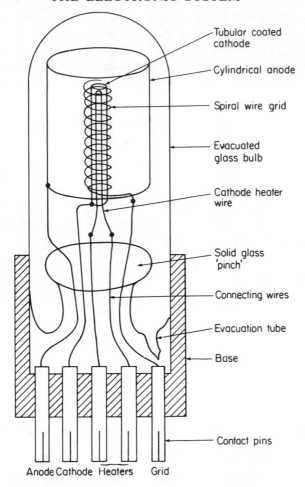

Figure 8.4   A diagram showing the construction of a
triode valve

control over the space charge of electrons than the anode does. The grid
is always made slightly negative with respect to the cathode, so that it
cannot collect any electrons. Therefore, no current can flow in the external
grid circuit, and thus no power is required to control the electron stream
flowing through the valve.

Let us now consider the operation of the valve together with the simplest
kind of external circuit, shown in Fig. 8.5. The valve is represented
symbolically by a circle containing symbols for the three electrodes and

the heater. The anode is connected to a battery B through a load resistor
R. The grid must be maintained at a potential slightly negative to the
cathode, so a further low voltage 'grid bias' battery G is connected between grid and cathode. Suppose the voltage of the anode battery is 100,
the value of the load resistor is 5,000 ohms, the grid is at −1 V, and that,

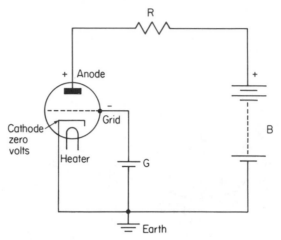

Figure 8.5   A diagram showing the external circuit
used in conjunction with a triode valve

under these conditions, a current of 10 mA (0·01 A) flows through the
valve and its associated load resistor. Then, by Ohm's Law, the voltage
drop across R will be the product of the current I and the resistance:

$$V = IR = 5{,}000 \times 0{\cdot}01 = 50 \text{ volts}$$

and the power dissipated in R will be:

$$W = I^2R = 5{,}000 \times 0{\cdot}01^2 = 0{\cdot}5 \text{ watts}$$

Next, let us suppose that changing the grid voltage from −1 V to
−2 V causes the current through the valve to drop to 5 mA. The grid is
capable of this degree of control because of its close proximity to the
cathode. The voltage drop across R will now be:

$$V = IR = 5{,}000 \times 0{\cdot}005 = 25 \text{ V}$$

and the power dissipated by R will be:

$$W = I^2R = 5{,}000 \times 0{\cdot}005^2 = 0{\cdot}125 \text{ W}$$

Therefore, a change in grid voltage of 1 V has given rise to a voltage
change in the external circuit of 25 V. The valve and its external circuit

are said to have an amplification or 'gain' of 25. The same change in grid voltage has controlled the power in the external circuit by a factor of 4. The 'sensitivity' or 'mutual conductance' of the valve is said to be 5 mA/V.

8.6 The Valve Stabilizer

Let us now see how the triode valve can be used in a stabilizer, either for maintaining lens current or H.T. constant. A simple stabilizer circuit diagram is shown in Fig. 8.6. The basic layout will be seen to correspond to the stabilizer block diagram shown in Fig. 8.3. The measuring device is a high-stability resistor r_m placed in series with the load, which is shown here as a lens coil, but could also be the load resistor of the high tension

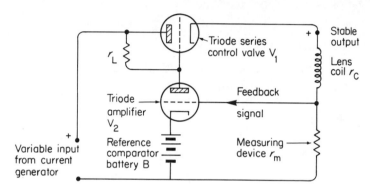

Figure 8.6   The basic circuit of an electronic negative feedback stabilizer using two triode valves

generator. When a current is passing through the lens coil, it also passes through r_m, and causes a voltage drop across it. This voltage drop is compared with the voltage of a battery B. If it differs from B, a signal is applied to the grid of the amplifier valve V_2, which controls the series control valve V_1 which is in series with the lens coil and measuring resistor.

The way in which the circuit works is as follows. Suppose the input voltage to the stabilizer drops, due to a sudden drop in the mains input to the current generator (which is not shown). This causes an immediate reduction in the current through the series circuit of V_1, the lens coil and r_m. The voltage drop across r_m becomes less. But the voltage of B remains constant; therefore the voltage on the grid of the amplifier valve V_2 becomes more negative. V_2 therefore passes less current, and so the voltage drop across its load resistor r_L also becomes less. But the voltage drop across r_L controls the voltage on the grid of V_1, which therefore becomes more positive (closer to anode potential). V_1 therefore passes more current. This is exactly the effect required to increase the voltage drop across r_m, which also increases the current through the lens coil, which is the object we wish to achieve. The stabilizer works correctly.

The basic principle of the circuit is to maintain the voltage drop across r_m equal to the voltage of the battery B. It does this effectively by altering the series resistance of the valve V_1, just as the man watching the meter in Fig. 8.2 did by moving his series resistance in response to the received signal. The greater the amplification given by the combination of V_2 and r_L, and the greater the sensitivity of V_1, the greater will be the overall stabilization factor of the stabilizer.

8.7 Stabilization Factor

The ratio of the change in output volts to the change in input volts in a stabilizing circuit is known as the 'regulation factor'. It can be shown that this parameter R, which measures the effectiveness of the circuit, is given by the expression:

$$R = \frac{dV_o}{dV_i} = \frac{1}{1 + G},$$

where G is the total gain of the circuit. G is the product of the amplification factors of the measuring circuit M and the servo valve S, i.e. $G = (G_{v_2} \times G_{v_1})$. Since the gain of the circuits used in practice is large, of the order of 1,000 or more, it is more convenient to consider the inverse of this relationship, which is called the 'stabilization factor', and which is therefore:

$$S = 1 + G \approx G, \quad \text{if } G \text{ is large}$$

Another factor, however, must be taken into consideration, which is the fraction of the output voltage which is used to generate the signal. In Fig. 8.6, if the resistance of the lens coil is r_c, then this fraction is given by the ratio of r_m to the combined resistances of r_m and r_c. The stabilization factor is therefore:

$$S = \frac{r_m}{r_c + r_m}(1 + G)$$

8.8 Practical Stabilizer Circuits

The simple two-triode circuit we have just been considering has a number of disadvantages. The main one is that the gain of a simple triode cannot be made very high; the figure we have suggested of a circuit gain of 25 is reasonable. This limitation is due to difficulties in valve electrode construction. They can be overcome by inserting more control electrodes or grids between the cathode and the anode. If a particular type of valve called a 'pentode', which has two extra grids and hence five electrodes, is used in place of the triode as an amplifier, a circuit gain of 1,000 or more can quite easily be achieved.

A stabilizer circuit using a pentode amplifier and triode series control valve is shown in Fig. 8.7. As well as the greatly increased amplification

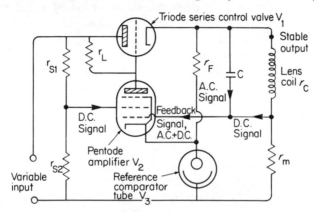

Figure 8.7 The basic circuit of the type of stabilizer using a pentode amplifier and triode control valve which is generally found in electron microscopes

obtained by using the pentode as an amplifier, a further advantage is gained by using the input voltage to control the signal on the series valve as well as the output signal from r_m. A potential divider r_{s_1} and r_{s_2} is connected across the input, so that the second or 'screen' grid acts as a secondary control. Input variations now affect V_2 directly, thus increasing the effectiveness of the stabilizer.

A further modification often made in practice is to replace the reference battery by a different type of reference source, the cold cathode gas discharge tube V_3. This consists of two electrodes sealed into a bulb containing an inert gas such as neon or argon at a reduced pressure. Its external appearance is identical with that of a valve, but when operating, it can be seen to glow with the characteristic colour of the gas it contains. The voltage drop across such a device, once it has 'fired', is almost independent of the current passing through it, and is dependent only on the physical dimensions of the electrodes, the temperature and the gas it contains. It is a 'non-ohmic' resistor. While its stability is not quite as good as that of a specially designed reference battery, it has the great advantage of requiring no maintenance. Failure of a gas discharge reference valve is very rare; batteries require frequent replacement. In the circuit shown in Fig. 8.7, the reference comparator valve V_3 is connected to the cathode of the pentode amplifier V_2 and acts in the same way as the battery in Fig. 8.6 to keep the cathode potential of V_2 constant. It is fed with an almost constant current from the stabilized output of the circuit through a resistor r_f.

A further refinement is the capacitor C, which is connected across the load. This capacitor is selected to have a considerably lower impedance (alternating current resistance) to ripple or 'mains hum' than the load itself, and thus the stabilization factor for ripple is made higher than that for slow D.C. variations. The resultant output can thus be made almost completely free from mains frequency ripple or hum.

The effective gain of such a circuit can be made fairly high—of the order of 2,000 times. The stabilization factor is reduced, however, in the ratio of r_m to the total output resistance, which is in practice about 1 to 5, or 0·2. The stabilization factor of the single amplifier circuit shown in Fig. 8.6 is therefore about 400. This can be increased by a further factor of 10 or more by introducing a more complex, two-stage amplifier in place of V_2. This is generally found in the stabilizer for the objective lens current, where the highest degree of stability is required.

8.9 High Voltage Stabilization

The achievement of a stability of the order of 10^6 in the 100,000 volt electron accelerating potential is not as simple as the achievement of comparable stability in the lens supplies. Three basic methods can be used to solve the problem. Firstly, the adaptation of a simple degenerative stabilizer such as is used for the lens circuits. Secondly, the mains input to the H.T. generator tank can be stabilized to the required degree. Thirdly, a low-voltage, high-frequency (50 kc/s) signal of the required stability can be generated, this being used to feed an oil-immersed step-up generator-transformer of simpler design than is necessary in the second case.

The simplest way of describing the high tension circuits used in practice is by means of block diagrams (Fig. 8.8). In all cases, the components which actually generate the high tension have to be immersed in a tank of transformer oil in order to prevent unwanted discharges, which would interfere with stability. The circuits which amplify the signal derived from the measuring resistor chain are mounted out of the oil tank, and must therefore be kept at a potential fairly close to earth. This means that the

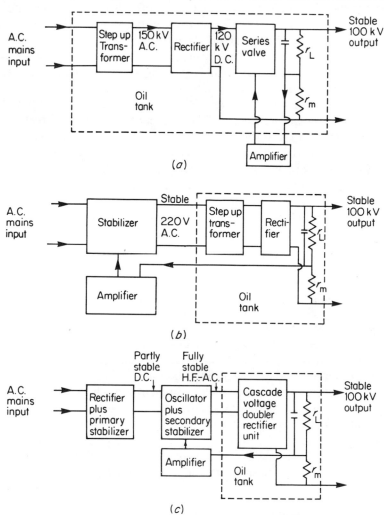

Figure 8.8   Block diagrams illustrating the three basic methods used for generating and stabilizing the high accelerating voltage used in electron microscopes

ratio of load resistor to measuring resistor has to be much smaller than in lens stabilizers; consequently, the stabilizer must have more amplification.

In the case of a simple degenerative stabilizer (Fig. 8.8(a)), a triode series valve of special construction must be mounted in the oil tank. This valve has to withstand the full high tension voltage across it, and in consequence is very expensive. The oil tank also contains the step-up transformer and the rectifier. Stabilization takes place in this case after the high voltage has been generated. In the other two classes of stabilizer, stabilization takes place before the high voltage is generated. In the second type (Fig. 8.8(b)), the monitoring signal passes out of the oil tank to an amplifier and thence to a stabilizer which maintains constant the A.C. mains input to the step-up transformer inside the tank. The resulting high alternating voltage is then rectified by means of solid-state rectifiers in the oil bath, and the stable rectified D.C. is passed to the electron gun. In the third type (Fig. 8.8(c)), the A.C. mains is first rectified to D.C. and stabilized; it then energizes a high frequency oscillator governed by the signal amplifier to give a constant voltage output at around 50 kc/s. This high frequency is then applied to a cascade voltage multiplier in the oil tank, which simultaneously generates and rectifies the high voltage.

8.10 Filament Voltage Supply

The filament requires a supply of about 1·5 volts at about 2 A. This supply does not need to be highly stabilized, because the electron gun is always run under 'saturated' conditions (see below). The supply can be either A.C. or D.C.; in the interests of highest stability, most high performance modern instruments use a D.C. supply. The filament supply has to be generated in the oil tank, since the filament and its associated electron gun are at the accelerating potential. In some designs, a part of the highly stabilized high frequency input to the high voltage generator is stepped down in voltage by means of a small transformer to feed the filament with high frequency A.C. This supply may subsequently be rectified and the filament fed with D.C. The filament supply can be at mains frequency, in which case it can be generated and stabilized by a separate transformer of the 'saturated reactor' variety contained in the oil tank.

8.11 Electron Gun Bias Supply

One other potential has to be generated in order to operate the electron gun—the bias potential applied to the control electrode or Wehnelt shield. Strictly speaking, the latter term is only applicable if the control electrode is connected directly to the filament or cathode. In this case, a two-electrode gun results, a type found only on very old instruments. If the control electrode is operated at a different potential to the filament and anode, the gun has three electrodes, and its operation is exactly analogous to that of a triode valve. In this case, the control electrode is usually referred to as the 'bias shield'.

As has already been described on p. 114, a negative potential of several hundred volts with respect to the filament is applied to the bias shield in order to control electron emission. The way in which this potential is derived is very ingenious, because it allows the gun to be made self-regulating and practically independent of filament temperature and hence filament voltage.

The method is to use the excess or waste beam current (that part of the beam which strikes the shield electrode and does not pass out towards the anode through the central aperture) to generate the shield potential. This is done simply by connecting the shield to the filament through a resistance R_B (Fig. 8.9) called the 'cathode bias resistor'. The presence of the resistor causes electrons to pile up on the shield, which becomes negatively charged with respect to the filament. The greater the value of R_B, the greater the negative potential on the shield, and the greater is the effect of the shield in repelling electrons back to the filament. The value of R_B can therefore be made to control gun emission. Varying the 'Emission' control on the microscope console alters the value of R_B, which, since it is at gun potential, has to be immersed in the oil tank along with the H.T. and filament generators. Control is effected either by means of a flexible cable or by a remotely controlled electric motor. The arrangement is called a 'self-biased gun'.

The effect of the bias resistor is to stabilize gun output. If the gun output tends to rise due to an increase in filament temperature, the shield potential becomes more negative and thus prevents it. It therefore maintains gun output independent of filament temperature. If gun emission is plotted against filament voltage (and hence filament temperature), increasing filament voltage will initially increase emission (see p. 112). A point will be reached where the value of the bias resistor prevents any further significant rise in emission with filament temperature. This is the so-called 'saturation' point of the filament, the implications of which must be fully understood by the operator. The term is something of a misnomer; the filament itself is not saturated in the sense that it cannot emit any more electrons; rather the gun is 'saturated' by the action of the bias resistor and the shield.

One effect of the saturation point in practice is that the filament temperature need not be raised above it. Any further increase will simply reduce filament life without improving the brightness of the gun. If, however, the value of the bias resistor is decreased, then emission will increase,

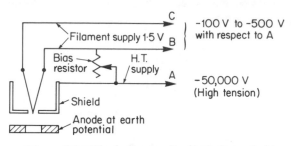

Figure 7.9   The basic circuit of a 3-electrode biased electron gun

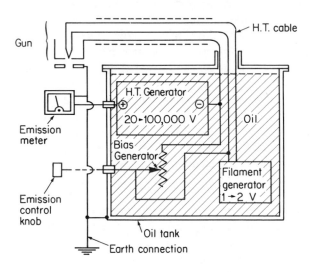

Figure 8.10   The arrangement of components in the oil-filled high-voltage generating tank of an electron microscope

illumination intensity will increase, and the filament temperature must be increased in order to return to the saturation point. Attempts to operate the gun below the saturation point will result in uneven and randomly fluctuating illumination due to the gun being no longer self-compensating with respect to filament temperature.

Fig. 8.10 shows the general arrangement of the H.T. bias and filament generators in the oil tank. On some instruments, two oil tanks are used, one to generate the high tension and a second, mounted close to the column, to generate the filament and bias voltages. The H.T. cable is of special construction, and is sometimes filled with transformer oil for additional insulation and flexibility. The cable is very delicate and susceptible to damage by bending or abrasion. It cannot be cut; surplus cable must be rolled to a stated radius and stored on a drum.

8.12 Solid State Electronics

Semiconductor diodes and triodes (transistors) have now virtually completely replaced the earlier thermionic valve circuits which have just been described. However, the general principles of the stabilizer and control circuits remain exactly the same. Semiconductor devices have enormous advantages over valves. They work at room temperature and hence do not require any cathode heater power. The physical volume of a semiconductor component can be at least 1,000 times smaller than that of a corresponding valve, since no vacuum-tight envelope is required and very little heat is generated. Transistors are electrically far more efficient than valves, requiring only 5–10 volts for operation instead of 150–300. They are physically far more robust and electrically far more reliable. These overwhelming advantages allow the physical size of a solid state circuit to be greatly reduced from that of a corresponding valve circuit. This, combined with the fact that almost no waste heat is generated, allows the whole solid state electronic circuitry of an EM to be built into the console surrounding the column.

The fundamental principle behind the valve is that a large amount of electrical power flowing from cathode to anode is controlled by a very small amount of power applied to the grid. In the transistor, a large amount of power flowing from the emitter to the collector is controlled by a much smaller amount of power applied to the base. Since solid state terminology differs from valve terminology, a brief description of the characteristics of these devices will be given.

In a triode valve, conduction is by means of electrons flowing across a vacuum, the flow rate being controlled by an electric field applied to a control electrode (grid) interposed between the electron source (cathode) and collector (anode). Current can only flow in one direction. If the grid is omitted, the valve becomes a diode or rectifier. In a transistor, however, conduction may take two forms. Firstly, it can be by electrons flowing through a single crystal of a semiconductor (either silicon or germanium) in the usual fashion of metallic conduction. Since neither of these quasi-metals contains many free electrons in the pure state, very little conduction can take place until the pure element has been 'doped' with foreign or 'impurity' atoms such as arsenic or antimony, which act as electron donors. A second form of conduction is possible, however, which makes three-electrode operation, and therefore amplifiers, feasible. If the pure semiconductor crystal is doped with an impurity such as indium, which captures electrons, 'positive conduction' can take place in the opposite direction to electronic conduction as a result of a deficiency of electrons. There is a flow of positive charges or 'holes'. These two forms of conduction are termed '*n*' (for negative) and '*p*' (for positive) respectively. A modern junction-type transistor is constructed of a sandwich of these two '*n*' and '*p*' materials (see Fig. 8.11). The two outer layers of the

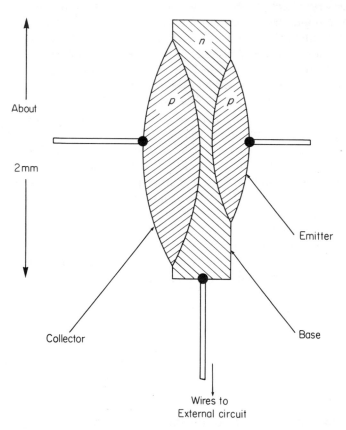

Figure 8.11   A section through a junction-type *p-n-p* transistor

sandwich are made of the same material, while the central layer (which must be extremely thin, 50 μm or less) is made of the opposite material. The three layers of the sandwich are the electrodes, termed 'emitter', 'base' and 'collector' respectively. Wires are firmly connected to each electrode so that external circuit elements can be connected. The emitter corresponds to the triode valve cathode, the base to the grid and the collector to the anode. An 'n-p-n' transistor works as follows. When the n-type collector is made positive with respect to the emitter, electrons will flow from the n-type emitter into the p-type base, which is deficient in electrons. The conduction electrons diffuse through the very thin base, and most will arrive at the collector and will pass from thence into the external circuit. Some conduction electrons, however, will be captured by the 'holes' in the p-type base, which therefore charges up and checks the flow. These electrons can be removed into the external circuit by biassing the base positively, and flow from emitter to collector can be restored. By suitable design of the sandwich, a small number of electrons can be made to control the flow of a far greater number; possibly 100 times as great. The device is therefore current-operated in contrast to the valve, which is voltage-operated (virtually no current flows in the grid circuit under normal conditions of operation). A current can be turned into a voltage by passing it through a resistor and, therefore, by small alterations to the circuitry, the transistor can replace the triode valve in all but the largest power applications. It will also be seen that if only the emitter and base electrodes are taken into consideration, the device acts as a diode or rectifier, allowing current to pass in one direction only. It can therefore replace the rectifier valve.

The original design of transistor developed at Bell Telephone Laboratories by William Shockley and his colleagues from 1948 onwards was the germanium 'point-contact' type which, although capable of amplification, was unable to handle appreciable amounts of power due to heating and melting of the point contact junctions. Early transistors could not be used in power stabilizer circuits. With the development of the junction-type transistor described above, transistors could be designed with much larger contact areas between the semiconductor slabs and power applications became possible. These power transistors and rectifiers have to be cooled, since their electrical characteristics alter markedly with temperature. They have to be mounted on 'heat sinks', which are massive pieces of finned aluminium to which the devices are bolted. Heat sinks in EM stabilizer circuits are frequently water-cooled, using the pump and lens water supply.

8.12.1 Transistor Circuits

A very basic transistor series stabilizer circuit corresponding to the valve circuit shown in Fig. 8.11 is shown in Fig. 8.12. The series control element is a large *p-n-p* silicon power transistor mounted on a water-cooled heat sink. A suitable transistor is about 2 cm in diameter and 0·5 cm in thickness, and will handle several amperes, which is the full current for one lens coil. In series with the lens coil is the stable measuring resistor, the voltage across which varies with lens current, which we wish

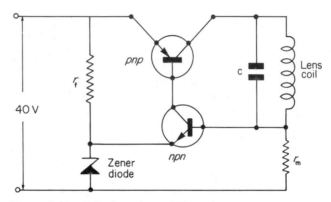

Figure 8.12   A basic series stabilizer circuit using an *n-p-n* amplifier transistor and a *p-n-p* series power transistor

to maintain constant. This voltage is fed back to the base of the *n-p-n* amplifier transistor, the emitter voltage of which is held constant by a device called a 'Zener diode' after Clarence Zener, who discovered it. This device corresponds exactly to the gas discharge reference tube. It is a non-ohmic resistor; a special semiconductor diode with 'breakdown' characteristics similar to those of a gas discharge tube. Above a certain current, 'avalanche conduction' takes place, and the voltage drop across the device becomes virtually independent of the current flowing through it. This voltage is the stable reference voltage. As the base voltage on the *n-p-n* transistor varies, the collector current is amplified, and the induced voltage on the collector, and hence on the base of the power transistor, causes the conductance of the latter to vary. This is a *p-n-p* device, and therefore the required 180° phase shift is automatically applied. The circuit acts to increase the conductance of the series power transistor when the current in the lens circuit drops, thus stabilizing the lens current. A feature of transistor stabilizers is the very much lower voltages at which they run compared with valve circuits. EM lenses are usually run off a 40-volt line, about 10 volts being the drop across the stabilizer. The lens coils are wound with fewer turns of thicker wire, and run at much higher currents than valve-stabilized coils.

The stabilization factor of the simple circuit described above is quite inadequate for a lens, therefore a cascade transistor amplifier of several stages, arranged so as to minimize thermal drift, is used between the measuring resistor and the series control element. A primary stabilizer followed by individual lens stabilizers is also used, following traditional valve practice.

The small physical size of individual transistors and diodes (other than power transistors) has encouraged the miniaturization of other circuit elements, notably resistors and capacitors. All the amplifier components for a stabilizer can be mounted on a 'printed circuit board' with integral wiring which is only 10 cm square or even less. Each individual printed circuit board plugs into the main circuitry by means of contacts formed along one edge. The complete board can be replaced as readily as a valve in a thermionic unit. This technique has revolutionized the servicing of EMs, since the service engineer has only to locate a fault to a particular circuit board and exchange it for a new, factory-tested one. Since the the boards are no larger than valves, he can be equipped with a full set of the different types used in a particular instrument. The components on a faulty board are often so cheap that it is not economic to trace the trouble to a particular component and replace it; the board is simply discarded. A minor disadvantage of this technique is the mass of wiring and plug-in joints required, which could reduce reliability. As components become more and more reliable, doubtless the plug-in circuit board technique will in its turn be replaced by instruments which never require any electronic servicing.

CHAPTER 9

Commercial Transmission Electron Microscopes

9.1 Introduction

Electron microscopes are very complex pieces of instrumentation combining design expertise in many different fields. They can, in general, only be manufactured by firms with vast capital resources behind them. The makers are therefore in general huge, internationally known firms or consortia with established departments of vacuum technology, electronics and general instrumentation as well as electron optics. Because electron microscopes tend to be spectacular and keep very much in the news, these firms seem to be happy to use their wares in this field partly as prestige advertising, and to take relatively small profits on them. Electron microscopes are therefore in general extremely good value for money in the field of scientific instrumentation. However, in this field as in any other, you get what you pay for. The prospective purchaser is very easily blinded by the frills while unaware of the design fundamentals.

In 1969, when the first edition of this book was written, no less than 14 different manufacturers in 11 different countries throughout the world were offering a total of at least 39 different models of TEM. In the intervening five years, many of these have fallen by the wayside, and the manufacture of TEMs in the Western world is virtually confined to seven major manufacturers, all of whom are large international corporations with subsidiaries and servicing facilities all over the world. These firms are, in alphabetical order, the following.

The British firm of G.E.C.–A.E.I. (General Electric Company merged with Associated Electrical Industries) has 135 principal subsidiary companies in Britain alone, one of which is the AEI Scientific Apparatus Division which manufactures TEMs. This subsidiary is descended from the scientific instrument division of Metropolitan–Vickers, which constructed the first British TEM at Imperial College, London, in 1935. The

192

two Japanese firms are Hitachi, who are amalgamated with Perkin–Elmer in the U.S.A., and the Japan Electron Optics Laboratory (JEOL) whose TEMs are given the code prefix 'JEM'. The Dutch electrical giant N.V. Philips' Gloeilampenfabrieken are the largest European manufacturers. In the U.S.A., the old-established firm of the Radio Corporation of America (RCA) continue to manufacture one model. Siemens of West Berlin were historically the first firm to market a commercial series-production TEM in 1939. Finally, Carl Zeiss of Oberkochen, the world-famous light microscope firm for whom Ernst Abbe worked, are also in the TEM field. A total of about 16 models is offered, including HVEMs.

The manufacture of TEMs is becoming increasingly specialized, and most manufacturers are now tending to concentrate on one standard model, either a relatively simple medium-performance model or a very complex high-performance instrument.

The complex models are in general a standard design suitable for plain transmission microscopy and designated 'biological' instruments, together with modifications to this standard design to supply demands for more specialized work such as high-angle specimen tilt, diffraction and possibly X-ray microanalysis of the specimen, together with specialist facilities for manipulating the specimen (rotating, tilting, straining, heating, cooling etc.) required either singly or together by various other specialist users such as metallurgists and physicists. The general pattern of design is to offer a TEM consisting of console, controls, vacuum system, electronic system, illuminating system, magnifying, viewing, and photographic system. This is modified to suit the user's requirements by fitting different types of objective lens, specimen stage and specimen manipulation facilities, generally as a single interchangeable unit. In theory, therefore, a purchaser can buy the standard instrument to begin with, and can fit any available modification at a later date simply by removing the objective lens unit and replacing it with another. Some manufacturers designate each modification with a different model number; others keep things simple and offer a single instrument with interchangeable accessories.

The available range of electron microscopes can conveniently be divided into three broad classifications. These are:

(a) Basic instruments
(b) Medium-performance instruments
(c) High-performance instruments.

Basic TEMs are here defined as those having a claimed resolving power of 50 Å or more, no condenser lenses and a top magnification of around

10,000 ×. Medium-performance TEMs have a claimed resolving power of around 5 Å, either a single or a double condenser, three imaging lenses and a top magnification of around 100,000 ×. High-performance TEMs have a claimed resolving power of 2 Å or less for line or lattice spacing and 3 Å for point-to-point (see p. 134), double condensers, four imaging lenses and a top magnification of 500,000 or more.

9.2 Basic Instruments

Few manufacturers offer basic instruments. They are of the simplest possible design, without condenser lenses and with a two-stage imaging system consisting simply of an objective and a projector, which may be combined as a double lens. Magnification is fixed at up to 4 selected values, the limit with such a simple and inefficient illuminating system being around × 10,000. The accelerating voltage is fixed at a fairly low value. There is no reason why the objective lens should not have quite a high resolving power; 20 Å can readily be achieved. At such low magnifications, however, micrograph resolution tends to be limited by photographic grain size and the ability to focus at low magnifications rather than by the performance of the objective lens. Basic instruments cost around £5,000—£10,000, and are intended mainly for use in quality control in industry where the size and shape of such things as small particles have to be checked routinely.

At this point, a fallacy must be exploded. The opinion seems to be held by some would-be electron microscopists that a basic instrument can be used for 'scanning' specimens and choosing the best ones for further, more rigorous examination on a high-performance instrument. This, it is claimed, would save time on the more expensive instrument. This concept is totally fallacious, for the following reasons. In the first place, a person who is experienced and competent in specimen preparation simply does not produce specimens that are so bad that they can be distinguished in a basic microscope. Really bad ones can be distinguished by the naked eye or the light microscope, and really good ones need a high-performance instrument to appreciate them. A basic instrument simply cannot give the required information. In the second place, even if the basic microscope could perform the job, the amount of time saved on the high-performance instrument would be minimal and would not justify the expenditure, space and maintenance necessary for the basic instrument. Some modern high-performance instruments can take up to 6 specimens at a time. In the third place, basic instruments, being necessarily simple, lack the means such as double condenser lenses and anticontamination

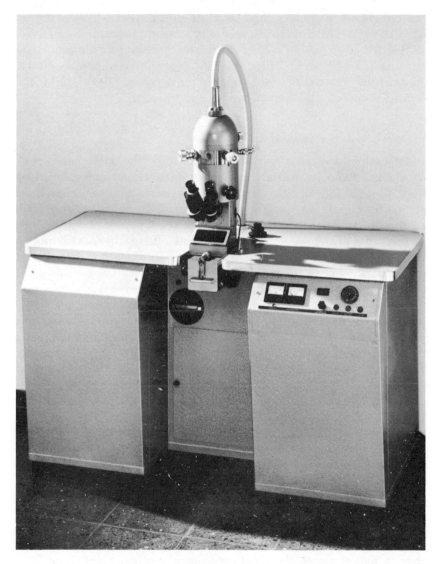

Figure 9.1 The Siemens Elmiskop 51 'basic' TEM with two permanent magnet imaging lenses. (Courtesy Siemens A.G.)

devices which are necessary to protect a good specimen and preserve its ultrastructure for later examination. The simplified optics of the illuminating system give rise to beam damage and contamination, and a delicate thin section or negatively stained virus preparation which has been 'scanned' in a basic microscope is very unlikely to be in a condition suitable for high quality work in a high-resolution instrument. Basic instruments are therefore of little value in a biological research laboratory, although they have their place elsewhere. A typical 'basic' TEM, the Siemens Elmiskop 51, is shown in Fig. 9.1.

9.3 Medium-Performance Instruments

Medium-performance instruments can offer the biologist, especially the cell biologist interested only in thin sections at magnifications below about × 50,000, everything he needs very much more cheaply than high-performance instruments. The biologist who works only with sections is limited at the present time to a resolution of about 20 Å by the thickness of the section. An instrument offering 10 Å resolution is therefore adequate for the work. 20 Å can be distinguished with the naked eye on a print at × 100,000, so an electron optical magnification of × 50,000 followed by an enlargement of × 2 makes full use of the available resolving power.

In addition, most medium-performance instruments offer considerable vacuum automation together with facilities such as anticontaminators and high-angle specimen tilt. Maintenance and alignment are generally cheaper and simpler on these microscopes, because factory prealignment is necessary for simplicity and cheapness in construction. As the cost of a medium-performance microscope is about one-half the cost of a high-performance one (£30,000—£50,000), the cell biologist working with thin sections who is choosing an electron microscope should examine very carefully any reason put forward for not buying one, and should be convinced that there really is a reason other than sheer prestige for wanting a high-resolution one.

Figs. 9.2 and 9.3 show two different types of medium-performance TEMs. The A.E.I. 'Corinth' (Fig. 9.2) is a breakaway from the conventional column-above-desk design shown in Fig. 9.3. The column is placed between the operator's knees, with a large, square (275 cm² area) transmission screen in the most convenient position for viewing. This instrument is the most compact on the market. Fig. 9.3 shows the more conventional Philips EM-201, which takes up approximately the same floor space as the 'Corinth' but is considerably taller.

9.4 High-Performance Instruments

These can cost up to £100,000, and offer the ultimate in electron micro-scopy. This necessarily includes ultra-high (better than 5 Å) resolving power. However, images obtained near the ultimate resolving power of a microscope are capable of conflicting interpretations, (see p. 271) due to phase contrast effects close to focus giving rise to image granularity. Resolving power at the 2 Å extreme is to a certain extent a shibboleth, and is only capable of somewhat uncertain measurement and dubious interpre-tation. This is, of course, an argument in favour of high resolution. If a microscope is limited to 2 Å by lens aberrations, then the information it

Figure 9.2   The AEI 'Corinth' compact inverted-column medium-performance TEM in the basic low-magnification 'clinical' form. (Courtesy GEC–AEI)

Figure 9.3   The Philips EM-201 medium-performance TEM showing the conventional column-above-console construction. (Courtesy Philips Eindhoven)

can give at the 6 Å level is much more likely to be correct than information at the 6 Å level obtained from a 5 Å instrument. However, such arguments tend to be rather academic.

Far more important points in favour of the high-performance instruments are the highly sophisticated facilities which they offer such as complex and accurately measureable forms of specimen manipulation while under observation; highly automated or even completely automatic, foolproof and fail safe operation of the vacuum system; extreme stability of all electrical supplies by the use of the latest electronic techniques; and facilities for attaching image intensifiers, SEM, STEM, EXMA, television displays and videotape information storage. Most users do not require all these highly sophisticated aids and frills; as they say, 'You pays your money and you takes your choice'.

Ultimate resolving power is mutually incompatible with maximum facilities for manipulating the specimen while under observation. Resolving power is a function of the focal length of the objective lens. The shorter the focal length, the less the spherical aberration and the higher the resolving power. A very strong lens of short focal length must have the polepieces as closely apposed as possible. Since the specimen is placed between the polepieces in a modern design of objective lens, there is little room for specimen movement in any plane other than the lateral traverse necessary to select the field of view. Tilting the specimen is very difficult. Ultra-high resolution immersion lenses have a focal length of about 1 mm, and point resolving powers of 3 Å or less are theoretically possible. It must be pointed out, however, that the possession of an ultra-high resolution electron microscope does not automatically confer upon its user the ability to take meaningful ultra-high resolution electron micrographs, any more than the possession of a very fast sports car confers upon its owner the ability to drive it to its limit. A very great deal of experience is necessary before the user has mastered the techniques of specimen preparation, microscopy and image interpretation sufficiently well to make genuine 5 Å micrographs a matter of routine.

Fig. 9.4 shows a typical modern high-performance TEM, the Philips EM 301, complete with full attachments for SEM, STEM, EXMA and TV display. The column, fitted with the eucentric goniometer tilting stage controlled by the foot pedals, stands on the control console in the centre of the photograph. To the left of the column is the liquid N_2 cooler for the ED X-ray detector. On the trolley to the left of the console is the TV monitor displaying the elemental X-ray peaks, operated by the spectrum generator and 'data improvement' computer on which it stands. A fibre optics TV display for the normal TEM image stands to the right of the column. On the trolley to the right of the console is the scan generator and TV display system for the SEM and STEM facilities. In this form the instrument is known as the 'TEAM 301', the acronym standing for Transmission Electron Analytical Microscope.

Figure 9.4   The Philips EM-301 high-performance TEM complete with ED–EXMA elemental spectrum display (on left) and SEM–STEM scan generator and display (on right). (Courtesy Philips Eindhoven)

Once it has been accepted that extreme resolving power is seldom either possible or necessary, and that instruments capable of extreme resolution are sought more as a matter of laboratory prestige than as a necessity, the choice of an instrument becomes at once more logical. Let us examine more closely some of the features which manufacturers tend to stress above others in the descriptions of their instruments.

9.5 Features of Modern Transmission Electron Microscopes

9.5.1 The Double Condenser

This is an essential feature for instrumental magnifications of over about × 50,000. The two condenser lens currents should be independently adjustable so that the size of the focused spot can be altered, or there is little point in having a double condenser. For magnifications below this figure, a perfectly adequate single condenser can be designed which is much cheaper. A double condenser should be fitted with an electrical stigmator and electrical beam shift controls.

9.5.2 Individual Lens Centration

With the increasing precision of modern manufacturing methods, it is becoming less necessary to provide individual adjustments for centration and tilt on each lens. The modern tendency is for factory prealignment of mechanical components, together with electrostatic or electromagnetic devices for tilting, shifting and deastigmating the beam. There is always a residual error in fixed factory alignment.

9.5.3 Column Jack

This is a very desirable feature for ease of maintenance. It enables the column to be easily split at any level without dismantling the whole column. With good management, it is seldom required, but its presence is very much appreciated when it is necessary to get at an otherwise inaccessible part of the column in a hurry.

9.5.4 Anticontamination Device

This is essential for work requiring high resolution, although if the column is clean and well-maintained, it should not be necessary to use it at magnifications below about × 50,000. The modern tendency is to combine it with a cryopump in the stage chamber, which is a large, cooled area acting as a condensing surface to improve the general vacuum around the specimen. The stage chamber should always be connected directly to the vacuum manifold by a wide tube.

9.5.5 Specimen Manipulation

It is becoming increasingly necessary for the biologist to be able to tilt the specimen through relatively large angles (30° or more) while examining the image. The provision of a tilting stage is a very worthwhile feature. Other forms of manipulation such as 360° specimen rotation, straining, heating and cooling are at present of little interest to the biologist, but may become so. Provision for the incorporation of such features at a later date is worth bearing in mind.

9.5.6 Selected-Area Diffraction

This is seldom required by the biologist, but it may well become more important in the future as means are found for extracting more information from the specimen by low-angle scatter. It is at present a desirable but non-essential feature.

9.5.7 X-ray Microanalysis

This is potentially a very powerful method for obtaining further information about the constituents of a specimen (see Chapter 17). The design of instruments which can be used both as electron microscopes and X-ray microanalysers is still in its infancy, but in principle all that is necessary is to modify the objective lens. This may be a good reason for getting an instrument with readily interchangeable objective lenses.

9.5.8 Scan Magnification

The provision of a low, fixed magnification of × 500 or less which is immediately available without going through a series of adjustments is a very desirable feature. It allows large areas of the specimen to be surveyed very rapidly in order to choose a suitable area for examination at high magnification.

9.5.9 Focusing Wobbler

This is a very desirable aid to image focusing at low magnifications, below about × 10,000, where the criterion of visual sharpness is not good enough to obtain perfect focus.

9.5.10 Alignment Modulator

This is a device for varying either the objective lens current or the high tension voltage over a small range at a frequency of about one cycle per second in order to align the current or voltage centre of the microscope to the screen centre. It is a worthwhile feature but not essential; it can be a great time-saver.

9.5.11 Fully Automatic Vacuum System

This can be regarded either as the latest marvel of instrumentation, or as yet another complication to go wrong. If a completely automatic vacuum system is provided, it must be utterly reliable and not prone to pressure sensor contamination, sticking valves and burnt-out motors. It is an expensive addition, regarded by some experienced microscopists as an unnecessary frill. To the beginner it could be a great blessing. As such systems become more reliable, they are incorporated on more instruments. It is now impossible to buy a high-performance microscope without an automatic vacuum system.

9.5.12 Multiple Specimen Holder

This is a great time-saver where a number of specimens has to be examined in order to select the really good ones. It is a great blessing for instance in selecting electron autoradiographs. It is also very useful for comparing a specimen with a standard. Multiple specimen holders can only be fitted to side-entry objective lenses, although several specimen holders can be placed into some top-entry specimen chambers.

9.5.13 Serial Section Specimen Holder

For biologists engaged mainly in stereology (the reconstruction of three-dimensional objects from two-dimensional images) the technique of serial sectioning can be useful. A multiple grid specimen holder is invaluable here, since ribbons of serial sections can be split into groups, each mounted on a separate grid. Of greater interest is a specimen holder with a slit 2·5 mm long, which can accommodate 250 or more sections of normal size in an unbroken ribbon. The ability to use such a specimen holder to its greatest advantage would be taxing the ability of the microtomist to its uttermost. It can only be applied to side-entry objective lenses.

9.5.14 Four-lens Imaging Systems

The advantages of this type of imaging system have already been discussed in Chapter 6. For magnification, only three lenses are used at any one time, the objective and projector plus one of the intermediate lenses. Of the two intermediate lenses provided, one is used for diffraction and as a demagnifying lens for low magnification work, while the other is used for high magnification work. The advantage of this arrangement is that there need be no compromise in design such as is essential in three-lens instruments, especially those in which the excitation of the projector lens is fixed. A further advantage in diffraction work is that by using all four lenses together, a focused selected area diffraction pattern can be obtained over a wide and continuously variable camera length with the specimen in its normal position. This is very useful for obtaining low-angle scatter diffraction patterns from relatively large objects such as regular arrays of biological molecules. More use of this facility will doubtless be made in the future.

9.5.15 Variable-Power Projector Lens

It is possible to obtain very low-power images with very low distortion by reducing the strength of the projector lens. This cannot be done simply by reducing the current in the lens; it must be done by increasing the bore of the projector polepiece. It is convenient to be able to do this from outside the column while operating the instrument. A turret of polepieces of different bores may be provided, which can be selected from outside the vacuum.

9.5.16 Electrical Shutter

This is almost essential if reasonably short exposures (less than 2 sec) are to be used. Short exposures are very desirable (see Chapter 14). All high-performance instruments are now fitted with them, and older models can usually be converted quite easily by an addition to the column between the projector lens and the viewing chamber. Electrical shutters are now appearing on medium-performance instruments.

9.5.17 Multiple Accelerating Voltages

Some medium-performance TEMs may be bought with one fixed accelerating voltage (generally 50 or 60 kV) or with a selection, generally 40–60–80–100 kV. For biological work involving sections, multiple accelerating voltage is seldom necessary. Generally, the 40 or 60 kV will be selected and used for all routine work. 100 kV is only of value if high-angle tilting of relatively thick (100–200 nm) specimens is contemplated.

9.5.18 Stereo-operation and High-angle Double Tilt

It is possible to get three-dimensional stereoscopic information from thick specimens if the specimen can be tilted on either side of the axis of the electron beam (see p. 268). Even with relatively thick sections, tilt angles of $\pm45°$ are necessary. This high-angle tilt also allows structures such as paired membranes to be aligned parallel with the electron beam and thus to be imaged more clearly, provided the specimen can be tilted in two mutually perpendicular axes (see Fig. 11.14). Frequently, the fitting of a high-angle double-tilt stage involves a loss in resolving power and magnification, since the objective lens polepieces must be spaced further apart, thus weakening the lens. This can be overcome if special very small ($1·5$ mm dia) grids are used. Ideally, the tilting device should be eucentric, i.e. the tilt axis always intersects the optical axis regardless of the position of the specimen. The methods are discussed in greater detail on p. 266.

9.5.19 Goniometer Stage

A goniometer stage enables the specimen to be turned through any required angle, the angle in degrees being indicated on a dial beside the stage. Such stages are almost indispensable to crystallographers and metallurgists, but are not in themselves of any great value to the cell biologist. When combined with a high-angle, single-tilt stage, the goniometer enables double-tilting to be carried out, since rotation of the specimen through $90°$ enables it to be tilted on an axis mutually perpendicular to the first. This mode of construction enables a eucentric stage to be constructed more easily, and is featured on the Philips high resolution goniometer (HRG) stage.

9.5.20 Scanning Facilities

It is now possible to buy some high-performance TEMs with three-lens probe-forming condensers giving a focused spot 250 Å or less in diameter, combined with a sweep system and a stage having detectors for EXMA, SEM and STEM. The cost of the modifications is almost as great as the cost of a separate scanning electron microscope which will do everything except TEM. In general, it may be said that two separate instruments, each designed for a specific purpose, are generally better than one multi-purpose instrument. With two instruments, one has the great added advantage that both may be used at the same time, although a combined instrument allows all four sets of information to be obtained either successively or simultaneously from the same specimen *in situ*, in the rather unlikely event of anyone wanting to do this.

9.5.21 'Zoom' Lens Systems

It is possible to link the currents in all the lenses of a TEM to the main magnification control in such a way that once the focus and screen brightness have been set to the optimum at any given magnification, the magnification can then be changed over a fairly wide range without a significant change in visual focus or image brightness. This rather complex luxury is available on two high-performance instruments, under the description of 'zoom' or 'proximity' focusing.

9.5.22 Self-cleaning Apertures

If the objective or condenser aperture is made by piercing a hole in a thin foil of molybdenum, the electrons it intercepts can be made to heat the metal around the hole to 200 °C or more. This greatly reduces the rate of contamination of such apertures, and allows them to be used for months at a time.

9.5.23 Other Features

Electron microscope manufacturers tend to vie with one another in the provision of various features which they hope will be unique and attract custom. These vary from the useful and important ones mainly listed above, down to gimmicks such as an automatic device to increase instrument-panel lighting intensity when the room lights are switched on. Such devices as automatic control of cooling water temperature, multiple interchangeable cameras, the ability to take several hundred micrographs at a sitting, the inclusion of a desiccator for photographic materials, meters to indicate magnification, devices to 'normalize' lens hysteresis to ensure reproducible magnification, foot switches for motorized drives, automatic exposure determination, airlocks between gun and column, very low (less than 30 kV) or very high (above 100 kV) accelerating voltages, exposure enumerators, automatic plate or film transport, whether or not the electronic circuits are built into the console and many of the other features which are described in the manufacturer's brochures are very much a matter of personal taste. They are mostly of marginal importance.

9.6 Second-Hand Instruments

It is not generally realized that a second-hand electron microscope can be an extremely good buy. This applies especially to a well-maintained older model by one of the top-ranking manufacturers. The commonest instruments which will have been in service for ten years or more are the

Siemens Elmiskop I and IA. More than 1,000 of these were manufactured in various guises since 1954. Electron optically, even the earliest
is superior to many of the medium-performance instruments available today. Modifications such as an improved gun and an airlock valve between
the column and camera chamber and an electrical shutter have been available for many years, and an early model fitted with these modifications is
well worth seeking by those with the ambition to possess a top-rank
instrument but without the depth of pocket to pay for a new one. Modern
operational facilities such as vacuum automation and large angle specimen
tilt will be lacking, and the instrument will require more skill and knowledge
to use to its fullest capabilities, but the results in normal (up to × 100,000)
transmission microscopy obtained by a skilled user will be almost indistinguishable from those obtainable from the most expensive instrument
available today. Modern instrumental improvements have been mainly in
the fields of automation, user facilities and specimen manipulation rather
than in the field of electron optics. Similar considerations apply to the
earlier models of the Philips 200 and the AEI–EM6 and EM–6B. Second-
hand instruments should become increasingly available as modern
specimen manipulation facilities become increasingly necessary in certain
fields of research, but plenty of scope will always remain for a first-class
transmission instrument of straight-forward design. A good well-maintained example may fetch one-third to one-half of its original cost ten
years ago. It still remains a very sound investment.

9.7 Prices

It is not possible to quote prices of new instruments, since these fluctuate
almost invariably upwards due to inflation and also rates of exchange vary
and import duty and taxes, if chargeable, vary according to governmental
whim. It must always be borne in mind that imported instruments are
liable in most countries to Customs duty, although educational establish-
ments are sometimes excused this. As a rough guide, the duty-free price
in Great Britain of basic instruments is between £5,000—£10,000; medium-
performance microscopes range from £30,000—£50,000; high performance
instruments cost between £75,000—£100,000. Erection on site and one
year's free maintenance are generally included in the price, plus free
tuition in the use of the instrument.

9.8 Specifications of Commercial Electron Microscopes

Summaries of the specifications of almost all of the transmission electron microscopes at present (1975) available commercially are given in Appendix I. Only instruments manufactured by firms having well-established international servicing networks have been included.

Special attention has been paid to features of direct interest to biologists; metallurgical instruments and modifications are mentioned in passing. The instruments are divided into 3 sections: basic, medium and high performance, and are listed in alphabetical order of the name of the manufacturer. This section should enable comparison to be made between various instruments for suitability for particular purposes. It must be stressed that electron microscopes are under continual development, and specifications are continually being changed and improved. The specifications have been obtained from the latest manufacturers' literature and by personal enquiry and are given in good faith, but the author accepts no responsibility for the accuracy of the statements made. Although the greatest care has been taken to ensure accuracy, the author would be grateful for any information about any errors which may have inadvertently crept in.

CHAPTER 10

Installation

10.1 General

An electron microscope is a very expensive and rather delicate piece of equipment which deserves care and thought in its housing. When properly installed, it is remarkably robust and trouble-free, taking into account its complexity. However, it must be remembered that it is inherently delicate, and will only prove reliable and comfortable to operate when installed in a suitable environment.

The keynote of the environment is stability and cleanliness. The microscope must be installed in a room by itself without any other apparatus such as vacuum evaporator or photographic equipment. The room must be as free as possible from disturbances due to mechanical vibrations and alternating magnetic or electrical fields; the main electricity supply must be free from sudden voltage changes; the cooling water supply must be clean and of even temperature, pH and pressure; and the air must be free from sudden temperature and humidity changes. The room must be capable of being completely blacked out, yet must be adequately ventilated and dust-free. It must be remembered that the operator and possibly several other people will sit for hours in front of it; they must be provided with comfortable seats giving adequate support. Fairly frequent routine servicing is required, so a suitable bench should be fitted in the microscope room on which components and tools can be laid out during dismantling and reassembly. Adequate space must be provided at the back and sides of the console for the removal of components and for working on them when they have been removed. A dustproof cupboard must be provided for the storage of spare parts. The walls and ceiling must reflect the maximum amount of light, and the floor must be plain and light in colour so that small parts or grids can easily be found again if accidentally dropped. The floor must be non-absorbent, or the oil which inevitably seeps from the rotary pumps will form a dust trap. The room must be easily cleanable and free from crevices which will harbour dust. A small hand vacuum cleaner should be provided to remove dust from beneath the instrument. Windows are best avoided; they leak dust into the room.

The modern tendency is to build electron microscopes as a single unit, but some, especially the older high-performance ones, may have up to four separate units—the column–console unit; the power supply cabinet; the rotary pump unit; and the plate desiccator. It is always best to instal the power supply cabinet outside the electron microscope room. The majority of the power consumption of the instrument is dissipated by the valves in the power supply cabinet; this can give rise to severe overheating and ventilation problems.

10.2 Vibration

In the early days of electron microscopy, it was considered essential to instal an instrument in an air-conditioned basement on a concrete raft in order to protect it from vibrations. Although such a situation is indeed ideal, modern instruments are not as sensitive to vibration as were their predecessors. The designers of modern instruments have to bear in mind the fact that many will have to be installed in environments which are far from ideal, and strengthen their columns accordingly. But, if an instrument is to be installed in a place where interference by mechanical vibrations may be troublesome, always test the model of choice for sensitivity to mechanical interference before finally becoming committed. This is best done by focusing a well-defined Fresnel fringe at high magnification, and then getting someone to jump up and down on the floor close to the console while watching the fringe through the screen magnifier. This is a very severe test, but several modern designs show no detectable movement of the fringe when it is applied. The other test is to focus the illumination spot at a low magnification, and to watch it while striking the gun casing gently with the side of the fist. On some instruments, this gives rise to an alarming amount of movement; others are virtually immune. In general, the thicker the column, the less susceptible the instrument is to vibration.

10.3 External Magnetic Fields

Since each lens is in effect a magnetic field, electron microscopes are peculiarly sensitive to stray external magnetic effects, especially if fluctuating. It is therefore essential to check the proposed site for stray alternating magnetic fields. It is first necessary to find out if neighbouring laboratories, especially those immediately above and below, have any large transformers, electric motors, chokes, mercury arc rectifiers, diathermy apparatus or high frequency heaters or other equipment which is known to produce magnetic fields, and also whether such equipment is turned on and off frequently during the working day. It is then necessary to test the site for stray magnetic fields under all conditions of operation of nearby equipment. This will generally be done by the manufacturer's representative when he comes to inspect the site. It is a very simple matter to test a site for magnetic interference. All that is needed is a coil of wire of many turns (an old electron microscope lens coil is ideal) and an oscilloscope. The coil is connected to the oscilloscope input at highest sensitivity, and the trace is watched while the coil is carried from place to place over the site. Stray alternating magnetic fields are picked up by the coil, and show as a waveform on the oscilloscope. The coil should be sensitive enough to pick up the field from the mains transformer of the oscilloscope. If a large waveform is picked up, the site is suspect. The maximum allowable is about 5 milligauss. The origin of the interference should be traced. It may or may not be possible to reduce it by resiting the cause, which will in general be a large transformer.

Because of the sensitivity of electron microscopes to stray fields, it is frequently necessary to mount the instrument's own power supply some distance from the column. It is necessary to find out about this from the manufacturer, as it may not be possible in the proposed space. Power supply cabinets take up valuable space in a laboratory, but can frequently be sited in a corner or in a corridor where no other apparatus can be used. The majority of modern instruments have the transistorized power supplies built into the console. In some instruments, the motor or motors for the rotary pumps may cause interference, both vibrational and magnetic, and must be sited a short distance away from the instrument itself.

10.4 Water Supply

It is essential to have a reliable water supply of constant pressure. A laboratory is known where, if both the ladies' toilets down the corridor are flushed together, both microscopes automatically turn themselves off. This situation is to be avoided if possible, and therefore the pressure of the water supply should be monitored with a pressure gauge throughout a working day. It may well be found to fluctuate from 80 p.s.i. to 20 p.s.i. Practically all instruments have some form of pressure switch incorporated in the water circuit, so that if the cooling water fails, the instrument turns off automatically. These safety switches generally require more pressure to operate them than the actual pressure required to force the water around the lenses and pumps. If only a very low pressure water supply exists (less than 10 p.s.i.) it may be possible to replace the microscope pressure switch by a flow switch, which requires very little pressure to operate it. In general, a water supply having a maximum pressure of about 80 p.s.i. and a minimum pressure of 40 p.s.i. will be found to be satisfactory. This sort of pressure is generally available in the basement of a building, but may be lacking high up, closer to the water storage tanks. In general, it is better to use water straight from the pressure main rather than to take it from a storage tank. It may be necessary to run a special pipe from the microscope to the place where the water main enters the building.

The quality and temperature of the water must also be investigated. Trouble may arise if the water is very soft, or if it is very hard. Very soft acid water from peat moors may cause corrosion of lens water passages; cases are known where the column has filled up with water overnight from this cause. On the other hand, very hard water may lead to furring-up, especially of the hot diffusion pump condenser. Excessively hard water may be softened by ion exchange. The temperature of the cooling water is very important; it must be neither too hot nor too cold. The range usually specified is 15–20 °C. If the water is too hot, the oil diffusion pump condenser will not trap all the oil vapour, and some will diffuse back into the column, contaminating it. Too cold cooling water can lead to condensation, especially in humid climates. Water may condense on the outside of the column, and run into the electrical plugs and contacts, causing instability or even damage to circuits.

It is generally necessary to filter the mains water supply. The best type of filter is the sintered ceramic candle type of about 20 micron porosity. It is necessary to have a tap between the mains and the filter, and there should be two pressure gauges, one on either side of the filter. The pressure drop across the filter can then be monitored, and the filter candle cleaned before it becomes completely choked. A needle valve is placed between the filter and the microscope, so that the correct water flow can be adjusted. The water outlet should be visible so that flow rate can be seen. It is generally necessary to allow the cooling water to flow for at least 10 minutes after turning the microscope off for the night, so that the oil vapour from the oil diffusion pump does not condense in the main vacuum valve and thus contaminate the column. It is therefore convenient to fit some form of clockwork-operated solenoid switch in the water supply before the filter, so that it can be set to turn off automatically after the specified time period. A suitable design of water circuit is shown in Fig. 10.1.

It is far better to instal a closed-circuit water system in which the cooling water is continuously recycled with a pump and cooled by a refrigerator. A microscope uses several hundred gallons of water each day, and in these times of shortage even water is becoming valuable and

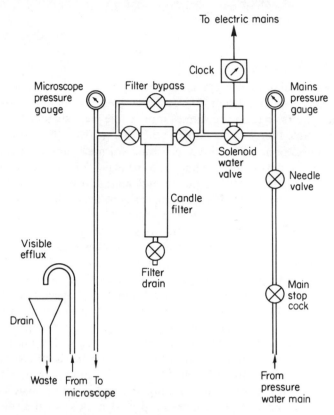

Figure 10.1   The cooling water supply circuit for an electron microscope. The difference in reading between the two pressure gauges gives the pressure drop across the filter, and is used to show when the filter needs cleaning. The filter bypass enables the instrument to be used while the filter is being cleaned, but is non-essential. The clock-operated solenoid valve enables the water to be turned off automatically after a preset time to enable the pumps to cool down after the nightly shut-down. The solenoid valve should be sprung open so that the water supply continues to run in the event of the failure of the mains electricity

in any case should be conserved for the sake of the environment. Distilled water is used in a closed circuit, and if recommended by the manufacturers of the cooling plant and the microscope, anticorrosion and antialgal compounds may be added. The cost of a suitable refrigeration plant is perhaps one-fiftieth of the cost of the microscope.

10.5 Electricity Supply

It is essential to have a mains supply which is free from sudden voltage surges, such as would occur if large, inductive pieces of apparatus such as electric motors sharing the same supply cable were switched on and off. It is advisable therefore to monitor the voltage from the outlet it is proposed to use, preferably with a recording voltmeter and over a period of some days. A slow voltage change is to be expected, especially during the winter, when supply authorities cannot maintain voltage in the face of large domestic heating loads. This can easily be dealt with by the electronic stabilizers. As has been explained in Chapter 7, stabilizers do not operate instantaneously, and sudden fluctuations in supply voltage will inevitably lead to momentary instability. It may be necessary to run a special lead to the mains input to the building, immediately after the supply meters. Practically all instruments require only single phase supplies, but some rotary pumps may have three-phase motors. It is of special importance to provide a very low resistance path to earth from the earthing point of the instrument. This conducts away any electrostatic interference or 'hum'. In general, the earth provided by the normal wiring will be unsatisfactory, and a heavy copper cable or strip should be run from the microscope earth point to the main earthing bus-bar of the mains supply, or to a main cold water pipe where it enters the building. A fused isolating switch breaking both live and neutral poles must be provided close to the microscope, which must be of a design which does not give rise to magnetic fields.

10.6 Air Conditioning

Considerable effort should be made to have the microscope room air-conditioned. Very often several people gather round an instrument for several hours at a time, and the atmosphere in a small room can become unpleasant. Apart from this, air conditioning is essential to prevent water condensation on the outside of the column, even in temperate climates.

10.6 Radiation Hazards

An electron microscope is a potential source of very penetrating and dangerous X-rays. If these are not properly shielded, they will endanger the health of the operator. X-rays are always produced when energetic electrons or other particles are slowed rapidly by a target. The major sources of X-rays are therefore the gun, the condenser apertures, the specimen and objective aperture, and the fluorescent screen. It is necessary to shield all these components from the operator and from other workers immediately above the instrument by using suitable shields made of lead, lead–bronze or other dense material of high atomic number. It is usual to place a lead shield around the gun and condensers, to make the viewing chamber of a lead-containing alloy, and to make the viewing window of thick lead glass.

It is most important for the operator to monitor the instrument immediately after it has been installed with a Geiger counter, while running the microscope at the maximum accelerating voltage and beam current. The counter head is held close to the gun, the pumping leads, the condenser lenses, the specimen chamber, the viewing chamber, the viewing window and especially under the camera. The position occupied by the operator must be carefully monitored, especially in the region of the gonads. It is possible for an intense sharply-defined X-ray beam to be emitted at a rubber vacuum seal if the shielding is not properly designed. It must be remembered that the operator will sit in front of the instrument for hours at a time, possibly over a period of years. The effect of harmful radiation is cumulative, and the maximum allowable dose is very small. A Geiger counter is not a particularly efficient indicator of X-radiation, so it is wise to be doubly assured by using photographic film radiation badges. These should be strapped to the instrument at the strategic positions suggested above, and a badge should be worn by the operator for the first 100 or so hours of use. The badges are then developed and the total dose calculated. All manufacturers are most meticulous over the shielding of their instruments, and no trouble should be encountered, but it is always possible for a shield to have been left out during assembly, and it is better to be safe than sorry.

It is most important never to run the instrument with any of the shielding removed. This can be a great temptation when testing an instrument after it has been dismantled. It is even more important never to run it with one of the lead glass viewing-windows replaced by an aluminium plate. Such blanking plates are provided by some manufacturers for testing. Do not run the microscope when using one of these blanking plates unless it has first been covered with thick sheet lead.

10.8 The Microscope Room

It must be remembered that a fair amount of space is essential around a microscope for servicing purposes. It cannot be pushed up against a wall. This is also true of the power supply cabinet. The amount of space necessary depends on the design; whether it is necessary to roll the pumps or the high-tension generator tank out on rails in order to get at it for servicing. If so, then further space should be allowed behind the pumps or tank when they are fully rolled out in order to work on them. It is a great mistake to attempt to jam an electron microscope into an inadequate space. If it is uncomfortable to work on to service it, then servicing will be neglected, leading in the long run to expensive trouble. The amount of space specified by the manufacturer should be looked upon as a minimum. The minimum requirement is about $10' \times 10'$, and $9'$ high.

An electron microscope is very heavy, and attention must be paid to permissible floor loadings. In general, a solid floor is to be preferred to a suspended floor. The room itself should be adequately ventilated, bearing in mind that several people may be sitting around the instrument possibly for hours at a time. An air conditioner is best, but care must be taken that it does not give rise to magnetic interference. The walls and ceiling should be painted with washable white paint in order to make the best use of the dim operating light necessary. The main room light should be very bright for servicing, but should not be of the fluorescent type, for the choke may give rise to magnetic interference. A dim light must be provided for general working; a large darkroom lamp with a yellow filter directed on the ceiling is best. Under no circumstances should photographic materials be processed in the microscope room. Spilled photographic solutions dry up to form a fine corrosive powder which will inevitably get into the column. A separate darkroom for developing plates is very convenient, since plate developing does not then interfere with the normal work of the main darkroom.

10.9 Delivery

Before the instrument is delivered, it is essential to have all the wiring and plumbing jobs finished, and the room decorated. It is necessary to make sure that all the doors along the route that the microscope is to take inside the building are large enough to admit the widest and tallest unit. The instrument will arrive in a number of wooden crates, the heaviest of which may weigh up to half a ton. It is therefore necessary to arrange for some form of crane or fork-lift truck to unload the crates from the road transporter. The crates must be opened under cover in a suitable place, and it may be necessary to have the contents examined by a Customs officer if the microscope has been imported and delivered straight from the docks. Some of these details will be seen to by the manufacturer, but some are the customer's responsibility. Inquiries must be made beforehand. The instrument will be assembled by the installation engineers on the site; this will take about a week. After the engineers are satisfied that the installation is satisfactory, the complete instrument is then generally put through its paces by an inspecting engineer or one of the designers. The customer should see that records are made of all lens currents, high tension voltages, monitoring meter readings and vacuum gauge readings for future reference. These should be measured monthly (see Chapter 14), and the results compared with those taken when the instrument was new. The engineer will take test micrographs showing resolution, contamination rate, magnification, distortion, etc. and the customer should see that he is furnished with copies of the prints for record and comparison purposes. After installation, free service is generally given over a period of one year, during which time the customer can make up his mind whether or not to take up a service contract (see Chapter 15), and if so, which one.

10.10 Instruction

Manufacturers generally arrange for some sort of instruction in the operation and maintenance of the instrument to be given to the person who will be in charge of it. Some manufacturers give excellent courses lasting a week or more; others are more perfunctory. These courses are free for one or two people, although the customer generally has to pay hotel expenses. Such instruction is essential for a laboratory without experience in electron microscopy, and should be given before the instrument is delivered.

All manufacturers give instruction manuals with the instrument. These, like the service courses, are very variable in quality. Some are vast tomes which amount virtually to textbooks on the practice of electron microscopy, containing working drawings and full information on voltages, currents, component values, etc. necessary to rebuild the electronic circuits completely. Others withhold such information and actively discourage the customer from attempting to service his own instrument. Anyone who is interested in the instrument for its own sake as well as a tool for obtaining results, and who wishes to undertake the maintenance work himself, would be very well advised to ask to see the instruction book before deciding on an instrument.

PART 2

USING THE ELECTRON MICROSCOPE

CHAPTER 11

Basic Operational Procedures

11.1 Introduction

It is not possible in a general book such as this to give a description of the precise way in which each model of commercial instrument is operated, since the design and layout of the controls vary so much in detail. Instrument manuals and instruction books give stepwise schedules for taking the microscope from one operating state to another, but seldom give any explanations for the procedures described. An attempt will therefore be made in this chapter and the following one to describe and explain the operational procedures applicable to all modern transmission electron microscopes. This should enable an operator to sit in front of any electron microscope and get the best results from it in the shortest possible time.

11.2 Microscope Controls

Before attempting to use an unfamiliar instrument, the positions of all the controls must be located. The operator should sit before the instrument with the instruction book, familiarizing himself with the general layout of the controls and the degree of automation to be expected. The controls can be divided into 3 groups: column, electrical and vacuum. Each group can be subdivided as follows: (controls not generally found are marked with an asterisk)

11.2.1 Column—Mechanical

1. Stage (specimen movement)
2. Apertures (changing and centration):
   a. *C_1
   b. C_2
   c. Objective
   d. *Anticontaminator blade aperture centration

221

      e. Intermediate (selected area diffraction)
3. Gun alignment
      a. Gun centration
      b. *Gun tilt
      c. Illuminating system centration
      d. *Illuminating system tilt
4. Lens alignment
      a. C_1 *tilt and centration
      b. C_2 *tilt and centration
      c. Objective tilt
      d. Diffraction *tilt and centration
      e. Intermediate *tilt and centration
      f. Projector centration
      g. *Projector polepiece selector
5. Specimen insertion mechanism and airlock
6. Fluorescent screen controls
      a. Raise or tilt main screen
      b. *Insert and withdraw focusing screen
      c. *Insert and withdraw beam current probe
7. Camera controls
      a. Plate transport mechanism
      b. Cassette box changing airlock
      c. Shutter
      d. Exposure meter
      e. Exposure counter
8. Image viewing
      a. Telescope translation
      b. Telescope focus
      c. Interpupillary distance adjustment
9. *Special methods
      a. Stereo tilt
      b. Diffraction spot interceptor
      c. Micron markers
      d. Half-frame exposure mask
      e. High-angle specimen tilt
      f. Others (e.g. goniometer) if fitted

11.2.2 Electrical
1. Mains on/off
2. Room and panel lights

3. Gun controls
   a. H.T. on/off
   b. H.T. kilovoltage selector
   c. Emission (beam current)
   d. Filament voltage
4. Beam controls
   a. Centration deflectors
   b. *Darkfield deflector
   c. *Focusing wobbler
5. Lens current controls
   a. C_1 coarse ($+$ *fine)
   b. C_2 coarse $+$ fine
   c. Objective (coarse, medium, fine, very fine)
   d. Diffraction ⎫ Stepwise magnification control $+$ individual con-
   e. Intermediate ⎭ tinuous controls
   f. Projector coarse ($+$ *fine)
6. Stigmator controls
   a. Condenser $+$ centration
   b. Objective $+$ centration $+$ *range
   c. *Intermediate lens stigmator
7. Alignment controls
   a. *Objective lens current modulator
   b. *H.T. kilovoltage modulator
   c. *Stigmator centration modulator
8. Monitoring meters (generally 2 or 3 multipurpose, switchable)
   a. Beam current
   b. *Filament volts
   c. Magnification (diffraction/intermediate lens current)
   d. High vacuum
   e. Backing/roughing vacuum
   f. Photographic exposure (screen or probe current)
   g. *Anticontaminator temperature
   h. Other lens currents
9. Pilot and warning lights
   a. State of vacuum
   b. Water pressure
   c. Diffusion pump heater
   d. Camera shutter
   e. Camera airlock
   f. Mains
   g. Power pack

10. Digital readouts
    a. Magnification
    b. Camera length
    c. Exposure countdown
    d. Negative number

11.2.3 Vacuum (Non-Automated System)

1. Roughing
    a. Column roughing valve
    b. Column air inlet valve
    c. Specimen chamber valve
    d. Camera airlock valve
    e. Camera air inlet valve
    f. Gun chamber valve
    g. Gun chamber air inlet valve
    h. Desiccator valve
    i. Desiccator air inlet valve
2. Backing
    a. Diffusion pump low vacuum valve
    b. Rotary pump air inlet valve
3. Main column high vacuum valve

11.3 Grouping of Controls

Most manufacturers make a careful ergonomic study of the positioning of controls for the maximum convenience of the seated operator. Vacuum valves in automated and automatic systems are operated by solenoids, electric motors or compressed air, and are controlled by manually operated push-buttons, which may light up when depressed. In manually operated systems, vacuum valves are generally linked mechanically and in correct sequence so that the least damage is done by foreseeable accidents. Mechanical controls must obviously be placed on the column; they are sometimes somewhat high and inconvenient to use. Electrical controls, however, being remote, can be grouped together for maximum convenience. The precise manner of grouping is a matter for the manufacturer, but the following definite pattern is beginning to evolve:

Right hand:
    a. Stage right (specimen movement)
    b. Focus (objective current)
    c. Magnification (diffraction/intermediate current)

    d. Beam shift right

    e. Stigmators right

Left hand:

    a. Stage left (specimen movement)

    b. Brightness (C_2 current)

    c. Beam shift left

    d. Stigmators left

    e. Emission

    f. Filament volts

    g. Kilovoltage

    h. Focused spot size (C_1 current)

The above list is given in order of frequency of usage of the various controls. Other auxiliary controls, such as lens on/off and reversal switches, modulator controls, etc., which are seldom used are generally grouped on a subpanel with a removable cover so that their settings are not interfered with unnecessarily.

11.4 Switching on from Cold

The aim is to turn on the vacuum pumps and electronic power supplies and allow them to warm up and settle down to a stable condition; to preevacuate the column; and finally to bring the column to high vacuum. The procedure is as follows:

1. Connect the instrument to the mains by closing the isolator switch.
2. Turn on the cooling water to pumps and lenses, checking that there is sufficient water pressure; that the water filter is not impeding flow; and that all the water pressure safety switches and cut-outs have operated.
3. Turn on the mains switch of the instrument and check that all indicators are showing correctly and that nothing (e.g. the filament) is on that ought to be set to the 'off' position.
4. Rough out the pumping system by:
    1. Close the rotary pump air admission valve;
    2. Turn on the pump motor;
    3. Wait until the pump is 'hard';
    4. Open the backing valve.
   These procedures are generally linked together.
5. Turn on the diffusion pump heaters.
6. Wait until the diffusion pumps are at full speed, and the high vacuum at the diffusion pump side of the high vacuum valve is sufficient.

This takes 5 to 15 minutes, depending on pump design.

7. Pump out the column with the rotary pump:
    1. Close the backing valve;
    2. Open the roughing valve.

These steps are generally linked together. Do not keep the rotary pump connected directly to the column for longer than necessary to reach the required pressure, otherwise contaminating vapours may diffuse back into the column from the rotary pump oil.

8. Open the column to the high vacuum pumps by:
    1. Close the roughing valve;
    2. Open the backing valve;
    3. Open the high vacuum valve.

These steps are generally linked. It is important to watch the Pirani (low vacuum) gauge during this procedure. The indication should rise as the diffusion pumps cope with the residual gas in the column, and then drop rapidly as the backing pump in turn deals with the gas. The rate of rise and fall of the low vacuum gauge is an important indication of the state of both pumps.

9. Check that the high vacuum and backing tank vacuum are within the prescribed limits.

10. In the case of an oil-mercury pumping system, the backing pump may now be switched off:
    1. Close the backing valve;
    2. Switch off the pump motor;
    3. Open the pump air inlet valve.

These steps are generally linked. It is most important to listen for the hiss of air entering the rotary pump.

The instrument is now ready for use. The total time taken should be between 15 and 30 minutes, depending on the design of the pumps.

With the latest type of fully automatic vacuum systems, it is only necessary to press a single button to initiate the complete procedure described above. Vacuum sensors placed in the system signal to servo-operated valves so that each succeeding stage follows automatically after the preceding stage has been completed. Each stage of the procedure is signalled to the operator by means of indicator lamps. With automated systems, the operator must wait for a signal lamp to light up and must then press the button initiating the succeeding stage himself. This relieves the operator from responsibility and also prevents damage, since the stages are arranged so that they can only be initiated in the correct sequence. The operator must however remain with the instrument while it is warming up, although he can obviously fit in other jobs during the warm-up time.

This is generally about 20 minutes, although diffusion pumps are now being designed which will reduce warm-up time to a few minutes.

11.5 Obtaining and Centring a Beam

The following instructions assume that the instrument is in reasonably correct alignment, i.e. that an image of the correctly adjusted electron source is capable of being focused on the final screen with all the lenses energized normally. If the instrument has been dismantled and re-assembled, it will need major realignment, the procedure for which is described later in this chapter.

The steps for obtaining a beam are as follows:
1. Switch on the H.T.;
2. Switch on the filament and bring it roughly to the correct operating temperature;
3. Focus the source on the screen and align it;
4. Centre and saturate the gun.

11.5.1 Checks

Before attempting to switch on the gun and find the beam, the following points must be checked:
1. The column vacuum is sufficiently good;
2. The required H.T. kilovoltage has been selected;
3. The filament is off or at minimum voltage setting;
4. The emission is set to minimum;
5. All lenses are switched on;
6. C_2 is fully underfocused (minimum current setting);
7. The specimen and objective apertures are withdrawn;
8. A low magnification ($\times$ 1,000–2,000) is selected;
9. The shutter is open and any camera or focusing screen is withdrawn from the beam.

11.6 Gun Controls

The gun generally has five controls:
1. High tension on/off (marked 'H.T.')
2. Filament on/off (may be combined with 1 or 3)
3. Filament heating (marked 'FIL')
4. Emission (marked 'EMISSION' or 'BIAS')
5. High tension selector (marked e.g. kV–40–60–80–100)

A meter to measure the value of emission current in microamps is almost invariably provided; a meter to measure filament volts is found on some instruments.

11.6.1 Switch on High Tension

It is good practice always to switch the high tension on with the filament off, or set to a low voltage. This prevents damage to the filament or specimen caused by surges of voltage. Press the 'H.T. ON' button. Watch the emission meter carefully and listen to the gun. The meter needle will probably jump about unpredictably, to the accompaniment of sharp clicks from the gun. The pressure in the column will rise slightly. The clicks are corona discharges taking place between the gun and the casing, due to the desorption of residual column gases, which are released by the discharge. The higher the kV selected, the longer the gun will take to clean itself and settle down. It is no use proceeding to the next step until the gun has settled.

It is good practice to select a kilovoltage higher than the one required when switching on. This gives the gun a chance to clean itself up. When it has become stable at the higher voltage, select the lower. The gun should then be completely stable. Some instruments are provided with an 'overvoltage' button, which increases the kilovoltage to a value higher than the maximum, so that the gun will be stable at maximum accelerating voltage.

11.6.2 Switch on the Filament

All modern electron microscopes have 'biased' guns (see p. 184). The advantage is that at the operating point, the gun emission and hence image brightness are almost independent of filament temperature.

Proceed as follows:
1. With the H.T. switched on and settled, darken the room.
2. Switch on the filament with its voltage control set at minimum.
3. Advance the filament voltage control while watching both final screen and beam current meter. At first, nothing may happen; the filament must be yellow-hot before it begins to emit. At the onset of emission, beam current rises very rapidly indeed (approximately as the fourth power of the filament temperature). The screen should then glow faintly and uniformly.
4. Darken the room so that the screen glow can be observed. Continue to advance the filament voltage control very slowly while watching the beam current meter and screen very carefully. Screen glow and emission current will rise to a maximum, after which further increase in filament voltage has little effect on either.

The operating point is now approximately set. It will probably be set a little too high, but not excessively so. Great care must be taken not to set the filament voltage too high, or filament life will be drastically reduced with consequent loss of operating time. On some instruments, it is possible to burn the filament out by overadvancing the filament voltage control. Some kind of stop, electrical or mechanical, should be provided on these microscopes.

The illumination generally takes the form of a sharply defined circle of light. The sharp edge is the image of the polepiece of the final projector lens and should be centred about the centre of the screen.

11.6.3 Focus the Source

Increase the current in C_2 by adjusting the coarse control (generally clockwise). The screen glow should become uniformly brighter, until a less sharply defined circle of light appears on the screen. This is the image of C_2 aperture. Continue to increase C_2 current using the fine control. The circle becomes smaller and the edges become less well defined. This is because the aperture of illumination is increasing (see p. 118). The circle diminishes to a spot about 1 cm in diameter, generally oval, with ill-defined edges. This is the image of the source. Further increase in C_2 current will overfocus the lens, and the image of the aperture will reappear, increasing in diameter again.

Focus the spot. If it is very small, increase the magnification. Centre the spot with the illumination centration controls (which may be mechanical or electrical). Defocus C_2 to obtain the aperture image, and centre this circle with the C_2 aperture centration controls. Refocus the illumination spot.

11.7 Centre and Saturate the Gun

Select the value of emission required. The higher the magnification to be used, the higher this will be. Each time the emission current is changed, the saturation point should be readjusted.

Reduce filament voltage very slowly, while observing the spot, which must be exactly focused, very closely. The spot will break up into a specific pattern, shown in Fig. 11.1(a). First the spot becomes dimmer in the centre, which then breaks up into an area of parallel lines. These are due to the 'wire-drawing lines' impressed into the metal of the filament during the drawing process used to thin it down during manufacture. The central area of lines becomes surrounded by a bright oval ring of light. This is called the 'halo' pattern. C_2 must be very accurately focused and deastigmated to obtain a sharp image of the draw lines. If the pattern is too small to see clearly, increase the instrumental magnification.

As the filament voltage is further reduced, the elliptical halo expands and breaks up into 4 spots. This is called the '5-spot' pattern shown in Fig. 11.1(b). With further voltage reduction, the pattern increases in size, becomes dimmer and finally disappears as the filament ceases entirely to emit electrons. The voltage at which emission ceases is only about 20 % less than the correct saturation voltage.

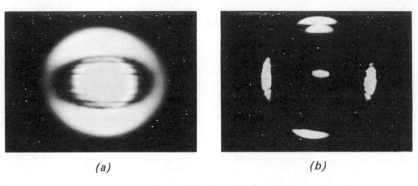

(a)   *(b)*

Figure 11.1   Aligning the electron gun

(a) The 'halo' figure (slightly underheated filament). The central area is the filament tip, showing the wire-drawing lines in the tungsten. The surrounding 'halo' is the bias shield. Philips EM 200

(b) The 5-spot figure (very underheated filament). The central area of the 'halo' figure breaks up into the 3 horizontal bars (the draw-lines are still visible). The halo breaks up into the upper and lower bright areas. AEI–EM6B

To align the gun, the figures are made exactly symmetrical by using the gun tilt and/or centration controls, or the anode centration controls

The gun must now be centred so that its axis coincides with that of the condenser lenses. Obtain the halo or 5-spot pattern (the design of gun determines which is the more convenient). Make the halo or the 4 outer spots symmetrically disposed about the central spot by using the gun centration control. This control may traverse the gun and insulator (anode fixed) or may traverse the anode (gun fixed). When the pattern is symmetrical, the gun is centred.

The gun must now be saturated so that its output is independent of filament temperature. Increase filament voltage so that the underheated cathode pattern disappears. The central spot increases in brightness (or, more correctly, 'luminance'). The lowest filament voltage giving maximum screen luminance is the correct saturation point. Screen luminance can be measured on some instruments by using the indication of the photoelectric exposure meter; on others by measuring the actual screen current, which may be the indication on the exposure meter. If neither aid is available, the correct saturation point is with the filament voltage a *very little* (1–2%) higher than the point at which the pattern disappears.

This saturation procedure must be followed each time the gun is switched on or 'fired up', e.g. after a specimen has been changed or the filament has been switched off for any reason. It must become second nature to the operator, or he will spend much of his time changing burnt-out filaments.

11.8 Adjusting Image Brightness

The brightness (strictly speaking the 'luminance') of the screen is a function of screen current per unit area, the kilovoltage and the efficiency of the phosphor. A very rough approximation is given in the following table:

Visual condition	*Screen current amp/cm^2*
Blindingly bright	10^{-9}
Very bright	10^{-10}
Comfortable viewing	10^{-11}
Photographic (1 second exposure)	10^{-12}
Just detectable by the dark adapted eye	10^{-13}

Screen current can be adjusted in either or both of two ways:
1. Adjusting gun emission
2. Adjusting condenser current (generally C_2).

The first method is inconvenient, since it involves two adjustments—bias volts and filament temperature. The second method has the twin advantages of convenience and the improvement of resolution and contrast, since the aperture of illumination is reduced as C_2 current is changed from focus (see p. 117). It is therefore usual to adjust the bias and filament voltage for a fixed emission sufficient for comfortable image viewing at the highest magnification to be used, and to reduce image brightness by adjusting C_2 current, generally by underfocusing (reducing lens current).

11.9 Adjusting the Illuminating System

Single and double condenser illuminating systems have been described in Chapter 6. A double condenser system can be turned into a single condenser system simply by switching one lens off. Either lens may in theory be switched off; in practice it is always C_2 which is used alone. This is because C_1, being closer to the source than the specimen, forms a magnified image of the source, which would be far too large even for the very lowest magnification operation.

11.9.1 Single Condenser Operation

The following applies equally to single condenser instruments and to double condenser instruments with C_1 switched off. The term 'condenser' now refers to C_2 only.

As we have seen (Figure 6.11), the size of the illuminating spot with the condenser at focus depends solely on the physical size of the electron source (which is dependent on the design and working conditions of the electron gun) and the magnification of the condenser (the ratio of image to object distances). At focus, the size of the condenser aperture controls the aperture of illumination, not the spot size. Reducing the physical aperture will make an image of the specimen on the final screen dimmer but sharper due to the Fresnel fringe effect described in Chapter 1. Increasing the aperture will make the image brighter but of poorer quality. The actual angular aperture of illumination of a single condenser lens fitted with a 200 μm diameter physical aperture is about 10^{-3} radians. A 50 μm aperture gives $2 \cdot 5 \times 10^{-4}$ and a 500 μm aperture gives $2 \cdot 5 \times 10^{-3}$ radians.

It is not possible, therefore, to alter the size of the focused illuminating spot with a single condenser system. If the source diameter is 50 μm and the condenser magnification is unity, the diameter of the spot focused on the specimen will also be 50 μm. This will fill the final screen at an imaging system magnification of only about $\times 1,000$ to $\times 2,000$. If the image is at a comfortable viewing brightness at $\times 1,000$, it will be 100 times as dim at $\times 10,000$. The gun emission must therefore be increased to restore image visibility, with the consequent possibility of heating and damage to the specimen over the whole illuminated area (which is far greater than that visible on the final screen). This is the disadvantage of working with a single condenser, and why it is correctly used only at relatively low magnification.

11.9.2 Centring the Beam and Condenser Aperture

If the focused spot is not at the centre of the screen (this will usually be the case), return it to centre either with the mechanical illumination shift controls (beam centration), or with the electrical beam shift controls if provided. Next defocus the condenser slightly (over or under). An unsharp image of the condenser physical aperture will appear on the screen, becoming larger as the condenser is further defocused. The lower the magnification, the sharper will be the image of the aperture and the more clearly will the effect be seen. Any dirt or contamination on the edge of the aperture will clearly be seen. If excessive, another aperture should be tried. Next centre the aperture image symmetrically about the screen centre with the aperture centration controls. The illuminating system is now correctly adjusted.

11.9.3 Correcting Condenser Astigmatism

If the focused image of the source is a drawn-out ellipse, and if the major axis rotates suddenly through 90° as the condenser focal setting is passed through, the condenser is astigmatic. The beam cannot be concentrated into an approximately circular spot, and therefore image brightness is lost and the illumination will be uneven. Correct the astigmatism to give an approximately circular spot by manipulating the two condenser stigmator controls (either X–Y correction or amplitude–azimuth) until the focused spot no longer shows a shift of axis as the focal point is passed through. The single condenser system is now correctly set for low magnification working.

11.9.4 Low-Power Double Condenser Operation

The upper lens (C_1) of the double condenser is used to control the diameter of the focused spot (see Figure 6.13). The physical aperture of C_1 is generally fixed and cannot be centred externally. For low-power working, the current through C_1 is set to a low value (one quarter to one half of maximum). This should give a spot size about one-third of that obtained with C_1 off. The focused spot should fill the final screen at a magnification of about × 5,000. This is a suitable general setting for work up to a plate magnification of about × 10,000. No hard and fast rules can be laid down; the precise settings of the double condenser depend on the design of the illuminating system. The procedure for the centration of the spot and C_2 aperture and the deastigmation of the illuminating system follow exactly that given for single condenser operation. For general low-power work, C_1 setting is left untouched. The operation of the double condenser at high magnifications (where it is most useful) was discussed on page 121.

11.10 Inserting a Specimen and Obtaining an Image

The precise means for inserting a specimen varies from instrument to instrument. Some have an airlock which admits a very small amount of air together with the specimen, this being dealt with in a reasonable time (30 seconds or so) by the high vacuum pump. Others admit a larger amount of air into a separate specimen chamber, which must be prepumped with the roughing pump before the specimen chamber is opened to the high vacuum. In all cases, it is a wise precaution to switch off the filament and turn off the H.T. before inserting a specimen. Air leaks will damage the filament by oxidation, and will overload the H.T. generator by causing

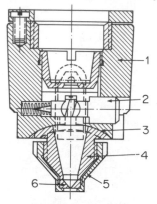

1 Cone	4 Segment
2 Transmitting pin	5 Holder for specimen diaphragm
3 Cylindrical slide face	6 Specimen diaphragm

(a)

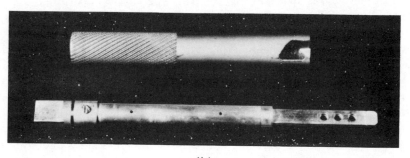

(b)

Figure 11.2   The two basic types of specimen holder

(a) Top entry bucket type, which is lowered into the lens from above. The conical taper fits accurately into the stage. The specimen grid or diaphragm is screwed tightly into the nose. Only one grid can be carried at a time. Specimen manipulation facilities are readily possible with this type; a stereo tilt holder is shown. (× 2 approx., by courtesy of Siemens Ltd.)

(b) Side entry rod type, with insertion tool above. This example carries 3 grids which can be selected very rapidly without stopping operation. Later types carry 6 grids (GEC/AEI EM6B)

heavy corona discharges. The gun is not turned on again until the high vacuum has been reestablished.

Before inserting the specimen on its copper grid into the microscope, the grid must first be placed in a carrier. If the instrument has an 'immersion' objective of very short focal length, the specimen carrier is a rod, Fig. 11.2(b), which is placed on the stage and lies between the objective polepieces. Other objectives, which carry the stage above the polepieces, generally hold the grid in a small 'bucket' cartridge, Fig. 11.2(a), which is lowered into the upper polepiece by means of a small crane. In whatever manner the grid is carried, it is essential to see that it is firmly clamped all round the edges, in order that heat may be conducted rapidly and evenly away from the specimen. If this is not done, the specimen may be damaged by overheating, and will in any case drift in the beam due to uneven thermal expansion. This will give rise to unsharp micrographs, even though a small amount of drift can be tolerated during visual examination.

11.11 Scanning the Specimen at Low Magnification

After the specimen has been inserted, the high vacuum reestablished and the gun switched on again, it is good practice to observe the whole grid at a very low magnification (around × 200) in order to judge which part of the section ribbon is likely to give most information. In this way, tears, knife scratches and other imperfections in the specimen can be avoided. 'Scan' micrographs are shown in Fig. 11.3. The best way of scanning the grid is to use the intermediate lens as an objective of very long focal length and low magnification, together with the projector, as a 2-lens microscope. The steps to be followed are:

1. Fully defocus C_2 with C_1 at low excitation;
2. Remove the objective aperture from the beam;
3. Switch off the objective;
4. Focus the image with the intermediate lens current. This gives one fixed focused magnification, generally about × 200.

The operator places the grid square judged to be most promising in the centre of the screen, and then reverts to normal operation by:

11.12 Inserting and Centring the Objective Aperture

1. Insert the objective aperture and place its image as close to screen centre as possible.
2. Switch on the objective.

3. Bring the intermediate to focus ('crossover').
4. Focus C_2. The aperture will now be seen as a sharp small circle with a sharp point of light at the centre. The point is generally known as the 'diffraction point' and the free lens control used to focus the point is often marked 'diffraction focus'. The aperture is now centred exactly by making its image symmetrical about the point of light at the screen centre.

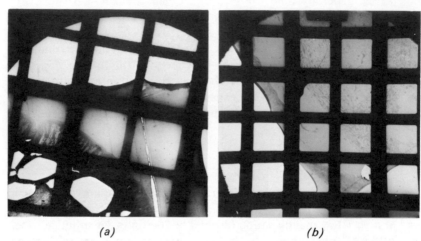

(a)   *(b)*

Figure 11.3   Very low magnification 'scan' micrographs. This image is used to judge which grid square is the most promising for higher power examination: (a) shows a very poor section with tears, holes, scratches and wrinkles. One square might be worth closer scrutiny; (b) shows a good section of large area almost completely free from imperfections; the histological detail is clearly visible and individual cells can easily be mated with the same cells seen under the light microscope, stained with a dye (see Chapter 18). Araldite sections approx. 500 Å thick, mounted unsupported on bare copper grids. GEC/AEI EM6B, approx. × 100

11.13 Changing Magnification

As has been explained on p. 129, there are three methods in use for changing magnification without introducing image distortion. These are:
1. The intermediate lens is always used as a magnifier, but the overall magnification range is changed by changing the bore of the projector polepiece.
2. Two intermediate lenses are provided; one (diffraction) is used to diminish the image size, while the other (intermediate) is used to magnify.

3. A single intermediate lens is used both to diminish and to magnify the final image, but the optical design is arranged so that the distortions introduced by the intermediate lens are exactly balanced by equal and opposite distortions introduced at low excitations (and hence low magnifications) in the projector lens.

The first method has the disadvantage of needing four separate ranges to cover magnifications between × 200 and × 200,000, but has the advantage that the image always stays the same way up, and distortion is always at a minimum. The other two methods have the advantage that the full range of magnifications is immediately available by rotating a single control knob, but at some point in the range, where the intermediate lens ceases to act as a diminisher and begins to act as a magnifier, the image suddenly inverts. This is because of the change from a 2-real-image system to a 3-real-image system (see Figs. 6.16 and 6.18). The intermediate and projector lens currents must in this case be very carefully balanced by the maker to ensure minimum low-magnification distortion and minimum chromatic change in magnification.

11.14 Choice of Electron Optical Magnification

When dealing with thin sections of embedded biological material in which a resolution of better than 25 Å is not expected, two factors influence the choice of magnification in the electron microscope: the ability of the operator to focus the image at low magnifications; and the ultimate resolution required. To deal with the latter point first, the overall magnification of an electron micrograph is determined as we have already seen by the ratio of the required resolution to the resolving power of the unaided eye. If it is agreed that the eye can readily resolve 0·25 mm at the standard distance of 25 cm then at an overall magnification of × 1,000, 2,500 Å can be resolved; at an overall magnification of × 10,000, 250 Å can be resolved and so on. To resolve 5 Å, × 500,000 is required. It must be remembered that the overall magnification of the final print is the product of the electron and the optical magnifications, since the plate or film (the negative or primary electron micrograph) is further magnified in an optical enlarger. If care is taken in developing the primary plate image (see Chapter 14), photographic grain can be made fine enough to allow optical magnifications of × 25 or more.

If a resolution of 25 Å is required, then the necessary overall magnification can be obtained either by an electron magnification of × 50,000 followed by an optical enlargement of × 2; or by an electron magnification of × 10,000 followed by an optical enlargement of × 10. The point here is that the original plate will have $5^2 = 25$ times as much information if the lower electron magnification is chosen, since the area of image included will be 25 times as great.

The problem in the latter case is one of focusing exactly at low magnifications, which requires skill and practice. If the original plate is not in exact focus, then the required resolution cannot be obtained by optical enlargement. This problem will be dealt with in the following section.

11.15 Focusing

When the instrumental magnification is altered by changing the current through the intermediate lens, the focal length of the latter changes, and therefore its object plane moves up and down the column. The image plane of the objective lens must then be made to coincide with the new intermediate lens object plane. This is done by adjusting the current in the objective lens while observing the final image; the procedure is called 'focusing'.

It might be thought that focusing merely entails obtaining the sharpest possible image, as with the light microscope. Unfortunately, there are other factors which make the issue rather more complicated. The basic dilemma is that image contrast is always at a minimum at the true or 'Gaussian' focus. The operator therefore has to choose between a micrograph which is in perfect focus and which may lose detail due to lack of contrast and one in which some detail is reinforced but other, smaller, detail may be lost due to loss in resolution.

The basic rule in focusing is: the higher the overall magnification of the final electron micrograph (the product of electron optical and enlarger magnifications) the more accurate must the focusing be. The resolution obtainable on an electron micrograph depends ultimately on the resolving power of the human eye. If we accept this as being 0·2 mm, then a micrograph at an overall magnification of × 10,000 cannot show to the unaided eye a resolution better than 0·2 mm divided by 10,000, or 200 Å. If this micrograph is examined with a × 5 hand lens, as electron micrographs frequently are, then the overall magnification will be × 50,000 and the required resolution will be 40 Å. To resolve 5 Å, an overall magnification of 400,000 is required. In theory, this should be possible by taking a micrograph at an electron optical magnification of × 10,000 and then enlarging parts of it 40 times in an enlarger. In practice, this is not possible, for two reasons. Firstly, the grain of the fluorescent screen is too large to enable perfect focus to be obtained at a magnification of 10,000; and secondly, even if the perfect focus setting were by chance to be selected, the grain of the photographic emulsion will be larger than the theoretical resolution of 5 Å. The limit of photographic enlargement in practice is about × 25; Fig. 12.8 shows (a) a comparatively low-power micrograph in almost perfect focus, together with (b) parts at high electron optical and normal enlarger magnification and (c) parts enlarged to the same overall magnification using a higher enlarger magnification. It will be seen that (b) and (c) are almost indistinguishable.

The grain of the fluorescent screen is about 0·1 mm in size, and can be seen very clearly when the screen is observed through the magnifying telescope, which generally gives an additional magnification of between × 5 and × 10. The first step in focusing the electron microscope, therefore, is to ensure that the telescope is in sharp focus for the operator's eyes, and that the grain of the screen can clearly be seen.

The grain of the fluorescent screen therefore limits the resolution of a micrograph taken at an electron optical magnification of × 10,000 to about 100 Å, although this can be reduced considerably by skilled use of the correct techniques of focusing. To bring this 0·1 mm detail up to the resolving power of the eye, an optical magnification of × 2 is all that is required, and in practice this is the degree of enlarger magnification generally used.

11.16 Methods of Focusing

Four methods are in general use for judging the point of true focus. These are, in increasing order of accuracy:

1. Observing image shift when the illuminating beam is tilted (the 'wobbler');
2. Obtaining the point of minimum contrast;
3. Obtaining the point of maximum contrast;
4. Observing the Fresnel fringe at a sharp edge.

Methods 1 and 2 are used at low magnifications; Method 3 is the method of choice for medium and fairly high magnifications; Method 4 is used when the very highest resolution is needed at the highest magnifications. Methods 3 and 4 will be discussed in the following chapter.

Contrary to expectation it is more difficult to focus an electron microscope at very low magnifications than it is at medium or fairly high magnifications. This is due to the combined effect of the resolving power of the eye and the resolving power of the fluorescent screen. Some aid for focusing at low magnifications is highly desirable.

11.16.1 Image Shift Method

If the axis of the illuminating system is tilted with respect to the axis of the objective lens, then image motion will take place when the current in the objective lens is changed. Advantage is taken of this in aligning the column, as will be explained later in this chapter. Conversely, if the objective lens is out of focus, then tilting the illuminating beam will cause image shift. Advantage can be taken of this effect in focusing if some means is provided on the instrument for introducing a sudden change illumination tilt. If the image is out of focus, then tilting the beam will give rise to a shift in image position; if the image is exactly in focus, no image shift should be detectable. The effect is equivalent to reducing the depth of focus of the objective lens.

This aid to focusing was first introduced by le Poole, and is now a feature of almost all instruments. The beam is tilted back and forth by applying an A.C. voltage at 50 c/s to a set of 'wobbler' coils mounted above the objective lens. This rapidly reversing induced tilt causes blurring of the image if the objective lens is out of focus, and gives rise to a greatly enhanced image unsharpness. The point of exact focus can thus be judged more accurately. The image movement effect can be made more sensitive if a sudden, single change in tilt can be applied while the image is being critically examined with the binocular magnifier. Some instruments have provision for energizing the wobbler coils with D.C. so that this effect can be obtained.

If the instrument is not equipped with a means for sudden tilting of the illumination, the image shift method of focusing cannot be applied.

11.16.2 Minimum Contrast Method

It is evident from a consideration of the Fresnel fringe phenomenon (see Chapter 4) that the point of true focus is the point of minimum contrast. This can be demonstrated very strikingly if a holey carbon film is examined at very low magnification with the objective aperture removed. The holes are easily visible on either side of focus, but disappear almost completely at the true focus point. The phenomenon becomes less and less marked as magnification is increased, or if the objective aperture is replaced. It is a reasonably accurate method at low magnifications, but necessitates the removal and replacement of the objective aperture. It is inadvisable to replace the objective aperture after the critical focus point has been determined, because it is very unlikely that it will go back into exactly the same position. If the aperture is dirty, the astigmation correction will then alter, and the resulting micrograph may show the worse fault of astigmatism. It is therefore unsuitable for high magnifications.

Methods of focusing at high magnifications will be discussed in the following chapter on high resolution operation.

11.17 Image Faults at Low Magnifications

After focusing, the low power image should be examined carefully for specimen drift, astigmatism, distortion and chromatic change in magnification before a micrograph is made.

11.17.1 Specimen Drift

This is generally due to thermal expansion arising from the heating of the specimen by the electron beam. It can also be due to the evaporation of material from the section, tears or holes in the section or substrate, or uneven clamping of the grid in the holder. Drift is minimized by preconditioning a grid square by scanning it with a defocused, low-intensity beam to 'cure' the resin section with heat before examining it at a higher magnification and beam intensity. This applies especially to unsupported Araldite sections. After focusing a suitable area to be photographed, watch a sharply defined feature closely through the telescope, checking for image movement against the grain of the screen or a speck of dust or flaw in the phosphor. Watch both at viewing intensity and also at the reduced photographic intensity; defocusing C_2 can cause a stable image to drift. If drift is present, find out the cause before taking a micrograph.

11.17.2 Astigmatism

This should not be detectable at magnifications below about × 20,000 if the instrument and especially the objective aperture are clean and well-aligned and the stigmator corrector controls are set to zero. The various procedures for correcting astigmatism at higher magnifications are discussed in the following chapter.

11.17.3 Image Distortion

This fault is especially noticeable at low magnifications (below × 5,000). It arises when the intermediate lens approaches the diffraction point and is working at increased aperture (see Fig. 4.8). As magnification is reduced, so pincushion distortion becomes more prominent. If it is desired to make a 'montage' of a number of overlapping fields taken at low magnification, the change in magnification between the centre and the periphery of the image must be less than about 5 %. This small amount of distortion is not immediately apparent in the image of an irregular object such as part of a cell. Distortion can be readily recognized if the image is moved rapidly back and forth across the centre of the screen with the stage controls. If distortion is present, any sharply defined feature can be seen to change in size as it passes across the screen.

If the instrument has a set of calibrated magnification steps selected by a single knob control, the manufacturers should have taken care of distortion by arranging the correct balance of currents in the intermediate and projector lenses at each step. If magnification is changed simply by altering the current in the intermediate lens, then the operator must himself check and if necessary balance out distortion by changing the projector polepiece or by reducing the projector current. Distortion is generally negligible at electron optical magnifications above × 5,000.

11.17.4 Chromatic Change of Magnification

This form of distortion causes the image to be out of focus at the periphery while in focus at the centre. It is caused by too thick a specimen or too low a kilovoltage or both (see Fig. 4.11). If the fault is noticeable, the probability is that the section is too thick and will in any case give an unsatisfactory micrograph of poor resolution. The only remedy is to increase kilovoltage; it is better however to change the specimen for a thinner one.

11.18 Taking a Micrograph

It is necessary to load a camera, desiccate the emulsion and insert the camera into the column before a micrograph can be taken. The exact procedure varies so widely from microscope to microscope that no general instructions can be given. The loaded camera should be desiccated for at least 2 hours over fresh P_2O_5 at a pressure of 1 torr or less before being inserted into the column. Film generally requires longer, especially roll film. After a freshly desiccated camera has been loaded into the column, the microscope will require up to 20 minutes to pump down to working vacuum, depending on the efficiency of the pumping system.

11.19 Exposure Procedure

This also varies in detail, but the basic steps are as follows:
1. Choose the area and magnification.
2. Focus at a comfortable viewing intensity, and examine the image closely for instability and other faults (correct if possible). Do not move the specimen after focusing.

3. Reduce screen luminosity to the correct value indicated by the exposure meter or comparator (corresponding to the preset shutter time) by reducing the current in C_2 (underfocusing).
4. Examine the image closely for drift again.
5. Lower the screen (or close the shutter).
6. Bring the plate smoothly and gently into the beam path, vibrating the column as little as possible.
7. Open and close the shutter (or raise and lower the screen) for the correct exposure time.
8. Remove the plate from the beam path.
9. Open the shutter and resume specimen examination.

If the microscope has a fully automatic camera, steps 5 to 9 follow in an automatic sequence following the signal from a single push-button. Other instruments have varying degrees of automation; simple instruments using the screen as a shutter require the sequence to be followed manually. Most instruments have at least a shutter giving a preset exposure time, which is often coupled to the screen, so that the act of raising the screen automatically closes and then opens the shutter. The output from the exposure meter may be coupled to the shutter timer, giving the correct exposure over a range of screen luminosities.

After the required number of exposures have been made, the camera is then removed from the column, taken to the darkroom, and the plates are then processed. Photographic procedure is discussed in Chapter 14.

11.20 Shutting Down

1. Check that the backing tank and desiccator are pumped out, and that the column is at a reasonable vacuum.
2. Check that H.T. and filament are off, and that emission and filament voltage controls are zeroed, and that C_2 is fully underfocused.
3. Shut off the column power supplies.
4. Close the high vacuum valve.
5. Shut off the high vacuum gauge (if recommended).
6. Shut off the diffusion pump heaters.
7. Shut off the rotary pump and check that the air admittance valve has opened.
8. Shut off the mains electricity.
9. Allow the cooling water to circulate through the pumps for 10–20 minutes before turning it off. This is to ensure that the diffusion pump oil does not boil up into the high vacuum valve and get into the column.

It is a convenience to have a water switch operated by a preset timer so that the microscope can be shut down without the operator having to wait around for the cooling water to run for the prescribed time. Some instruments include such a device. A mains-operated solenoid valve and clockwork timer can easily be built into the water system (see Fig. 10.1).

11.21 Column Alignment

11.21.1 Why Accurate Alignment is Necessary

Before the electron microscope can be operated, it must first be correctly aligned. The necessity for correct alignment cannot be too highly stressed. A badly aligned instrument is not only tedious and frustrating to operate, but cannot be expected to perform properly. Resolution, image brightness and filament life will all be adversely affected.

To obtain the optimum performance from an electron microscope, it is essential that all the optical components are disposed symmetrically about the axis of the column. Misalignment gives rise to movement either of the image or of the illuminating spot when lens currents or high tension voltage are altered. This means that when any lens control is altered, in order to change focus, magnification, spot size or brightness, either the image or the illuminating spot moves away from the screen centre, and has to be recentred. This effect is particularly annoying at high magnification. Quite apart from the annoyance, the instrument cannot give maximum resolution on an electron micrograph. The reason for this is not immediately apparent. It is due to the fact, as we have seen in Chapter 8, that the stabilizing circuits for the lens and high tension power supplies cannot be made perfect. There is an inevitable residual fluctuation on both supplies. The residual instability gives rise to two effects: rotation of the image and change in image magnification. If the alignment of the column is not within the accuracy specified by the manufacturer, image movement on the screen may be greater than the resolving power of the instrument. This movement will blur the final micrograph. The longer the exposure time, the more likely is the effect to show.

The first effect causes the image to rotate when the current in any imaging lens is altered. The second effect causes the image to increase or diminish in size as the high tension varies. If the column is correctly aligned, large mains supply variations of $\pm 20\%$ will have a negligible effect at the centre of the micrograph, and an effect less than the resolving power over the central half of the plate (see Fig. 11.4(a)). If the column is misaligned, the rotation centre will no longer fall at the screen centre.

If it is a long way off (see Fig. 11.4(*b*)) the effect of supply variations may well exceed the instrumental resolution over the whole area of the plate.

It is an interesting exercise to misalign the objective lens (the lens most sensitive to misalignment) deliberately, so that the rotation centre falls a foot or more away from the screen centre. If a small object is watched closely through the binocular magnifier at high magnification, the random shift due to supply instability can be seen very clearly, and is a salutory reminder of the necessity for correct alignment.

11.21.2 The Alignment Modulator

Two methods can be used to give the information necessary to align the optical components symmetrically about the microscope axis. These are (*a*) current centration and (*b*) voltage centration. In (*a*), lens current is

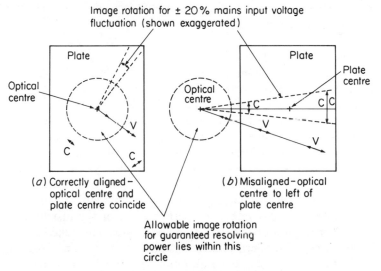

Image rotation for ± 20% mains input voltage fluctuation (shown exaggerated)

(*a*) Correctly aligned – optical centre and plate centre coincide

(*b*) Misaligned – optical centre to left of plate centre

Allowable image rotation for guaranteed resolving power lies within this circle

Figure 11.4   The importance of correct column alignment. Fluctuations in the mains input voltage cause (*a*) rotation of the image due to objective lens current fluctuation (the arrows C), and (*b*) radial change in magnification due to H.T. voltage fluctuation (the arrows V). The designer can only guarantee resolving power for a given fluctuation if the current and voltage centres lie within a given distance of the plate centre (the area shown within the dotted circle). Optimum conditions for maximum resolving power occur only if the optical axis coincides with the plate centre (left-hand diagram). If the axis lies away from the plate centre (right-hand diagram), image movement due to current and voltage fluctuations will exceed the resolving power. The operator must maintain the alignment of the column or resolving power will suffer

displaced on either side of the mean; in (*b*) the high tension is varied about the mean.

The task of aligning the tilt of the objective lens is greatly simplified if a means is provided for swinging the lens current at will by a small fraction (up to about $\pm 10 \%$) at a low frequency (about 2 c/s), so that the image revolves rhythmically through about $\pm 45°$. The operator then has both hands free to centre the lens and make the image rotation centre coincide with the screen centre. This device is called a 'modulator', and can be applied to modulate either the lens current or the H.T. voltage. The latter form of modulation gives the 'voltage centre', and aligning this with the screen centre is an alternative method of alignment favoured by a few manufacturers. The voltage and current centres seldom coincide exactly on any particular instrument, due to various axial asymmetries which are very difficult to eliminate during manufacture. The closeness of the current centre to the voltage centre is one measure of the accuracy of manufacture of a microscope column.

11.21.3 *Adjustments Provided*

Ideally, each optical component should be capable of being traversed perpendicular to the axis, and also tilted with respect to it. This requirement demands tilt and traverse adjustments for gun, C_1, C_2, objective, intermediate and projector; a total of 12 adjustments. In addition, traverse adjustments must be provided for all the movable apertures— C_1, C_2, objective and diffraction. Finally, centrable stigmators may be provided for both objective lens and illuminating system. This makes a somewhat formidable theoretical total of 18 separate adjustments.

Experience has shown that many of these alignments can be dispensed with, since they do not affect the performance of the instrument very critically. The less important lenses can be adjusted and locked in position at the factory. The modern tendency is to reduce the number of adjustments as far as possible, since although in skilled hands a particular alignment can be made to improve performance, in unskilled hands it is more likely to affect performance adversely.

The basic minimum adjustments are:

1. Traverse gun
2. Traverse illuminating system
3. Tilt objective lens
4. Traverse intermediate lens

In addition, some instruments provide:

5. Traverse and tilt C_1

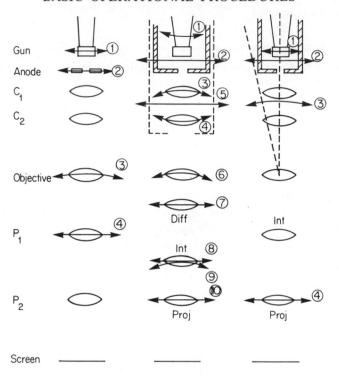

Gun
Anode
C_1
C_2
Objective
P_1
Diff
Int
P_2
Int
Proj
Proj
Screen

(a) GEC/AEI EM6B   (b) Philips 200   (c) Siemens Elmiskop IA

Figure 11.5   The alignment adjustments provided by 3 representative makers

(a) GEC/AEI EM6B

1. Centre gun (preset)
2. Centre anode
3. Tilt objective
4. Centre P_1 (preset)

(b) Philips EM 200

1. Tilt gun
2. Centre gun + anode
3. Tilt C_1
4. Tilt C_2
5. Centre whole illuminating system
6. Tilt objective
7. Centre diffraction
8. Centre intermediate
9. Tilt intermediate
10. Centre projector

(c) Siemens Elmiskop I

1. Centre gun
2. Centre gun + anode
3. Tilt the whole of the illuminating system about the objective lens centre
4. Centre projector

GEC/AEI and Siemens endeavour to keep the number of operator alignment adjustments to the minimum for simplicity in operation. This necessitates very accurate factory prealignment of components. Philips supply all requisite adjustments for perfect alignment, but this presupposes a high degree of skill and knowledge in the operator

6. Traverse and tilt C_2
7. Tilt intermediate lens
8. Traverse projector lens

The alignments given by three representative manufacturers are shown in Fig. 11.5. All manufacturers follow one or other of these designs, or a combination of them.

11.21.4 Perfect Alignment

When the column is perfectly aligned, the following criteria will be satisfied:

1. The image will rotate about the centre of the final screen when the current in any imaging lens is changed.
2. The image will stay centred when the high tension is changed.
3. The image will grow or diminish evenly from or towards the screen centre when magnification is changed.
4. When the illuminating spot is focused, it will form a circle at the screen centre the same size as the area of specimen under investigation.
5. When the illuminating spot is defocused, it will expand evenly about the screen centre, and the illumination will remain perfectly even over the screen.
6. When the intermediate (or diffraction) lens is at focus, the diffraction point will be at the screen centre.
7. The resolution as measured with a Fresnel fringe at the screen centre will be equal to or better than the manufacturer's specification.

Some alignments are more important than others. Misalignment of the image-forming system gives rise to loss of resolving power; misalignment of the illuminating system gives rise to loss of contrast, image brightness and filament life. The closer a lens is to the specimen, the more critical is its alignment. The most important alignment as far as resolution is concerned is to make the axes of the illuminating system and the objective lens coincide. The way this is done varies with column design; some tilt the illuminating axis into the fixed objective lens, and some the other way around.

11.21.5 Alignment Procedure

The basic objects of the alignment procedure are: (1) to bring all the components of the illuminating system about a common axis; (2) to make the axis of the illuminating system coincide with the axis of the objective lens; and (3) to make the other components of the imaging system (intermediate, diffraction and projector lenses) coaxial with the common axis.

The exact procedure for alignment depends on the mode of construction of the column, and it is obviously impractical here to deal with the precise methods used for each individual instrument. The full procedure to be described shows the basic principles of alignment, and is intended to provide explanations for the manipulations described in the microscope instruction manual. It should only be necessary to follow the full procedure after the whole column has been dismantled and reassembled. In general, day-to-day alignment should only require a small adjustment to the gun for maximum brightness and aperture centration; a possible small adjustment to the illumination or objective lens tilt; and a possible small adjustment to the intermediate lens centration. The gun requires realignment after a filament change.

Let us assume, then, that the column has been completely dismantled and reassembled. The components will be anything up to a millimetre or more off axis, and the gun will be incorrectly set. The first procedure is to check the gun setting. Fill the column or gun chamber with air, open the gun, tilt back the insulator and examine the position of the filament tip with respect to the shield aperture. It must be exactly centred, and the height of the filament tip with respect to the underside of the shield must be exactly that quoted by the manufacturer. These adjustments are very important, because the brightness and stability of the gun as well as the life of the filament depend on having the filament position correct to within 100 μm or less. Having checked and centred the filament, close the gun chamber and pump down to working vacuum.

In the instructions which follow, each stage follows the preceding one. The illuminating system is aligned initially with the objective off, the intermediate being used as a low magnification, long-focus objective.

11.22 Align Illuminating System (Single Condenser Operation)

The first steps in the complete alignment procedure are to remove the specimen, specimen holder and all movable apertures from the beam. For instruments having interchangeable projector polepieces, select the one giving minimum magnification. Turn all lenses off. Set gun emission to minimum. Turn on H.T. and filament. Increase filament heating very carefully. A small, dim circle of light 2–3 mm in diameter should be seen close to the centre of the screen. The centre of this circle can be taken for all practical purposes as the intersection of the microscope axis with the screen; it does not necessarily coincide with the centre of the screen. The exact position must be noted with respect to some marking on the screen. The point will be referred to as the 'screen centre'.

11.22.1 Stage 1—Centre Gun

Adjust the controls which traverse the gun or the gun anode until the circle of light suddenly becomes intensely bright. Patience is generally needed here. Obtain maximum brightness as quickly as possible, or the screen may become burnt at the centre. Turn on the projector lens immediately. The whole screen should now be evenly, dimly lit.

11.22.2 Stage 2—Align Projector Approximately

If a centration control is provided for the projector lens, adjust projector current until a sharply defined circle of light is projected on the screen. Adjust this circle with the projector centration controls until it is disposed symmetrically about the screen centre, then set the normal projector working current. Otherwise ignore this stage.

11.22.3 Stage 3—Saturate the Filament

Increase filament heating while watching the beam current meter. When the beam current ceases to rise, do not increase filament heating any further. This is a rough adjustment; the setting is made more accurately at Stage 5.

11.22.4 Stage 4—Align C_2

Switch on C_2. Objective off; low magnification selected. Focus spot, keeping it centred with the illumination centration controls. Defocus C_2 slightly; insert and centre aperture. Note: it is bad practice to run an instrument for any length of time with the condenser apertures removed, due to (*a*) column contamination by the excessive beam current, and (*b*) excessive X-ray generation in the illuminating system, which may be dangerous, and may also cause the rubber vacuum seals to deteriorate.

11.22.5 Stage 5—Adjust Gun Tilt

Focus C_2. Reduce filament heating until the 'halo' or '5-spot' pattern (see Fig. 11.10) appears on the screen. Adjust C_2 focus to get the pattern critically sharp. Adjust gun tilt or anode centration to get a symmetrical pattern. This adjustment is very critical; incorrect setting causes a rapid falling-off in gun brightness. For greatest accuracy, resaturate the filament and measure either screen current or screen brightness while adjusting gun tilt, the aim being to obtain maximum screen brightness for a given beam current. This method ensures optimum setting under normal working conditions.

11.23 Align Illumination System (Double Condenser Operation)

All high-performance electron microscopes are provided with a two-lens condenser system ('double condenser'). It is used to keep the area of specimen illuminated the same as the area under observation. In this way, the whole of the specimen away from the area under examination is protected from damage by the beam. Keeping the illuminated area small also reduces specimen heating, which in turn minimizes image movement due to thermal drift.

In many column designs, the two condenser lenses are constructed in a single block, and it is not possible for the operator to adjust the alignment of C_1 with respect to C_2. In this case, the alignment of the double condenser system is simple. First, focus C_2 with C_1 off. Next, turn on C_1 at its lowest current. The spot will defocus. Bring it to focus with C_2. It will now probably be found that the spot has shifted from screen centre. The amount of shift is a measure of the accuracy of initial alignment of the two lenses during manufacture. In a perfectly aligned system, the spot should not shift, only defocus. Therefore, bring the spot back to screen centre, preferably by using the mechanical adjustment with the electrical beam shifts switched off. If there is no mechanical means for centration, the electrical beam shifts must be used. Next, increase the current in C_1, refocus with C_2 and recentre the spot. The diameter of the focused spot will have decreased. Continue the process until the spot is of the desired diameter for the magnification it is intended to use.

It will be noticed that as the spot size is decreased, so spot brightness decreases also. This is an unexpected result; the explanation is as follows.

In a single condenser system, since the relative axial positions of source, condenser and specimen are fixed, the single condenser can only project a focused spot of a fixed size on to the specimen. A second condenser lens must therefore be introduced if it is desired to vary the size of the focused spot. This is the function of the first condenser lens, C_1. It is used to form a diminished image of the source formed at the electron gun. The size of the image of the source formed by C_1 is made smaller as magnification is increased, by increasing C_1 current. But as the image of the source is made smaller, it moves closer to the gun, and therefore C_2 must be refocused. It will therefore be seen that the strengths of C_1 and C_2 are mutually interdependent, and ideally both should be changed each time magnification is altered. The ray paths for the condenser lenses in the positions of low and high magnification are shown in Fig. 6.13, (b) and (c). It will be seen that as the strength of C_1 is increased, the angle of the beam at the image plane of C_1 becomes much greater than the acceptance angle of C_2. This means that most of the electrons which pass through the aperture of C_1 will be thrown outside the aperture of C_2, and will not contribute to the beam passing through the microscope. Therefore, as C_1 excitation is increased and the spot diameter decreases, the spot also becomes dimmer. To compensate for this, the gun emission must be increased for high magnification working. This in turn leads to reduced filament life. The operator must therefore make a choice between contaminating large areas of his specimen unnecessarily, and the life of the filament. The general rule is: the higher the magnification, the stronger is C_1 and the weaker is C_2; the lower the magnification, the weaker is C_1 and the stronger is C_2.

An exception to this rule is when C_1 is switched off for low magnification working. This is not necessary provided the magnification of C_1 can be made close to unity, but is occasionally preferred by some operators, since alignment is simplified. The ray diagram (Fig. 6.13(a)) shows that C_2 must now be made very weak in order to focus the gun crossover directly.

Table 11.1

Working condition	C_1 current mA	C_2 current mA	Spot size at specimen, μm (200 μm aperture in C_2)
very low mag. (scan)	15	65	45
low mag. $\times$ 1,000	25	53	13
low mag. $\times$ 5,000	35	50	6·0
medium mag. $\times$ 10,000	45	48	4·8
medium mag. $\times$ 50,000	55	47	3·0
high mag. $\times$ 100,000	85	46	1·2
very high mag. $\times$ 200,000	125	45	0·6
C_1 off	0	35	20

Table 11.1 shows the relative currents in C_1 and C_2 in relation to the focused spot size at the specimen. The measurements were made on a Philips 200 instrument. There is no reason other than ease of alignment for working with C_1 off at any magnification, even the lowest; in general, the image quality is much improved when the double condenser is in use.

11.23.1 Procedure for Aligning an Adjustable Double Condenser

Some instruments are provided with adjustments which enable the operator to adjust the gun centration; the tilt of C_1; the tilt of C_2; and the centration of the whole illumination system with respect to the axis of the imaging system. These adjustments, when correctly made, enable the currents in C_1 and C_2 to be changed independently without the illumination spot moving off the screen. This is a convenience to the operator, but is not essential, because the instrument is generally run with the excitation of C_1 fixed, a low current being used for low magnification work and a high current for high magnification.

Lens tilt is generally adjusted by moving one polepiece radially with respect to the other (fixed) polepiece. In the Philips system, the upper polepiece of C_1 and the lower polepiece of C_2 are each radially adjustable by means of screws operated by Allen keys in one direction, and a spring-loaded return mechanism in the opposite direction. The procedure for adjustment is as follows. First align C_2 as described above.

1 Align C_1 Switch off C_2 and switch on C_1. Switch off objective. Remove both condenser apertures. Use lowest beam current and magnification. Focus C_1. Centre the spot with the gun centration adjustment. Reverse the current in C_1. In general, the spot will jump to another position, due to the lens axis being tilted. The tilt adjustment (which centres the upper polepiece of C_1) is therefore manipulated so as to bring the spot halfway between the two extremes. In practice, one simply keeps on reversing C_1 current and making fine adjustments to the C_1 upper polepiece centration controls until the spot does not move on reversal. The spot, which will not in general be at screen centre, is then brought to screen centre by centring the electron gun. Now check reversal again, adjust if necessary, insert C_1 aperture, defocus C_1 and centre the aperture. C_1 and the gun are now aligned to the microscope axis.

2 Align C_2 Switch off C_1 and switch on C_2, using the same conditions as above. Focus C_2. Reverse C_2 current and 'split the jump' as above. This time centre the spot by traversing the whole illuminating system. This will obviously disturb the centration of the electron gun. Therefore, repeat the centration of C_1. It should now be possible to select any value of C_1 current and focus the spot with C_2 without having to recentre the spot with the beam shifts.

11.24 Alignment of the Imaging System

Most microscopes with a 3-lens imaging system provide only two alignments: a means for tilting the axis of the objective lens so that it coincides with the axis of the illuminating system (which has already been aligned); and a means for centring either the intermediate lens or the final projector lens. Some instruments, built on the Siemens principle, have the objective axis fixed in manufacture, but allow the whole illuminating system to be tilted about the centre of the objective lens. The 4-lens Philips microscopes provide means for tilting the objective; centring the diffraction lens; tilting and centring the intermediate lens; and centring the projector lens.

11.24.1 Stage 1—Adjust Objective Tilt

The current centre method of objective alignment will be described, since it is universally applicable to all instruments having means for varying the lens currents—in effect, all electron microscopes. For voltage centration, a modulator to vary the H.T. over a small range is necessary; few instruments have this device.

Switch off filament and H.T. Insert a specimen (preferably a holey carbon film—see Chapter 13); switch on filament and H.T. again. Focus illumination; switch on objective; focus specimen. The intermediate lens remains off; overall magnification should be × 5,000 to × 10,000. Find a small, sharply defined object such as a small hole or piece of dirt on the film and bring it to the screen centre with the stage controls. Focus exactly. Now defocus the objective by over- and underfocusing rapidly with the coarse focus control. The object should rotate slightly but stay at the exact screen centre. If it does not, the axis of the objective lens is tilted with respect to the illumination/projector axis. Either the objective lens must be tilted to conform to the illuminating axis or *vice versa*, according to the construction of the column. In either case the procedure is the same.

1. With small object at screen centre, focus exactly.
2. Underfocus the objective by about 5% of the lens current. The object moves.
3. Bring the object back to screen centre with the tilt (objective or illuminating system) controls. Keep the light spot at the screen centre with the mechanical illumination shift all the time. If the objective lens is badly tilted, this procedure will require much patience.
4. Refocus the image. Bring the spot back to screen centre with the stage controls.
5. Overfocus the objective by about 10 % of the lens current. Repeat step 3.
6. Keep repeating steps 2, 3, 4 and 5, defocusing by increasing amounts each time, until the objective lens can be defocused by as much as ± 25 % or more without perceptible movement of the image at the screen centre. The image should rotate about the exact screen centre as the objective current is varied.

The tilt is now correctly adjusted. This alignment is the most important of all for high resolution work. If the rotation centre is more than about 2 cm from the screen centre at an overall magnification of × 10,000, resolution at the edges of the plate at × 100,000 will almost certainly be diminished.

11.24.2 Stage 2—Centre Intermediate and/or Diffraction Lenses

With the image in focus, turn on the intermediate lens and slowly increase the current through it. The image will become smaller and smaller, finally shrinking to a spot of intense brilliance about 1 mm in diameter. This is the 'diffraction spot'. Bring it to the screen centre with the intermediate lens centration control. If a diffraction lens is provided in addition, repeat the procedure on this lens.

11.24.3 Stage 3—Centre Projector Lens Accurately

Few microscopes provide an adjustment to centre the projector if the intermediate and/or diffraction lenses are centrable. The axis of the microscope is generally defined by the line passing through the centre of the (fixed) projector and the (fixed) lower polepiece of the objective. The best method of centring the projector is to reverse the current through the objective and adjust the projector centration so that the centre point of the image (which should coincide with the centre of the screen) does not move on current reversal. Few instruments provide this facility; the Philips is a notable example. Microscopes such as the older Siemens which are provided with a turret of projector polepieces have to provide a means for centring the polepiece in use. The method is to have a circle scribed at the periphery of the screen centred on the screen centre, and to adjust the current in the projector and the projector centration controls so that the projected image of the polepiece coincides with the scribed circle. This ensures that the projector lens always operates at a known fixed magnification.

The microscope should now be completely aligned and capable of maximum screen brilliance for minimum beam current, together with maximum resolving power and operating convenience. If the instrument was previously completely dismantled and reassembled, it may be necessary to repeat the whole procedure.

For anything other than low-power work, it will be necessary to compensate for astigmatism, both in the objective and in the illuminating system. This procedure is described in the following chapter.

11.25 Quick Alignment Check

This takes less than 2 minutes. It should be run through before starting any session on the microscope.

1. Focus illumination spot at low magnification; set gun centration and filament saturation with screen current or brightness meter. Centre beam. Centre C_2 aperture.
2. Swing $C_2 \pm 20\%$ about focus; the spot should stay on centre and grow evenly across the screen.
3. Insert specimen and swing objective current $\pm 20\%$ about focus; the image should rotate about the screen centre.
4. Bring to zero magnification with intermediate lens current and check that diffraction spot is at screen centre.
5. Increase magnification to full; check that image and illumination spot stay centred.

11.26 Accuracy Required

The lower the magnification to be used, the less critical the alignment becomes. A good instrument will give good low power (up to $\times$ 20,000) micrographs even when badly misaligned. For high resolution work, accurate alignment is absolutely essential. The operator should always try to maintain good alignment as a matter of principle and pride. The instrument is far easier to work when properly set up. Neglecting alignment may save a little time at one stage, but far more time will be wasted in the end.

11.27 Other Methods of Operation

By far the greatest amount of biological information is obtained from electron microscopes by the straightforward examination of transmission images, which has already been described in detail. There are, however, some other methods which are sometimes of use, and which can give further information about suitable specimens. These methods can be used to give more information about the crystalline structure of the specimen, or about its structure in the third dimension.

The most important ancillary methods are:

(a) Dark-field operation
(b) Electron diffraction
(c) Stereo operation
(d) Double specimen tilting.

11.28 Dark-Field or 'Strioscopy'

This is a method of image formation which uses only the electrons scattered by the specimen, and which are removed by the objective aperture in normal transmission operation. The main beam passing through the specimen must be deflected or intercepted so that it cannot reach the final screen; the scattered electrons are directed into the imaging system and form the final image. The method suffers from the disadvantage that the scattered electrons cover a broad spectrum of energies, and the final image must therefore lose resolution due to chromatic aberration. It has the advantage of throwing holes, dirt and other contamination on a thin section into sharp relief, thus enabling them to be distinguished clearly from the biological detail embedded in the section. The method is however of little other use to the biologist, although it is of great use to the metallurgist examining defects in crystalline structures; details are therefore included for the sake of completeness.

There are three methods for obtaining dark-field images. These are: (1) to tilt the illuminating beam so that it does not pass along the axis of the objective lens; (2) to move the objective aperture to one side so that it intercepts the main beam but allows the scattered electrons to enter the imaging system asymmetrically; and (3) to use a special objective aperture having a central stop which intercepts the main beam. The ray diagrams for the three methods are shown in Figs. 11.6(b), (c) and (d). The advantage of the first method is that the scattered beam passes symmetrically into the imaging system, and chromatic aberration is reduced to the minimum. This method is sometimes called 'high-resolution dark-field operation', but the term merely means that greater resolution can be obtained by this method than by the others. It requires the provision of a tilt adjustment to the illuminating system, possessed by relatively few instruments. The best way of tilting the beam is to incorporate a double deflector system just below the second condenser lens. The first pair of coils deflects the beam to give the initial tilt of about $2°$, and the second pair deflects the beam back to give the required tilt. This method has the advantage that the deflector coils can be switched on and off, giving immediately available dark-field without any centration adjustments.

The second method is the simplest, and dark-field images can be obtained in a few seconds simply by decentring the objective aperture so that it is just out of the field of view. The third method requires a special objective aperture, but has been found to improve the contrast of biological specimens in the million volt electron microscope. Dupouy and his coworkers, who devised the method, use a very fine wire stretched across the objective aperture and which they call a 'contrast screen'. The term 'strioscopy' is sometimes applied to this method. Fig. 11.7 shows the improvement in contrast in a bacillus using this method, taken with 1 million volts accelerating potential. The method can be improved by using an annular aperture in C_2. Both annular and beam-stop apertures can now be obtained commercially.

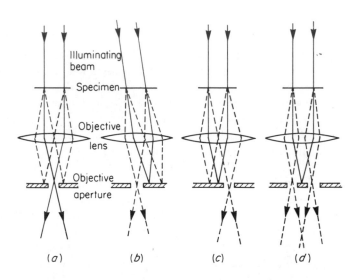

Figure 11.6   Methods for obtaining dark-field images

(a) Normal bright-field transmission microscopy, in which the central pencil (full lines) passes through the objective aperture, but the scattered pencils (dotted lines) are intercepted by the objective lens aperture

(b) Beam-tilt method. The illuminating beam is tilted so that the direct pencil is intercepted by the central objective aperture. One scattered pencil enters the imaging system

(c) Off-centred aperture method. The objective aperture is moved to one side so as just to intercept the direct pencil. The image is formed by the extreme peripheral scattered pencil, and is necessarily of low resolution

(d) Beam-stop aperture ('contrast screen') method. A fine wire is placed across the centre of the objective aperture, which intercepts the direct pencil. Almost all of the peripheral scattered pencils enter the imaging system, giving a brighter image than is obtained in (c)

11.29 Electron Diffraction

If a specimen is made up of regularly spaced arrays or lattices of atoms such as are found in a crystal, a beam of electrons passing through it will be scattered preferentially into discrete rays by being reflected at the lattice structures. An amorphous specimen does not possess this property. If the illuminating beam is coherent the reflected rays will interfere with one another, forming a pattern of spots or rings of electrons. This pattern can be imaged at the final screen by suitably energizing the microscope lenses, and can be photographed. The resulting micrograph is called a 'diffraction pattern', and the dimensions and spacings of the spots and rings can be used to calculate the lattice dimensions of the crystalline specimen. The smaller the lattice spacing, the more widely spaced will be the component parts of the pattern. Crystalline substances such as metals give rise to useful, easily formed diffraction patterns, but in general biological specimens do not have this property. Diffraction patterns can be used to identify

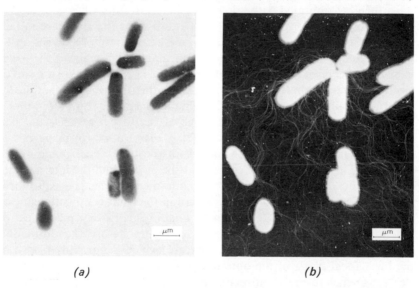

(a) (b)

Figure 11.7 Bright-field (left) and dark-field (right) electron micrographs of the same field, made by the 'contrast screen' method (Fig. 10.6(d)). An application of the million-volt electron microscope. The bacteria (*B. proteus*) (left) are almost transparent to the very penetrating direct beam, and show some internal detail. The bacterial flagellae are lost. The dark-field image shows detail of the bacterial cell walls and flagellae, lost in the bright-field image. (Micrographs by courtesy of Drs. Dupouy, Perrier and Verdier, C.N.R.S. Toulouse; and *Journal de Microscopie*)

crystalline substances. In general, the spacing of the lattices in protein crystals is too large to form a useful diffraction pattern. These 'low-angle scattering' patterns are currently being investigated, and may be of use to the biologist in the future.

A diffraction pattern can be formed on the final screen without using any lenses. If an image of the source is focused on the screen, and a crystalline substance is placed in the path of the beam, as shown in Fig. 11.8(a), the central image of the source will be surrounded by a diffraction pattern of dots or circles, as shown in Fig. 11.8(b). The spacing clearly depends both on the lattice dimensions and the distance from the specimen to the screen. This distance is called the 'camera length'.

The size of the final pattern tends to be very small without lenses, since the scatter angles are also very small. It is therefore convenient to magnify the pattern, which can be done very easily by using the imaging system of the microscope. Fig. 11.9 shows the ray diagram for this method of diffraction operation. It effectively increases the camera length, which must then be calibrated with known specimens. It must always be clearly understood when forming diffraction patterns that the spots are in fact images of the source. The imaging system must therefore be set up to focus an image of the source on the screen, and not an image of the specimen. The specimen must be illuminated with as nearly parallel a beam of electrons as possible (the smallest possible aperture of illumination) in order to obtain a coherent beam. The first image of the source will therefore be formed in the back focal plane of the objective lens, and if a diffracting specimen is in the path of the beam, a sharp image of the diffraction pattern will also be formed in the back focal plane of the objective. This therefore now constitutes the object, and the intermediate lens must now be focused on to it. The intermediate lens is used to form a focused image of the diffraction pattern in the front conjugate plane of the projector lens, which then forms a final image, further magnified, of the diffraction pattern on the screen or plate. Before the diffraction pattern can be seen, the objective aperture, which is normally placed in the back focal plane of the objective in order to intercept the scattered electrons which form the diffraction pattern, must be withdrawn from the beam.

This method of diffraction produces a pattern from the whole illuminated area of the specimen, which may be far greater than the single crystal to be examined. It is necessary therefore to screen off all that part of the specimen except the area from which the diffraction pattern is required. This method is called 'selected area diffraction'. It would clearly be very difficult to select the required area at the specimen itself,

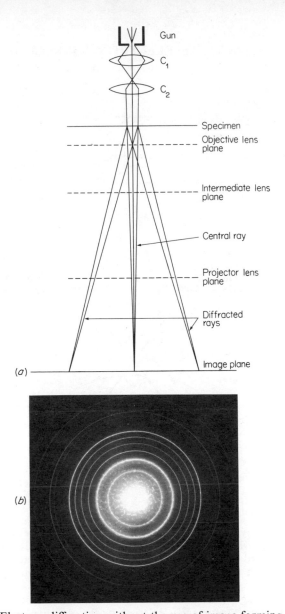

Gun

C_1

C_2

Specimen

Objective lens
plane

Intermediate lens
plane

Central ray

Projector lens
plane

Diffracted
rays

Image plane

(a)

(b)

Figure 11.8   Electron diffraction without the use of image-forming lenses. The
double condenser illuminates a selected area of the specimen, and the scattered
pencils are focused on the screen by the action of C_2. The resulting diffraction
pattern (b) is an image of the electron source after interference with the specimen.
The method is only of use for very, close (interatomic) spacings; the diffraction
pattern is normally enlarged by the action of the image-forming lenses (see
Fig. 11.9). (Diffraction pattern of thallium chloride crystals at 100 kV by courtesy
of Siemens Ltd.)

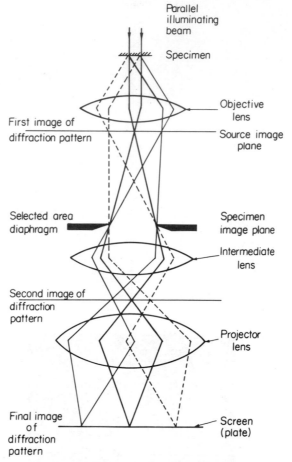

Figure 11.9   A ray diagram for selected area diffraction. The specimen is illuminated with as nearly parallel a beam of electrons as possible. The objective lens
forms an enlarged image of the diffraction pattern at its back focal plane (the
objective aperture diaphragm, which normally intercepts this pattern, must be
withdrawn). An image of the specimen will be formed in the lower conjugate plane,
so that a diaphragm placed in this lower plane will also be in focus with the
specimen. This primary image is enlarged by the objective, so that a relatively
large diaphragm will intercept all but the rays from a very small area of the
specimen. The intermediate lens strength is now adjusted so as to form an image
of the diffraction pattern from this selected area in the upper conjugate plane of the
projector lens, which in turn forms a magnified image of the diffraction pattern
on the screen or plate

so advantage is taken of the fact that a normal focused image of the specimen is formed below the back focal plane of the objective, as shown in Fig. 11.9. This image is magnified by the objective, so a relatively large stop or diaphragm can be placed in this primary image plane in order to select the required area. The diaphragm can be a circular aperture identical with those used in the objective and condenser lenses, or it can be four movable blades which can be moved independently in order to screen off a square area of any required size. The conditions for forming a sharply focused diffraction pattern are therefore (1) the objective must focus the specimen at the plane of the selector diaphragm; and (2) the intermediate lens must image the selector diaphragm into the projector lens, which in turn will image it sharply on the screen. Clearly, these conditions can only be obtained at one fixed magnification. This is a disadvantage of a three-lens imaging system when used for diffraction. In order to be able to obtain diffraction patterns at any desired magnification, a fourth lens must be placed between the intermediate lens and the projector. The lens we have been referring to as the 'intermediate' is now called the 'diffraction lens', the lens below it being called the 'intermediate'. This optical system is now found on all modern high-performance TEMs.

The procedure for obtaining diffraction patterns is as follows. First, the specimen is inserted and examined by transmission to find a suitable object for diffraction (Fig. 11.10(a)). Next, insert the area selector diaphragm and blades, and focus them sharply with the intermediate lens current. Then focus the specimen sharply with the objective lens current, which must be now be left fixed (Fig. 11.10(b)). Next, obtain parallel illumination by fully under- or overfocusing C_2. The image will now, of course, be very dim. Increase the intermediate lens current to obtain a sharp spot on the screen; this is the image of the source. Finally, remove the objective aperture, and a sharply focused diffraction pattern will appear on the screen provided the specimen is crystalline. The central spot will be far brighter than the spots forming the pattern, and it is usual to reduce the glare from this central spot by intercepting it with a small disc or cup before it reaches the final screen. A diffraction cup is provided on all instruments which allow selected area diffraction operation. The spots making the pattern are very dim, and exposures of the order of minutes are required to record them as a micrograph. Figs. 11.10(c) and (d) show two diffraction patterns. The circles are due to the graphite structure of the supporting carbon film, and the spots are due to the crystalline structure of the specimen.

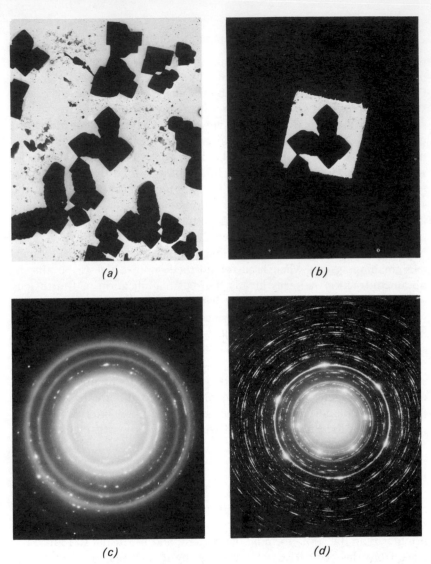

Figure 11.10   Obtaining a selected area diffraction pattern (*a*) A suitable area of specimen including obviously crystalline material is found. (*b*) A small area is selected by suitably manipulating the four area selection blades (or by inserting a diffraction aperture of suitable diameter). (*c*) The source is focused on the screen with the intermediate (diffraction) lens, the condenser is defocused fully and the objective aperture is withdrawn. The pattern of dots and circles is the diffraction pattern of the specimen. The sharply-defined pattern (*d*) is of unsupported aluminium oxide. The poor definition of (*c*) is due to the thickness of the specimen (sodium chloride crystals) and the presence of the carbon and Araldite supporting films. (*a*), (*b*) and (*c*) Philips EM 200. (Micrograph (*d*) by courtesy of Dr. Grasenick and Siemens Ltd.)

11.30 Stereo Operation

Information about the third dimension of depth is obtained in part (by persons who are normally sighted in both eyes) by the fusion in the visual cortex of the brain of the two images formed from the slightly different viewpoints of the two eyes. The interocular distance gives rise to a lateral separation between corresponding points in the two images which is called 'parallax'. A similar effect can be obtained in the following way. If a thick object is photographed twice from the same viewpoint, but is turned through a small angle perpendicular to the optical axis of the camera between the two successive exposures, there will be a lateral separation (the parallax) between corresponding points on the two

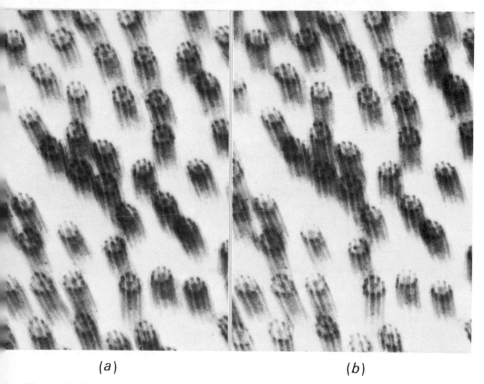

(a)   (b)

Figure 11.11   A stereo pair of a thick (0·5 μm) section of frog cilia tilted through ±6° at a magnification of × 40,000. The axial inclination of the cilia within the section becomes clearly visible after the pair have been optically fused. (Philips 200, 100 kV)

photographs. If now the two photographs are viewed simultaneously, each with one eye, in such a way as to cause a fusion of the two separate visual images in the brain, the two photographs will, if the amount of parallax is right, appear to 'lock' and form a very realistic appearance which is immediately interpreted by the brain as 'depth' or 'thickness'. The effect is the same as if the thick object had actually been viewed from a distance using both eyes.

If the two photographs are now superimposed, either physically (if they are transparencies) or optically by means of a stereoscope (the instrument normally used for fusing the images), the parallax between corresponding points on the stereo pair may be measured with a ruler. It has been found that for most people maximum realism is obtained when the measured parallax is 3 to 5 mm at the normal viewing distance of 25 cm. With more parallax, the brain finds it impossible to 'lock' the fused images; less gives a poorer impression of depth.

The principle is readily applicable to electron microscopy, and, provided the specimen is sufficiently thick, it can often provide useful information about the third dimension. The method is very simple. The specimen is first tilted through an angle on one side of the beam axis, and the required area is found, focused and photographed. The specimen is then tilted to the same angle on the other side of the beam axis, and the identical area is recentred, focused and photographed again. The two micrographs are then printed (contact prints are generally made) and the prints are placed with their centres about 10 cm apart. The prints must be exactly parallel, set at the same vertical height and, of course, at the same magnification. They are then viewed through a pair of narrow-angle prisms or lenses arranged so as to converge the optical axes of the eyes inwards by about 14°. The images of the two members of the stereo pair are then superimposed, and should fuse, allowing the brain to interpret the three-dimensional structure of the specimen. Experienced viewers can fuse the two images of a stereo pair if they are the correct angular distance apart, without the use of any optical aid. The eyes are deliberately squinted inwards to superimpose the pair, and are then refocused. Most people are able to acquire this facility with practice. It is particularly useful when stereo pairs are projected onto a screen, otherwise polarized light or light of two different colours must be used in conjunction with special spectacles.

The four parameters involved in making a successful stereo pair are the full angle of tilt θ, the parallax P, the specimen thickness t and the overall viewing magnification M (the product of the magnifications of the microscope, the enlarger and the stereo viewer). They are related by the expression:

$$\sin \tfrac{1}{2} \theta = \frac{P}{2tM}$$

Families of curves have been published in the literature (see bibliography p. 510) giving the optimum tilt angle for given magnifications and specimen thicknesses.

Most TEMs are fitted with a stereo specimen holder giving a tilt angle around $\pm 7°$, this being the average angular separation between the eyes at a viewing distance of 25 cm. Assuming an overall magnification of $\times 50,000$ (the highest at which stereoscopic 'depth perception' seems to work) this small tilt angle necessitates a section thickness of 0·3 μm–six times as thick as a normal ultrathin section. As magnification is reduced, so specimen thickness must be increased. The stereo effect obtainable with the normal biological ultrathin section, even using a high-angle tilt holder (see the following section) giving tilts of $\pm 60°$, is barely perceptible, and gives little or no worthwhile extra information. In the special case of a well-dispersed and geometrically highly ordered specimen such as the cilia shown in the stereo pair of Fig. 11.11, a reasonable stereo effect can be obtained from a 0·5 μm thick section at 100 kV. Even better results can be obtained at 1 MV, using 2 to 5 μm thick sections of even more highly specialized specimens such as the Golgi apparatus sectioned after the classical silver or osmium impregnation techniques. In general, the clutter of detail superimposed in such a thick section completely obscures any three-dimensional effect which might be present. Stereo is, however, useful in the usual 50–100 kV TEM for examining shadowed surface replicas such as those obtained from freeze etching (see p. 484). These are themselves thin enough to transmit the beam, but the thin replica is itself three-dimensional. The great depth of field of the TEM ensures that the images are uniformly sharp throughout.

11.31 High-angle Double Tilting

This technique has considerable potential interest for biologists, especially those interested in the three-dimensional reconstruction of intracellular features. A 75 Å thick 'unit membrane', for example, is between one-fifth and one-tenth the thickness of the section in which it is embedded, and unless it is almost exactly perpendicular to the plane of the section, its image will be diffuse and unrecognizable, as shown

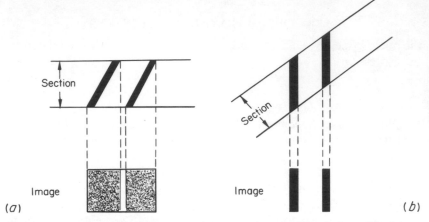

Figure 11.12   The advantage of the high-angle double-tilt technique. The image of a pair of membranes lying at an angle in a section will appear unsharp (*a*) when the section is perpendicular to the electron optical axis, as is normally the case. If the section can be tilted in two mutually perpendicular planes so as to bring the membranes themselves parallel to the electron optical axis, the image will appear maximally sharp, as in (*b*)

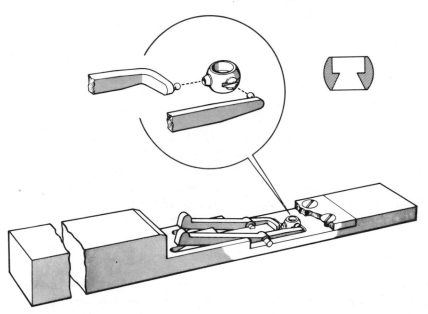

Figure 11.13   A drawing of the high-angle double-tilt holder which fits in the AEI 'Corinth' TEM specimen stage in place of the standard 4-grid holder. A small (1·5 mm dia.) grid is carried in the semi-spherical ball, which can be tilted orthogonally ±30° by means of the two levers. These are operated by cams driven by two reversible electric motors controlled by foot pedals. (Drawing Courtesy GEC/AEI Ltd.)

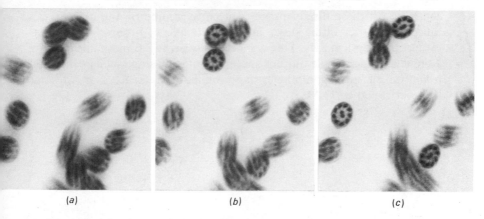

(a)   (b)   (c)

Figure 11.14   An example of the use of the double-tilt stage shown in the previous Figure. The specimen is a 0·2 μm-thick section of frog cilia. In (a), the section lies perpendicular to the beam axis. No cilium lies with its longitudinal axis parallel to the beam, and the '9 + 2' filament structure is not clear in any transverse section. In (b), the section has been tilted − 10° on the y-axis and + 5° on the x-axis. Two cilia now lie parallel to the beam, clearly showing the '9 + 2' structure. In (c), the section has been tilted − 5° on the y-axis and + 20° on the x-axis. Three other cilia are now lying roughly parallel to the beam. (AEI 'Corinth', 50 kv, × 5,000)

diagrammatically in Fig. 11.12(a). If the section is tilted so that the paired membranes lie parallel to the electron optical axis of the TEM, as shown in Fig. 11.12(b), the blurring due to superimposition is minimized and the true geometrical structure is revealed. Fig. 11.13 shows a mechanical double-tilt specimen holder, and Fig. 11.14 shows how the method can be applied to demonstrate the classical '9 + 2' filament structure in a thick section of cilia.

Stereo operation and double tilting both present certain difficulties. Unless the specimen area being imaged lies exactly on the electron optical axis and on the exact axis of tilt, in other words, at the exact centre of the grid, it will move up and down with respect to the objective lens object plane while being tilted. The objective must then be refocused after tilting, which will cause the image to change in magnification and to rotate. It may be impossible to fuse 'straight' prints of the resulting pair of micrographs optically, therefore the enlarger magnification and print angle must be changed to compensate during printing. Also, if the object lies away from the grid centre, the image will move out of the field of view while being tilted, and it must be kept centred on the screen with the stage controls. All these disadvantages can be overcome by the use of a so-called 'eucentric' tilting stage, in which the axis of tilt always intersects the electron optical axis. This is most easily arranged mechanically with a single axis high-angle tilt plus specimen rotation stage such as the high-resolution eucentric goniometer stage obtainable as an accessory on some high-performance TEMs.

CHAPTER 12

High Resolution Operation

12.1 Introduction

An adequate definition of the term 'high resolution' is possibly the most controversial subject in the field of electron microscopy. The difficulty lies mainly in the interpretation of images which appear to show detail of 5 Å or less. The main reason for this difficulty is the enhancement of fine structure by Fresnel fringes, which give rise to spurious detail when the objective lens is slightly out of focus. The presence of the fringe surrounding an object can cause an apparent doubling or tripling of fine detail. For instance, the image of a single line can give the impression of three very closely spaced lines, due to the image of the actual line object being surrounded on either side by a single Fresnel fringe. An extended line object 10 Å wide can appear as three parallel lines spaced 5 Å apart. In a similar way, the periodicity of a regularly spaced lattice structure can be doubled by the formation of Fresnel fringes between the images of the lines. This spurious 'doubling' of the periodicity of a grating has been recognized in light microscopy for decades; it is due to interference, and is often called a 'phase-contrast' effect. An identical effect arises in all slightly defocused electron images. The spurious contrast effect is called 'defocus granularity'. The high magnification micrographs in Fig. 12.1 of a through-focus series of the structure of a thin carbon film coated with evaporated platinum show defocus granularity increasing on either side of focus. Any slightly out-of-focus inhomogeneous object, even the thinnest carbon film, shows defocus granularity. It is very easy to be misled into interpreting this apparent structure as 'resolution'.

True resolving power is very difficult to measure objectively; the methods used will be discussed in the following chapter. Theoretical resolving power can be calculated if the spherical aberration coefficient and the focal length of the lens are known. The problem in high resolution microscopy is to achieve this theoretical performance in practice.

271

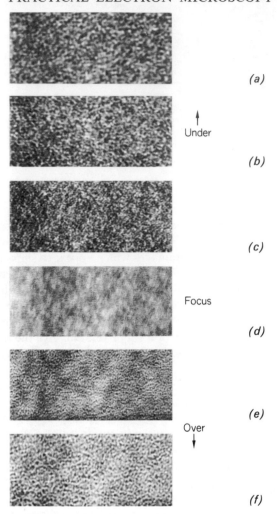

Figure 12.1 A through-focus series of electron micrographs of the same area of a thin carbon film lightly shadowed with platinum, to show 'defocus granularity'. Each micrograph differs by one step or 'click' of the finest objective lens current control. (a) 2·1 μm under focus; (b) 1·4 μm under focus; (c) 0·7 μm under focus; (d) exact focus to within the resolving power of the instrument; (e) 0·7 μm over focus; (f) 1·4 μm over focus. Magnification: electron optical × 250,000, enlarger × 2; total 5 × 10⁵. In (d), particles can be resolved on the originals which are about 0·15 mm apart, indicating a point resolution of 3 Å, as claimed by the makers of the instrument (AEI–EM6B). This fine detail can be seen to be enlarged and made more contrasty and hence more easily visible to the eye on either side of focus, but resolution falls off rapidly. Maximum contrast lies somewhere between (b) and (c), 1 to 2 clicks of the fine focus control under focus

12.2 Practical High Resolution

It would probably be agreed among most practising electron microscopists that the general term 'high resolution' means the demonstration of true point or lattice spacing closer than 5 Å. All high-performance microscopes claim to be able to do this; medium resolution instruments claim 8–15 Å. The term 'high-resolution operation' will be used in this book to mean the attainment of the maximum resolving power out of the particular instrument in use, for it can be just as difficult a feat to get 18 Å from a medium performance instrument as it is to get 3 Å from a high performance one. To get the ultimate in performance out of a particular instrument, the operator must develop to the highest degree the basic skills of alignment, deastigmatism and focusing. Real proficiency in these skills is a matter of long and continual practice plus a full understanding of the principles involved.

It must always be borne in mind that high-resolution work is simply not possible unless the specimen is suitable. Proficiency in specimen preparation is just as important as proficiency in operation. The preparation of specimens suitable for high resolution work is discussed in the following chapter.

It should hardly be necessary to point out that, before even thinking about undertaking high-resolution operation, the microscope itself must be in as perfect condition as it is possible to make it. It is a waste of time to sit down in front of a high-performance microscope which has been used for nothing but low magnification work for months and expect good results at high magnifications. It is essential first to check on all the points listed below.

12.2.1 Column Cleanliness

The objective and condenser apertures, specimen holder and anti-contaminator blades must be freshly cleaned. The objective lens must be tested for residual astigmatism to make sure that it has not deviated significantly from the basic value measured when the instrument was first commissioned. If there is any significant difference, the probable cause is contamination of the objective lens polepiece bores and of the specimen chamber. In this case, there is no alternative but to strip the column down and clean these vital parts, following the maker's instructions with scrupulous care.

12.2.2 Faultless Vacuum

The ultimate vacuum in the column as measured by the Penning gauge must be as good as it was when the instrument was new, and must be reached just as rapidly. Any vacuum leaks or excessive contamination due to dirty pumps make high-resolution work impossible.

12.2.3 No Contamination or Stripping

Contamination of the specimen is the major bugbear of high-resolution work. It can be reduced to a negligible amount by the use of the anti-contaminator. This must be filled with liquid nitrogen at least half an hour before a high-resolution session, or temperature gradients may cause stage instability and consequent specimen drift. Stripping is caused by poor vacuum and the presence of water vapour in the specimen chamber; ensuring well-desiccated plates and absence of vacuum leaks should take care of this. There is often an optimum temperature for the anticontaminator which gives a minimum of both contamination and stripping.

12.2.4 Stage Stability

The microscope must be checked at the highest magnification for any sign of specimen drift, which will probably be due to mechanical strain in the stage control mechanism or to temperature gradients arising from the anticontaminator. It is a waste of time to attempt high-resolution work unless the control of the movement of the specimen stage is completely reproducible and stable.

12.2.5 Stable Electronics

The stabilities of the high tension voltage and lens currents must be within the limits specified by the manufacturer, generally a few parts in a million. Absolute stability is very difficult to measure, but a very sensitive test for general electronic stability using the microscope as its own detector is described further on in this chapter. All reference batteries must be fresh and in good condition; all valves must be correctly 'aged' and stable (not brand new and not more than 10 % down on emission); and there must be no dirty or intermittent contacts in any of the switches and potentiometers used to control lens currents and high tension. If unstable electronics are suspected (and this must be expected in an instrument more than about 3 years old) it is wise to have the whole electronic system overhauled by a service engineer.

12.2.6 Perfect Alignment

All the foregoing counsels of perfection will be unavailing if the column is not correctly aligned. The alignment of the double condenser and the final alignment procedure are described in the following section.

12.2.7 Warm Up

All the circuits must have been running with the H.T. on (preferably on the kilovoltage above the one which is to be used) and the filament off for at least an hour before a high resolution session. The microscope must be left untouched during this time.

12.2.8 Choose the Time

It is not only essential for the instrument itself to be in perfect adjustment. The operator must also be in the right mood of almost limitless patience, the conditions must be right and the specimen must be right. The operator must see to it that he is undisturbed by interruptions to deal with irrelevant problems, and he must be prepared to work far into the night, stopping only when the desired result is achieved, or when he is finally convinced that he is attempting the impossible. It is always advisable to have high-resolution sessions at night, since the mains electricity is generally more stable, the water pressure is more constant, and there is less likelihood of external magnetic or vibrational disturbances. It is essential to prepare everything beforehand; a supply of suitable specimens, a plentiful supply of desiccated plates, and plenty of freshly-made photographic solutions for developing the negatives.

12.2.9 Select the Specimen

The very highest resolution can only be demonstrated on specimens which have been specially prepared for the purpose. Special specimens for the measurement of performance are described in the following chapter. The specimen must be as thin as possible, and be supported by the thinnest possible clean carbon film. Very high resolution cannot be expected from ordinary 'silver' (c. 50 nm thick) epoxy resin sections. Parts of such sections where particularly well-stained objects are in particularly well-orientated positions may in favourable cases show resolution of 15 Å in a high-performance instrument, provided they are well removed from any other structure which might interfere with the image. For instance, the plasma membrane of the animal cell can be resolved as a 'trilaminar' structure consisting of two dense lines approximately 25 Å in width separated by a space of approximately 30 Å (Fig. 12.2). This represents about the best performance in resolution which can normally be expected from sections of embedded biological material. Smaller objects embedded in a section, such as particles of colloidal gold ingested by phagocytic cells, can be imaged even when they are 10 Å or less in diameter, but such preparations can hardly be described as normal biological specimens. To realize the full resolving power of a high-performance microscope on biological material, it is necessary to use a preparation of particles dried down on a very thin carbon film and contrasted with negative or positive stain such as sodium phosphotungstate or uranyl acetate (see Chapter 19). The most suitable objects so far found in the biological field have been virus particles. Resolution of the order of 5 Å has been shown on certain virus structures, but, as always in high resolution microscopy, the interpretation of the image is often open to criticism.

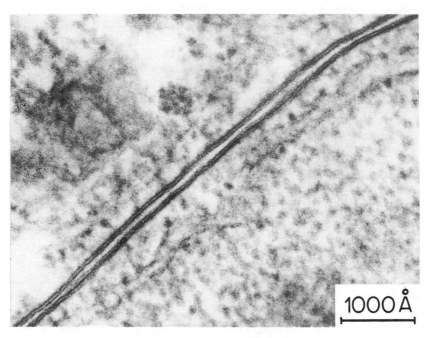

Figure 12.2   A high-resolution micrograph of a thin section of the borders of two cells. Each cell membrane is clearly resolved as two dense lines approximately 25 Å wide, separated by a space of about 30 Å (the trilaminar 'unit membrane' structure). The two cells are separated by a space 100–150 Å wide. This micrograph shows the highest resolution that can normally be expected from an ultrathin (approximately 50 nm thick) section of biological material, fixed in glutaraldehyde/ osmium tetroxide, embedded in Araldite, and stained with lead. Plate magnification × 94,000, enlarger magnification × 2. Philips EM–200

12.3 Test for Electronic Stability

Before attempting work at high resolving power, it is necessary to check the stability of the electronic supplies of the microscope. A very useful,

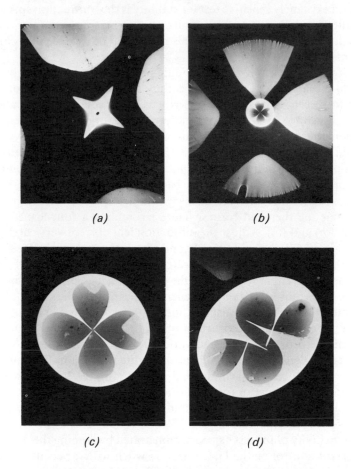

Figure 12.3   The 'diffraction spike' test for stability in the lens and H.T. electronic circuits. (*a*) The diffraction spot (normally intensely bright, but indicated here by a black spot in the centre) surrounded by a highly pincushion distorted image of a grid square. (*b*) The image on the overfocused side of the diffraction point. The pincushion figure is turned back upon itself, resulting in the four-leaf-clover spike figure in the centre of the field, shown enlarged in (*c*). To detect instability in either H.T. or diffraction (intermediate or Projector 1) lens current, watch the spacing between opposite points of the spikes. Any detectable movement corresponds to an instability on either diffraction lens current or H.T. supply of a few parts per million. The test does not indicate instability on the other lenses, but if the diffraction lens supply is perfectly stable, the other lenses are likely to be stable also. The test shows the presence of astigmatism in the diffraction (or inter-mediate) lens, shown in the asymmetrical figure (*d*). This is of no consequence except for high resolution diffraction work. Philips EM–200

rapid and extremely sensitive test for stability is the 'diffraction spike' test. Instability in H.T. and lens current of a few parts in a million can be detected. If the instrument passes this test, it will be capable of its highest resolving power.

First focus the specimen, which should preferably be a holey carbon film on a 200-mesh grid, at low magnification, and bring the centre of a grid square to screen centre. Fully underfocus C_2. Reduce magnification to zero by adjusting the current in the intermediate (or diffraction) lens so as to obtain the diffraction spot. This will appear surrounded by a very distorted image of the grid (Fig. 12.3(a)). Continue to increase the diffraction lens current, passing right through the diffraction point. The image of the grid will then 'turn back' on itself, and will form four 'spikes' which, if the diffraction lens is completely free from astigmatism, appose one another and form a figure resembling a four-leaved clover (Figs 12.3(b) and (c)). Adjust the diffraction lens current very carefully so that a pair of spikes almost touch, and watch the spacing between them closely through the screen magnifier. If any random change in this spacing can be detected, either the H.T. or the diffraction lens current is unstable. If one lens current is unstable, it generally means that the current supply to the other lenses is also unstable, and high resolution work will not be possible. The size of the 'diffraction spike' figure depends on the magnification of the projector lens, and varies from instrument to instrument. The test also demonstrates astigmatism in the diffraction lens; an astigmatic lens produces an asymmetrical pattern (Fig. 12.3(d)). Few instruments provide a stigmator for the diffraction lens; astigmatism in this lens is only significant when high-resolution diffraction work at long camera length is being attempted.

The sensitivity of the test can be demonstrated by tapping the diffraction lens current control or the H.T. selector switch with a pencil. The slight change in contact resistance generally shows up as a change in spike spacing, and is a salutary reminder of the necessity for keeping all control contacts clean.

12.4 Astigmatism

The success or otherwise of a high-resolution session depends more on the ability of the operator to compensate for astigmatism than any other single factor, provided the specimen is suitable and all other conditions are at their most favourable.

Astigmatism becomes more noticeable as magnification is increased. It is due, as has been discussed in Chapter 4, to an axial asymmetry in a lens which cannot be corrected to a sufficiently high degree during manufacture. The only lenses needing correction are the objective and, in high performance instruments, the condenser. The other image-forming lenses have less effect on the final image, since their errors are progressively less magnified. Because of this, they can be manufactured with a sufficiently high degree of tolerance, and do not need correcting. The limiting factor in instrumental resolution in the present state of the art is the residual uncorrectable astigmatism of the objective lens, which in turn depends on the skill of the operator in correcting it.

The effect of astigmatism is to make the focal length of the lens differ in two mutually perpendicular planes. It can be corrected as we have seen in Chapter 4 by introducing a cylindrical lens of correct strength and in the correct plane to introduce and equal and opposite compensating asymmetry. The device which is used to introduce and orientate the compensating field is called the 'stigmator'.

12.5 The Stigmator

Stigmators have been briefly discussed in Chapter 5. They are always found on the objective lens, and generally on the second condenser lens. Three types are commonly used:

1. Mechanical magnetic
2. Electromagnetic
3. Electrostatic

1. The mechanical magnetic stigmator was the earliest type to be used. The method of compensation is to place four small soft iron slugs around one of the polepieces of the lens. The slugs are connected in opposing pairs (see Fig. 6.21), and can be rotated concentrically about the polepiece. When the four slugs are equally spaced, each cancels the effect of the others, and no correcting field is applied. If one pair is moved relative to the other pair, an axial asymmetry or artificial astigmatism is introduced. The strength of the correcting field is controlled by altering the spacing between the pairs of slugs, and the direction is controlled by rotating all four slugs together around the polepieces. The mechanical controls of the stigmator generally take the form of two concentric knobs mounted on the lens casing. The inner or 'amplitude' knob controls the distance apart of the pairs of slugs; the outer or 'orientation' knob rotates the slugs together around the polepieces.

2. In the electromagnetic type, eight small coils are wound on iron cores and are energized by currents supplied from a stabilized source (Figs. 6.19

and 6.20). The currents can be applied in a logical fashion, with separate controls for amplitude and orientation, or can be applied by two controls connected so that one knob always applies a correcting field in one direction, the amplitude being varied by the amount the control knob is turned; and the other knob applies a similar field but at 90° to the first. Correction is thus obtained by balancing the applied fields so that they cancel the lens astigmatism. The latter system of stigmator operation is called 'X–Y correction'.

3. The electrostatic type of stigmator generally consists of a set of 8 insulated rod electrodes or pins mounted coaxially beneath the lens. Each opposing pair can be charged, either positively or negatively, and the resulting field acts as a cylindrical electrostatic lens. The voltages are applied to each pair of electrodes in such a way that the field strength or 'amplitude' is controlled by one knob, and the field distortion or 'orientation' is controlled by another knob.

The correcting field applied by any stigmator must be exactly coaxial, or the image of the hole which is being used to correct the astigmatism will move out of the field of the binocular telescope as the stigmator strength is increased, and may even move right off the screen. This makes correction very tedious. It is relatively easy to provide electrical adjustments to centre electromagnetic and electrostatic stigmators, but mechanical stigmators have to be centred during manufacture. The centration controls for the stigmators are generally mounted on a subpanel away from the main controls of the microscope, since they should only require adjustment after the column has been completely dismantled and reassembled.

12.6 Deastigmating the Objective Lens

All objective lenses have a certain amount of basic astigmatism. The maker sets a certain maximum acceptable value, and all lenses are rigorously tested to conform to this after assembly. In any batch of lenses, some will have a lower basic astigmatism than others, but all should be capable of giving the claimed resolving power. The stigmator setting necessary to compensate for the basic lens astigmatism must be carefully recorded when the instrument is first commissioned. Some microscopes have two sets of stigmator controls, one of which is on a subpanel out of normal reach and is used to compensate basic lens astigmatism. The basic compensation must be checked with the specimen holder and objective aperture out of the column. The cause of any large deviation from this basic value must be investigated before beginning a high-resolution session. Dirty lens bores, polepieces, stage, specimen chamber, etc. are indicated. Astigmatism due to dirt often fluctuates with the beam current, and must be removed even if it can be compensated with the stigmator.

There are two methods in general use for detecting, measuring and correcting objective astigmatism. These are:

1. To observe the behaviour of fine particulate detail (of 50 Å or less in diameter) in the image as the objective lens is taken through the point of focus (the 'line-drawing' method); and

2. To observe the behaviour of the Fresnel fringes around an area of specimen where there is a sudden change in contrast.

The procedures followed when correcting astigmatism are given below.

12.6.1 The 'Line-Drawing' Method

When a specimen with a well-defined residual background granularity such as a lead-stained section is just underfocused, the background grain becomes very sharp to the eye. If the lens is astigmatic, the grains will not appear to be circular, but will appear to be drawn out into lines. When the objective lens current is varied rapidly about the point of true focus, by rotating the fine focus control rapidly back and forth, the astigmatic lines will appear to jump through 90° (Figs. 12.4(a) and (b)). The more astigmatic the lens, the more 'line-drawing' will be present, and the more pronounced will be the effect.

The method of correction is as follows:

1. Guess how much astigmatism is present. This requires experience.
2. Set the 'amplitude' knob of the stigmator to the amount shown by experience to give the required amount of correction.
3. Rotate the 'orientation' knob while taking the objective lens back and forth through focus until the background granularity does not jump through 90° and the image is maximally sharp. It may be necessary to repeat stages 2 and 3.

The accuracy of compensation of the method depends on the contrast of the specimen and the skill and experience of the operator. It can be very rapid, and will give good enough correction for electron optical magnifications of up to × 50,000, and resolution of about 10 Å. The X–Y correction system is much easier to use than the more logical amplitude-orientation method, and is now found almost universally on all high-performance instruments.

12.6.2 The Fresnel Fringe Method

The second method, which is always used for high-resolution work,

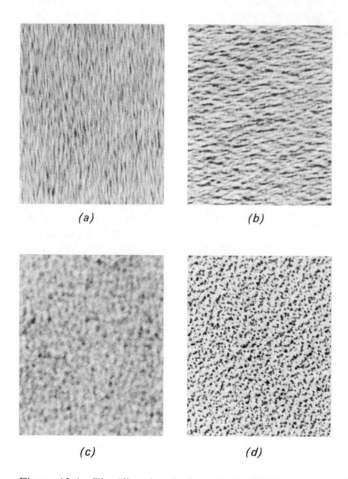

Figure 12.4   The 'line drawing' method of objective lens deastigmation. Specimen—a highly contrasty, rather heavy coating of Pt–Ir on carbon substrate. (*a*) Astigmatic lens, underfocused. Astigmatic lines run N–S. (*b*) The same area, slight overfocused. Astigmatic lines run E–W at 90° to (*a*). (*c*) Same area at 'jump-over point' of best focus with the astigmatic lens. (*d*) Same area at best focus with astigmatism corrected to better than 10 Å. Plate × 71,000, enlarger × 5 = × 355,000. Philips EM–200

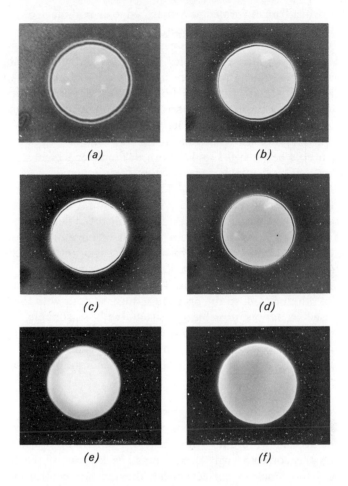

Figure 12.5   The steps in correcting objective lens astigmatism, using a Fresnel
fringe. (*a*) Obtain a round hole at × 100,000 (instrumental) easily encompassed
by the field of the viewing binocular. Overfocus to obtain a broad, blurred fringe
around the edge. The multiple fringes inside and outside the hole can be seen.
(*b*) Reduce C_2 current to bare visibility and reduce objective current to obtain a
sharp, crisp, just overfocused fringe. (*c*) Obtain direction of astigmatism by just
touching fringe to hole edge at two points. (*d*) Apply maximum amplitude
correcting field, and adjust azimuth (orientation) control to bring the excess
correcting field at 90° to the astigmatic field. (*e*) Reduce objective current, reduce
amplitude and adjust orientation slightly to bring the fringe exactly coaxial with
the edge of the hole, until the fringe just disappears equally all round the periphery (*f*). For maximum accuracy, find a smaller hole, increase magnification to
maximum and repeat the procedure. (AEI–EM6B)

uses the Fresnel fringes formed at a sharp edge or boundary of contrast. Only one fringe will in general be seen, which takes the form of a bright line surrounding the boundary. The best test object is a sharp-edged hole in a carbon film, but the small holes which frequently occur along the slight scratch lines which are present in nearly all sections are almost as good. The fringe formed around a hole is a very sensitive indicator of astigmatism, since it displays both the amount and the orientation of any radial change in focal length of the lens.

The steps to be taken when using this method are shown in Fig. 12.5. First find a hole between 0·1 and 0·5 μm in diameter which can readily be encompassed by the field of view of the screen magnifier at an overall magnification of at least × 100,000. The hole should be circular as seen through the magnifier; such holes are difficult to find when the screen is tilted. Focus the hole with the coarsest focus control so that a rather wide, blurred fringe stands well away from the blurred edge of the hole (Fig. 12.5(a)). Now reduce the lens current with the next or medium sensitivity focus control. As the fringe approaches the edge of the hole, it becomes sharper and 'crisper' (Fig. 12.5(b)). The fringe is sharpest at minimum aperture of illumination, so the procedure must be done with the C_2 lens underfocused as far as possible consistent with reasonable visibility. Maximum contrast is necessary, so it is usual to correct for astigmatism with the objective aperture in place, unless a check is being made on basic lens astigmatism. When a fringe has been found the procedure is as follows:

1. Switch the stigmator off, or reduce amplitude to zero.
2. Reduce objective lens current until the fringe just touches the edge of the hole at two points (Fig. 12.5(c)). The amount of astigmatism which the corrector will handle must be known; if the fringe is judged too much to correct, check the aperture centration and if necessary change the aperture for a clean one. Only relatively small amounts of astigmatism can be corrected satisfactorily at high resolution.
3. Draw an imaginary line connecting the two points 180° apart at which the fringe touches the hole. It is best if a rotatable crosswire is incorporated in one of the eyepieces of the screen magnifier; this enables the orientation direction to be recorded more precisely.
4. Set the amplitude control to maximum, and switch the stigmator on and off rapidly (or turn the amplitude control rapidly from zero to full strength). The fringe will jump from its original position to another position.

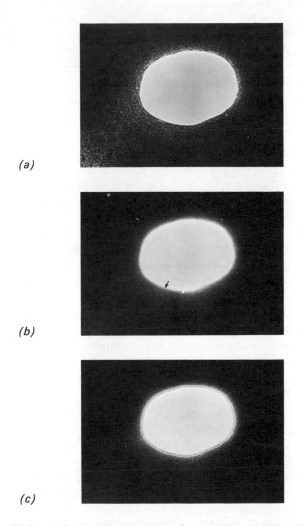

(a)

(b)

(c)

Figure 12.6 A through-focus series of a hole in a carbon film, as perfectly corrected as possible for astigmatism: (a) underfocused: note the defocus granularity shown by the carbon substrate surrounding the hole; (b) at closest focus: note the slight fringe (arrowed) which is not balanced by a similar fringe on the other side of the hole. This residual uncorrectable astigmatism represents the limiting resolving power of the lens, in this case about 3 Å; (c) overfocused: note the reappearance of the defocus granularity. × 425,000. AEI–EM6B

5. Now slowly turn the orientation control until the fringe jumps exactly 90° when the stigmator is switched on. The correcting field will now be in the correct direction, but will be too strong (Fig. 12.5(d)).

6. Now reduce the strength of the stigmator and at the same time alter the fine focus setting to bring the fringe as close to the hole as possible. As correction is established, it is necessary to alter orientation, amplitude and focus simultaneously in order to obtain as perfect correction as possible (Figs. 12.5(e) and (f)).

With practice, the fringe method can be used to remove all the first order astigmatism from the objective lens. It will frequently be found that there is a residual trace of 'one-sided' astigmatism (Fig. 12.6(b)) which cannot be removed. If this can be detected, the lens is compensated as completely as possible, and will be able to yield its highest resolving power. This perfect compensation will not last for long; contamination of the aperture or specimen will probably reintroduce detectable astigmatism after half-an-hour or less, depending on the cleanliness of the column.

The method requires a great deal of practice in order to ensure a really high degree of correction. It is however essential to use it when resolutions better than 10 Å are required. Once again, the X–Y system of stigmator control is easier and more rapid to use, although the final correction is no more accurate than that obtained with the amplitude-azimuth system.

12.7 Deastigmating the Condenser

The effect of astigmatism in the condenser system is to make it impossible to obtain an illuminating beam of even brightness and circular cross-section. As C_2 is taken through focus, the illuminating spot will have the appearance of an ellipse. The major axis of the ellipse will turn through 90° at the point of exact C_2 focus. Astigmatism in the condenser wastes illumination, because the electrons cannot be concentrated into a circular spot. It has no effect on instrumental resolution in the lower magnification ranges, although it may give rise to unevenly illuminated micrographs which have to be 'dodged' in printing (Chapter 14). At the highest magnifications, condenser astigmatism can give rise to a slight loss of image resolution due to the aperture of illumination differing in two mutually perpendicular planes.

There are three methods in common use for correcting astigmatism in the illuminating system. These are, in increasing order of accuracy:

1. To make the image of the C_2 aperture circular;

2. To make the underheated filament pattern as sharp as possible; and
3. To set the underfocused caustic pattern for maximum symmetry.

The procedures to be followed are given below. The specimen should always be withdrawn when adjusting condenser astigmatism.

12.7.1 Circular Aperture Method

Pass C_2 control rapidly back and forth through focus at low magnification. If the illuminating spot is obviously elliptical, underfocus C_2 to obtain an image of C_2 aperture which is about 2–5 cm diameter on the screen. Manipulate the C_2 stigmator until the aperture image is circular. If the aperture is contaminated, it will not be circular in the overfocused position, but this does not in general matter, since C_2 is always underfocused in order to reduce illumination before a micrograph is taken.

12.7.2 Underheated Filament Pattern Method

Focus C_2 at low magnification, and then reduce filament heating. The pattern of 'draw-lines' of the filament tip (Fig. 11.1(a)) will appear on the screen, if the magnification is correct for the gun in use. To deastigmate the condenser, manipulate C_2 stigmator in direction and amplitude until the 'draw-lines' appear sharpest. Resaturate the filament and check that the image of C_2 aperture at slight C_2 underfocus is circular. If it is not, then the aperture is contaminated or damaged and should be exchanged.

12.7.3 Condenser Caustic Method

This method is the most accurate. First make sure that the gun is set to minimum beam and that the filament is saturated. Focus C_2 at low magnification and withdraw C_2 aperture. Care must be taken here, since the number of electrons passing down the column will be greatly increased, with the resultant dangers of excessive X-ray output from the microscope, possible damage to the fluorescent screen and column contamination. The correction for astigmatism must therefore be made as rapidly as possible. Slightly underfocus C_2 and observe the 'caustic figure' thus obtained. This figure is a section of the envelope of rays emerging from a lens with spherical aberration. Since C_2 with the aperture withdrawn is a very wide-aperture uncorrected lens, this caustic figure is readily obtainable. If C_2 is astigmatic, a complex figure such as is shown in Fig. 12.7(a) will be obtained. As the condenser stigmator is operated in the correct direction, the figure becomes more circular (Fig. 12.7(b)), until a roughly circular

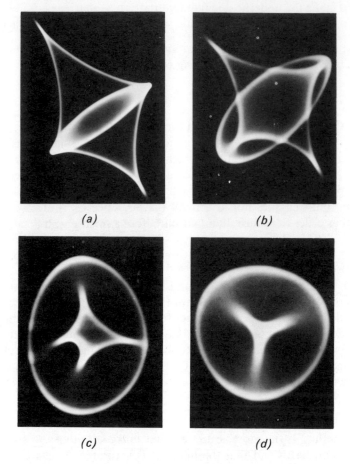

Figure 12.7 The steps in correcting condenser lens astigmatism, using the caustic figure method

(*a*) The caustic figure from a slightly underfocused, very astigmatic condenser

(*b*) The caustic figure from a less astigmatic condenser closer to focus

(*c*) The caustic figure from a very slightly astigmatic lens

(*d*) The '3-legged' caustic figure from a condenser as perfectly corrected as possible. The shape arises from uncorrectable third-order astigmatism

spot containing a 4-, 6- or 8-legged pattern (Fig. 12.7(c)) can be obtained. The fine focus control of C_2 must be adjusted at the same time. Very careful adjustment of stigmator and focus controls should now produce a 3-legged pattern (Fig. 12.7(d)), which represents the residual uncorrectable third-order astigmatism of the lens. The condenser is now as perfectly corrected as it is possible to make it. If it is impossible to obtain a 3-legged figure, the condenser polepieces are probably contaminated or damaged.

Method 3 has the drawback that C_2 aperture contamination cannot be corrected, since the caustic pattern is only obtainable when it is removed. Methods 1 or 2 are therefore more generally used. Method 3 should be used in conjunction with a perfectly clean, circular C_2 aperture when the very highest resolution is required from the microscope.

12.8 Causes of Astigmatism

The basic cause of astigmatism is the presence of an asymmetrical electrical or magnetic field strong enough to affect the electron beam. A bar magnet held close to the outer casing of the objective lens will cause considerable astigmatism. The commonest cause is the presence of so-called 'contamination' close to the electron beam, e.g. on apertures, anticontaminators, lens bores, specimens and specimen carriers. Contamination (see p. 159) is a brown, tarry substance formed from the break-down products of diffusion pump oil, which settles on surfaces close to the beam. The contamination never settles in a symmetrical fashion and, being an electrical insulator, becomes charged, and thus produces a weak electrostatic lens close to the beam. This lens is stronger in one direction than in another, and therefore introduces astigmatism. The best and simplest way of curing astigmatism caused by contamination is by cleaning the offending components. The commonest place for contamination to give rise to detectable astigmatism is for it to settle on the objective aperture. The smaller the aperture, the closer are its edges to the beam, and thus the greater the effect of contamination. The cleanliness of the objective aperture must be checked frequently by the operator. Microscopes generally have 'sticks' of 3 apertures which can be changed and centred within the vacuum, so inserting a clean aperture presents no problem.

Damage to lens polepieces will also give rise to astigmatism, so very great care must be taken when cleaning these components. External magnetic fields can also cause astigmatism, and can be detected by a blurring of the Fresnel fringe in one direction if the interfering field is an alternating one, as is most likely.

12.9 Focusing for High Resolution

The eye is the final judge of the focus of the image on the fluorescent screen. At low magnifications—up to × 5,000 or so—apparent image 'sharpness' is a sufficient criterion of focus. At higher magnifications, however, the enhancement of contrast by the Fresnel fringe effect (see Fig. 12.1) becomes increasingly noticeable. The eye chooses the point of maximum contrast as the 'sharpest' setting of the focus control. This, unfortunately, does not coincide with the point of exact focus, which is the setting for maximum resolving power.

Another factor has to be taken into consideration in focusing the image visually on the fluorescent screen. This is the inherent graininess of the screen phosphor itself. This depends on the fluorescent mixture used, but in general the grains have a diameter of 20–50 μm. It is therefore necessary to magnify the grains to bring them up to the resolving power of the eye, which under best conditions of illumination is about 200 μm. This is the purpose of the focusing telescope, which usually has a magnification of between × 5 and × 10.

If we wish the screen to resolve 5 Å, we must magnify this distance electron optically at least up to the screen grain size. The minimum instrumental magnification must therefore be × 100,000. If a × 7 binocular magnifier is in use, the overall specimen-to-eye magnification is × 700,000, and the eye will have to resolve 350 μm. But under maximum-resolution conditions, C_2 will be defocused for a small aperture of illumination and high coherence, and screen brightness will be low. So the eye is being taxed to the uttermost in attempting to resolve 350 μm under these conditions, and it is more usual to increase the instrumental magnification to × 250,000 or more, giving an overall magnification of at least 2×10^6.

Under these conditions of very high magnification, the specimen must be very contrasty to enable the eye to judge accurately the point of exact focus. If the specimen is lacking in contrast, it is common practice to take a series of micrographs, beginning on one side of focus and ending on the other side, hoping that one of these plates will show exact focus. A series of 6 plates each differing in focal setting by 3 Å would be taken for an expected resolution of 5 Å. The micrographs taken by this technique are called a 'through-focal series' (Fig. 12.1).

In general, the higher the resolution desired, the greater must be the electron optical magnification used. This is not always strictly so, because it is possible by using the technique of 'maximum-contrast' focusing to be described below to take comparatively low power micrographs which are so close to true focus that the resolution is limited only by the photographic grain, which is at least ten times smaller than the grain of the screen. Fig. 12.8(a) shows a low-magnification micrograph of a cell in a thin Araldite section. A small area of this micrograph has been enlarged optically in an enlarger × 24 in in Fig. 12.8(b), and can be compared with the direct electron micrograph of same area of the same section, which was enlarged × 24 electron optically, in Fig. 12.8(c). The two high magnification micrographs are difficult to distinguish from one another. The advantage here is that $24 \times 24 = 576$ times as much information is included on Fig. 12.8(a) as is included in Fig. 12.8(c), and this information can be extracted by simple optical enlargement.

12.10 Maximum Contrast Focusing

We have already seen (see p. 100) that Fresnel fringes give rise to an increase in contrast on either side of (over and under) true focus. The point of maximum contrast is a little under true focus, that is, the lens current is a little less than it is at true focus. The point of maximum contrast varies with instrumental magnification; the lower the magnification, the greater the required amount of underfocus. The eye can, with practice, judge the point of maximum contrast with fairly high accuracy. What is required therefore is a curve showing the amount of underfocus present at any given magnification at the maximum contrast point. It is then possible to read off the correction to apply to obtain a micrograph close to true focus. This is the principle of the 'maximum-contrast' method of focusing.

The steps in the construction of such a curve are as follows. First, we need to know the magnification to apply in order to bring an object of a given size up to the resolving power of the eye. This is shown diagrammatically in Fig. 12.9. This diagram is based simply on the ratio of the resolving power of the eye (200 μm) to the resolution required on the micrograph, and is of course applicable to any instrument. Next, we must construct a curve correlating magnification and the maximum tolerable resolution loss with the amount of defocus of the objective lens (Table

Table 12.1.

Defocus, μm	Max. magnification	Resolution = fringe width
0·01	10^6	2 Å
0·1	10^5	6·4 Å
1·0	10^4	20 Å
10·0	10^3	64 Å

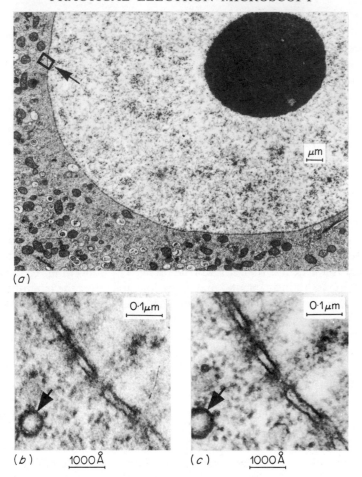

Figure 12.8   Light optical versus electron optical magnification

(a) A low power (× 4,200) electron micrograph of a section of part of a very large cell (gastropod oocyte) embedded in Araldite and stained with lead, taken very close to focus

(b) An optical (× 24) enlargement of the small area shown arrowed in (a). Total magnification × 100,000

(c) An electron micrograph of the same area of the same section taken at a direct electron optical magnification of × 100,000

The resolution shown by (c) is slightly better than that shown by (b), as demonstrated by the presence of the trilaminar membrane structure in the small vesicle (arrowed) on the left. The amount of biological structural information contained in (c) is very little more than that in (b). On the other hand, (a) contains 24 × 24 = 576 times as much structural information as the higher magnification micrograph, which can be extracted by simple optical enlargement

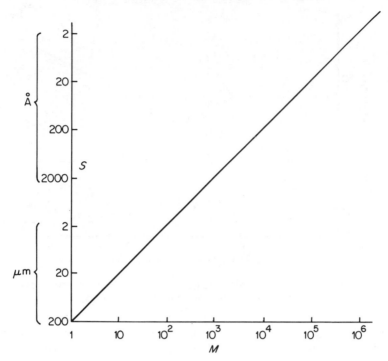

Figure 12.9   Curve showing the size S of a feature plotted against the magnification M necessary to bring it up to the resolving power of the eye, i.e. to make it visible

12.1). This curve (Fig. 12.10) is calculated from the expression for the width d of the Fresnel fringe:

$$d = \sqrt{\lambda.\Delta f.10^4}$$

where λ is the electron wavelength and Δf is the amount of defocus of the objective lens, measured in microns of focal length. Since λ is a function of accelerating voltage, the expression can be further simplified to:

$$d \approx 20\sqrt{\Delta f}$$

which applies to within $\pm 10\,\%$ for the accelerating voltages between 50 and 80 kV normally used in biological thin section work.

Finally, we must convert Δf into increments ('clicks') of objective lens current, and plot this against the instrumental magnification required, obtained from Fig. 12.9. This gives us a curve, applicable to only one particular instrument, giving the amount of underfocus which will be

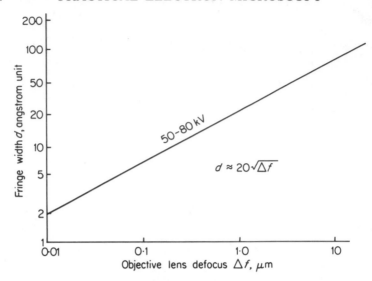

Figure 12.10   The amount of defocus in μm, Δf, of the objective lens
for a given fringe width d in Ångstrom units

applied at any given instrumental magnification at the maximum contrast
point. From this, the number of fine clicks of overfocus (increase of
objective current) to obtain true focus can be easily read off, and the
correction is then applied before taking the micrograph. This curve should
be hung by the microscope for easy reference. Fig. 12.11 shows a specimen
curve constructed for the AEI–EM6B instrument.

The technique of focusing by using the maximum contrast method is as
follows. First focus the grain of the fluorescent screen accurately with the
binocular telescope. Next, select the field required, and increase the
magnification until it fills the whole plate area. Examine the central area
of the field with the telescope, and swing the fine focus control rapidly
back and forth through focus. Provided the specimen is contrasty and
contains fine detail, the point of maximum contrast can be accurately
judged after a certain amount of practice. Next, use the curve for the
instrument (e.g. Fig. 12.11) to read off the number of fine focus clicks between maximum contrast and true focus at the particular magnification in
use. Apply this correction so as to *increase* objective lens current (overfocus), and expose the plate. The higher the magnification, the less will be
the correction to be applied. With experience, the correct amount of overfocus to apply can be judged without reference to the curve, and correctly
focused micrographs will be obtained every time.

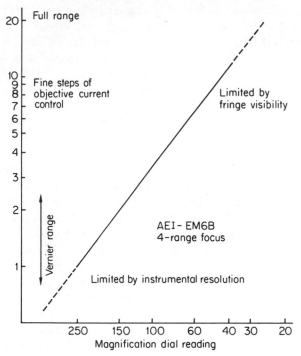

Figure 12.11   A curve constructed for the GEC/AEI
EM6B electron microscope, showing the number of
fine steps of objective current to be applied as an overfocus correction to obtain true focus from the point of
maximum contrast at any given step of the magnification dial

The range of defocus over which this technique is useful is not very
great; the effectiveness of the method falls off as magnification decreases
and defocus increases. This is because the fringe and the corresponding
bright underfocus line becomes more blurred and less contrasty as the
magnification is reduced and the amount of defocus increases. The maximum amount of defocus which is useful in practice is about 20–30 Å,
corresponding to a defocus of 2 μm at 50 kV at an instrumental magnification of about 25,000. At the high magnification end of the scale, the
usefulness of curve is limited by instrumental resolving power.

12.11 Fresnel Fringe Focusing

This is the most accurate method of all, and must always be used when the highest resolutions are required. It is necessary to observe the behaviour of a well-defined Fresnel fringe at a sharp contrast boundary as the specimen passes through focus. This means that either a small hole of the required size or a small scrap of dirt such as a particle of lead stain must lie within the required field. The edge is observed, and the fine focus control is simply adjusted to the point where there is neither fringe nor bright line surrounding the edge. Needless to say, these ideal conditions are very rarely found. It is no use whatever to find a hole, focus it critically and then move the specimen with the stage. Focus will inevitably alter, due to change in specimen height arising from imperfections in the stage mechanism or a non-planar specimen or both. If no hole or suitable dirt particle is present in the required field, then a through-focus series of micrographs must be taken. The focus setting for maximum contrast (with the objective lens slightly underfocused) is selected, and then a series of plates is exposed as rapidly as possible at increasing focus increments which are the minimum available on the instrument. Minimum steps on high-resolution instruments are generally about 3–5 Å. If 6 plates are taken, one of them should give the required resolution, provided all other conditions remained perfect. It is essential to take the series as rapidly as possible to avoid circuit drift and contamination, but great care must be taken at the same time not to jolt the column. The method is very tedious, expensive and frequently disappointing, but it is sometimes the only appropriate one.

12.12 A High-Resolution Session

To summarize, the preliminaries and procedures for a successful high-resolution session can briefly be stated as follows.

First, make sure the instrument is in the peak of condition as regards column cleanliness, vacuum system and reference batteries. Make sure the filament has a reasonable life expectation, and will not blow in the middle of the session. Prepare your specimens beforehand, and make sure that the essential test specimens (holey carbon films and evaporated metal specimens) are to hand. Make sure you have a supply of clean forceps and freshly cleaned specimen holders. Clean up the microscope room, and tidy away any unwanted apparatus and specimens; make sure there is adequate bench space for writing notes and notebooks to write them in. Make sure the specimen record book is lined up, and that there are plentiful supplies of freshly-desiccated plates and made-up processing solutions at the right temperature to hand.

Choose a time when you will be undisturbed for some hours, preferably in the evening, early morning or at night. Warm the microscope up for at least an hour. Fill the anticontaminator with liquid nitrogen and leave it to settle. On sitting down before it, first run through the quick alignment check (see page 280) and make any fine adjustments that will almost certainly be necessary. Deastigmate the condenser system. Insert the holey carbon test film, and check instrumental stability with the diffraction spike test. Correct astigmatism at the highest magnification, and if possible take a couple of plates and develop them quickly to make sure that your fringe is truly symmetrical around your hole. While you are deastigmating on the hole in the carbon film, keep a watchful eye on the specimen and make sure the film is neither contaminating nor stripping.

Finally, insert your specimen and fire away. Good Ångstrom hunting!

CHAPTER 13

Test Specimen Preparation and Performance Measurement

13.1 Introduction

In order to keep an electron microscope at the peak of its performance, two parameters must be monitored at a minimum frequency of once per month and preferably more often. These are firstly, resolving power; and secondly, magnification. Magnification is affected by slow, unidirectional drifts in the electronic supplies (notably the voltage of the reference batteries) which affect the values of the high-tension voltage and the lens currents. Resolving power is affected by dirt and contamination collecting on sensitive parts of the column such as the objective lens polepiece bores, and also by short-term fluctuations in the electronic supplies. It is necessary to keep a record of these changes in order to have the instrument properly serviced (see the following chapter). To measure magnification and resolution accurately, suitable test objects must be kept on hand. These should give results independent of the methods used to make them, so that results are strictly comparable from laboratory to laboratory.

Instrumental resolving power is best measured by determining the minimum distance apart of two points of suitable size which the microscope can separate. This demands reasonably accurate knowledge of the value of the instrumental magnification used to determine this minimum distance. Resolving power may also be measured, although rather less satisfactorily, by the Fresnel fringe method, which also requires knowledge of the magnification used. It can be measured directly, using suitable crystals, the lattice spacing of which is known accurately by the completely independent method of X-ray crystallography. This method measures unidirectional line spacing, which is not as absolute as point-to-point spacing, but has the great advantage of providing a built-in magnification standard.

There is, unfortunately, no single standard suitable for calibrating the magnification of an electron microscope over its full range. Metal, graphite or organometallic crystals of accurately known spacing are best for the highest magnification ranges; negatively stained protein crystals or other objects of known size such as polystyrene latex spheres serve for the medium magnification ranges; and metal shadowed carbon replicas of ruled diffraction gratings are best for the low magnification ranges. Suitable test objects may be bought ready prepared from electron microscope suppliers (see list on p. 504) by those without the time or facilities to prepare them themselves. For those who wish to make and mount test objects, brief instructions are given in the following sections.

Test specimens—and, for that matter, all specimens used in electron microscopy—have to be prepared in a form suitable for insertion into the column vacuum. They must be able to withstand mechanical handling and irradiation by a high-energy electron beam. The most convenient and most frequently used test specimen therefore is the 'selfless object' which we have already discussed at length, the Fresnel fringe. This is always formed at any sharp contrast boundary, so the most suitable test object is one having well-defined, sharp edges. Suitable objects are of two kinds: small electron-opaque particles mounted on a very thin, electron-transparent film; and the converse object, which consists of holes in a relatively thick, electron-opaque film. Of the two, the latter is preferable, being less fragile. In either case, the film must be mounted on a suitable support. The support almost invariably used nowadays is the so-called 'grid'.

13.2 The Grid

We have briefly mentioned grids on p. 123. They form the basic support for the specimen supporting film or substrate. This very thin film is the electron microscope analogue of the familiar transparent $3'' \times 1''$ glass slide of the light microscopist. Grids consist basically of a metal disc punched or etched with holes which are small enough to allow an extremely thin film (200 Å thick or less) of a suitable electron-transparent substance (a tough plastic film or a film of evaporated carbon) to be stretched across them without breaking when gently handled or, more important, when irradiated by the electron beam. The electrons which pass through the supporting film pass on through the objective lens and form the image; those which strike the metal parts of the grid are conducted to earth and take no part in image formation.

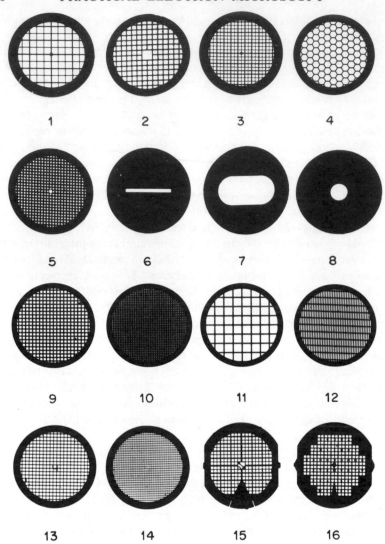

Figure 13.1   A representative selection of commonly used 3·05 mm ($\frac{1}{8}''$) diameter specimen grids. (1)–(8) respectively: 'New 100', 'Gap 150', 'New 200', 'Hexagon', 'Hole 483', 'Slot', 'Big Slot' and '1-Hole' ('Athene', made by Smethurst Highlight Ltd., Bolton, Lancs.); (9), (10): VECO 200 and 400 (VECO N.V., Eerbeek, Holland); (11), (12): LKB 4829A/1 (75 mesh) and LKB 4829A/7 (75/300 mesh) (LKB, Stockholm, Sweden); (13), (14): 'Micron' 200 (Blue) and 400 (Green) (Micron Grids, Brentford, Middx.); (15), (16): Graticules '200 H3' and 'Finder H2' (Graticules Ltd., London). Some designs are obtainable in other diameters, e.g. 2·3 and 1·5 mm

The first specimen carriers were relatively thick platinum discs 2 mm in diameter, drilled with very fine holes. These had to be cleaned each time after use. This tedious procedure was soon superseded by the use of discs punched out of sheets of very fine, closely woven (200 holes per linear inch) copper or brass wire mesh such as is used in very fine sieves and filters. These discs, being introduced by American workers, were standardized at $\frac{1}{8}''$ (3·05 mm) diameter. They had the great merit of being cheap enough to throw away after a single usage. The grids punched from wire mesh sheet were in turn superseded by perforated copper discs made by a photographic electro-deposition process. This made it simple to produce grids cheaply and in any desired pattern or size of holes. In consequence, a variety of meshwork patterns became available, each being designed by a particular worker to suit his own particular needs. Some of the most useful patterns have now become standardized, and those in greatest demand are readily available from a number of suppliers. Some of the most commonly used patterns are shown in Fig. 13.1.

Grids with square holes can be obtained in meshes of 50, 100, 150, 200 and 400 holes per linear inch. The more open the mesh, the smaller is the proportion of the specimen which is obscured by the grid bars. The transmission ratio varies from about 95 % for square 50 mesh to about 35 % for 400 mesh. While it is a great advantage to be able to see a large proportion of the specimen, the degree of support offered by the more openwork pattern falls off drastically; 100 mesh and below will only support thick sections or films suitable for relatively low resolution work. The 200 mesh grid is the most commonly used compromise; the best patterns have about 75 % transmission. It is convenient to have the centre marked in some way, and it is a further improvement if the centre marker is completely asymmetrical, so that the orientation of the specimen can be checked.

Very high resolution work demands extreme specimen stability, which can only be obtained by the use of fine mesh grids (more than 200 perforations/linear inch), preferably with round holes. These have the advantage over square holes in that there is no point of strain such as occurs at the corners of a square, and also that there is plenty of metal to carry away the heat and thus to minimize specimen movement due to thermal drift.

Grids are by their nature very fragile, being made of soft copper (they can also be obtained in inert metals such as gold for special purposes). They are less than 50 μm thick, so they must always be handled with the finest (No. 4 or No. 5) watchmakers' balance forceps.

13.3 Filmed Grids

Grids may be used either bare or filmed; this depends on the nature of the specimen to be examined. Thin sections of biological material embedded in epoxy resins will withstand the electron beam on a bare grid without a supporting film if they are thicker than about 600 Å, but a resolution better than about 25 Å cannot then be expected. For higher resolution work, thinner sections must be mounted on filmed grids, or, better, carbon–Formvar nets (see Fig. 13.2(*d*)).

The film may be of a plain plastic such as collodion or Formvar; a plastic film coated with evaporated carbon; or a plain carbon film. The plain carbon film is the best for high-resolution work, since it is tough and withstands the electron beam; it conducts electricity and hence reduces specimen drift due to charge effects; and it conducts heat from the specimen to the copper grid bars so that drift due to thermal effects is also reduced. The preparation of grids filmed with plastic is reasonably simple, but the preparation of good, thin plain carbon films, essential for high resolution work, is a tedious, time-consuming and somewhat irreproducible procedure.

13.4 Specimen Support Films

Before any particulate material can be examined, grids have to be coated with suitable substrate or support films. These may either be of plastic or carbon. The plastics in general use are collodion and polyvinyl formal, the trade name of which is 'Formvar'. Plastic films are easier to make than carbon films, and are used for work where the highest resolution is not called for. For really high resolution, carbon films are essential.

The general principle in making plastic films is first to make up a very dilute (0·5 % or less) solution of the plastic in a non-polar solvent, such as ethylene dichloride, chloroform or amyl acetate. Two methods are then available for making the film. For the first, a microscope slide is dipped into the solution, drained and then allowed to dry. The very thin (200 Å thick or less) film of plastic thus formed is then floated from the surface of the glass slide by dipping it under water. The film floats off if the upper side is not wetted, and appears like a large, dark grey ultrathin section by reflection. The thinnest plastic films are made in this way. The second method is to place a drop of the plastic solution on the surface of clean water, allow it to spread out and then to dry. A thin film of plastic is left on the water surface. This method tends to produce thicker films of somewhat uneven thickness. The method recommended for making plain Formvar grids is as follows:

1. Make up a very dilute solution (0·2 to 0·5 %) of polyvinyl formal (Formvar) in redistilled ethylene dichloride. The solvent should be freshly redistilled and grease-free.
2. Dip a previously cleaned 3″ × 1″ glass slide into the solution, leave it for a few seconds and transfer it immediately to a Coplin jar containing a little solvent so that it is filled with saturated solvent vapour. Close the jar with a lid, and allow the slide to drain in the saturated vapour for 5–30 minutes, depending on the desired film thickness. It is best to handle the slide with forceps; do not allow the forceps to dip into the solution. All apparatus must be scrupulously clean and dust-free.
3. After draining (the draining time determines thickness and evenness of the film) allow the film to dry thoroughly on edge in a dust-free place.
4. When the film is completely dry, scratch it through with a mounted needle about 2 mm from the edge of the slide, thus freeing an area of film about 2 cm by 5 cm.
5. Hold the slide almost parallel to the surface of distilled water contained in a glass dish about 15 cm in diameter, and gently lower it below the surface. The Formvar film should detach at the scratches and float on the water surface. It may be necessary to start it parting from the glass by gently lifting it with the needle. The water surface must be made completely dust free by wiping it with clean, lint-free paper.
6. When the film is floating on the water surface, it should be possible to see it only by the interference colour it gives by reflected light, which should be a dark grey. If it has a yellowish tinge, it is too thick. The film should show an even colour without patchiness. Cover the dish immediately to prevent dust from falling on the film.
7. The film is now transferred to the grids by laying them, rough side down, on the film as it floats on the water. It may be necessary to give a very slight bend to each grid so that the film does not fall through and adhere to the glass slide during the next stage of the procedure. About 20 grids can be laid on a good film.
8. Pick up the grids plus film by laying a clean 3″ × 1″ slide on top of the grids and sweeping them together with the film through the water, bringing the slide out film side up. This must be done very quickly, and success requires practice. The technique is best demonstrated.
9. Allow the slide plus grids plus film to dry thoroughly in a dust-free place. The film, if sufficiently thin, will adhere tenaciously to the rough surface of each grid, which can then be picked off the slide with forceps. The filmed grids are then laid on clean filter paper in a plastic Petri dish. They should be used as soon as possible, as the film tends to rupture on keeping.

These Formvar film coated grids can be used as specimen supports without any further preparation. When examined in the electron microscope, they should appear structureless at magnifications below × 100,000 and free from dust particles, tears, ripple, uneven thickness, holes and other defects. They should be thin enough to be almost completely electron transparent; if they absorb more than about 5% of 60 kV incident electrons, they are too thick. The absorption can easily be measured with a microscope equipped to measure screen current or screen luminous output accurately. Thick films do not adhere properly to the grids, apart from their deleterious effect on image contrast, and cause specimen drift.

If the film is too thin, it will not stand up to the electron beam, and may burst when the condenser lens is focused on it. Very thin films must be strengthened with a layer of carbon evaporated on the Formvar film, giving a 'Formvar–carbon' film. The technique of carbon evaporation is described on p. 482. These films have the great advantage of being electrically conducting, so that they do not charge up when irradiated. This reduces the possibility of specimen drift due to mutual charge repulsion, which can very easily occur with plain Formvar films. The method of preparation is the same as for plain Formvar films, but the filmed grids are not removed from the slide when dry. The slide plus filmed grids is placed in a vacuum evaporator, and a layer of carbon so thin that it gives only the barest visible tan colour on a white indicator plate is evaporated on top of the Formvar (see page 475). The grids are then picked off and stored as before; they have a virtually indefinite life, and can be made in large batches. The addition of the carbon layer should not increase the electron absorbance by more than about 2%. Good carbon–Formvar films have a total absorbance of less than 5%. They are the best all-round supporting film.

For the very highest resolution work, it is necessary to make plain carbon films. This is done simply by immersing the Formvar–carbon films in a Formvar solvent, leaving them undisturbed until the Formvar has dissolved completely, and then removing the solvent with great care so that the carbon films settle down on the surface of the grid and dry, firmly adherent. For this purpose, .the apparatus shown in Fig. 13.2 has been found very convenient. A tap is glass-blown into the base of the bottom member of a 15 cm Petri dish, and a stainless steel mesh grid carrier is made to fit into it. The apparatus is held in a retort stand or other firm support, and the mesh is then covered to a depth of about 2 mm with fresh redistilled ethylene dichloride or chloroform. The Formvar–carbon grids are then laid, carbon side up, on the mesh below the surface of the solvent. If the operation is carried out quickly and carefully, the films remain adherent to the grids. The dish is then covered, great care being taken not to give the apparatus a sideways knock, and the apparatus is left undisturbed for at least 6 hours to allow the Formvar to dissolve completely. The tap is then opened slightly with great care, allowing the solvent to run out very slowly. The grids are left to dry *in situ*, each being completely covered by a disc of carbon. If the apparatus is knocked sideways as the solvent is being run out, the carbon films will not remain

Figure 13.2 The apparatus used for preparing carbon substrate films on specimen grids. (Photograph Michael Turton)

centred on the grids. When dry, the films adhere very tenaciously to the grid bars. The grids should be laid on clean filter paper in a Petri dish, and are then examined by reflected light through a low-power stereo microscope. Those having more than about 10 % of torn or damaged squares are rejected. Good carbon grids are very time-consuming to make, and should be stored with care in a covered dish in a dust-free place. The carbon films are extremely brittle, and the grids may not be bent or stretched. They also tend to be water-repellent.

13.5 Test Objects

Having prepared suitable support films, test objects for determining the magnification and resolution of the electron microscope can now be prepared on them. The use of the test objects in measuring instrumental performance will be discussed later in this chapter. Details will first be given for the preparation of suitable test objects. These can be classified into 4 types:

1. Specially made holes in support films, generally carbon–Formvar;
2. Electron dense particles laid on very thin carbon films;
3. Crystals having suitable lattice spacings and characteristics; and
4. Replicas of ruled diffraction gratings.

13.5.1 Preparing Holey Films

The most satisfactory all-round test object is a carbon–Formvar film of moderate, even thickness which is pierced with a large number of circular holes with sharp edges having diameters ranging from about 0·1 to 1·0 μm. A low-power electron micrograph of a well-prepared film is shown in Fig. 13.2(a). The preparation of suitable 'holey' films is a very chancy and irreproducible procedure. Basically, the method is to prepare a film of Formvar or other plastic pierced with suitable holes. This holey Formvar film is then floated on to grids and a thin layer of carbon is evaporated on top. The difficulty lies in the production of a Formvar film which is completely pierced with well-separated circular holes of a suitable range of sizes and in sufficient number, and which retain their circularity without running into one another. There are two methods for achieving this end. The first is to condense minute droplets of water on to the surface of a drying Formvar film. As the film dries, the water droplets pierce the film, leaving circular holes. The second method is to dissolve the Formvar in a solvent containing water. This must necessarily be a mixture; a non-polar solvent for the Formvar, plus a partly polar solvent which is miscible both with water and with the Formvar solvent. In practice, mixtures of ethylene dichloride, dioxane and water are used. Glycerol may be used instead of dioxane. As the Formvar film formed from this mixture dries, the more volatile solvents evaporate first, leaving minute water droplets which pierce the film.

The first method is used most commonly, and was the one used to prepare the film shown in Figs. 13.2(a) and (b). A very thin Formvar film is formed on a 3″ × 1″ microscope slide by dipping the slide in a 0·3 % solution of Formvar in redistilled ethylene dichloride. This is allowed to drain in solvent vapour in a Coplin jar for at least 5 minutes to give a very thin film. Water droplets are then condensed on the film. The traditional method for doing this is for the operator to hold the slide in his widely opened mouth, and to breathe out on to it for a second or so. The

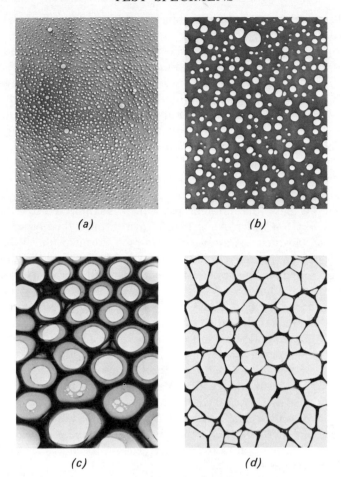

(a)   (b)

(c)   (d)

Figure 13.3 The 'holey carbon film' test object. (a) is a low-power micrograph of a well-prepared film with plentiful holes in the right size range ($\times$ 1,000); (b) is a higher power view ($\times$ 3,600), showing almost circular holes with clearly demarcated edges between 1·0 and 0·1 μm in diameter. The thickness of the film is even, right up to the hole edges; (c) is an example of a very poor film with partly perforated holes of too large a diameter, with a few ragged small holes in the septa; (d) is the result of carrying (c) to the extreme—a carbon–Formvar 'net', which is sometimes used as a support for extremely thin epoxy resin sections when the very highest resolution is required from them. (a) also shows the effect of chromatic change in magnification (c.c.m.); the Fresnel fringes appear on opposite sides of the holes as the field moves radially outwards from the centre, where no c.c.m. can be detected. A well-prepared holey film or net is a useful test object for showing up distortion and c.c.m. qualitatively

film will then be seen to have become 'milky', due to condensation of water from the breath. The degree of milkiness must be judged very finely by the operator; needless to say, the results are highly capricious. It is essential to have an extremely thin Formvar-solvent layer on the surface of the slide, or the water droplets will not pierce the dry Formvar film completely. Instead, the prepared film will be found to be studded with 'pseudo-holes' (Fig. 13.2(c)) which appear at low power to be suitable, but will be found at higher power to have a septum of Formvar across them. These films are useless. A more reproducible way of condensing water vapour on the slides is to hold them over the surface of hot water, or, better, to pass water vapour into the Coplin jar filled with ethylene dichloride vapour in which the slides are draining. This can be done by introducing a glass tube into the jar, which is connected to a bubbler filled with very hot water. Air is then blown through the hot water, and the saturated air then passes over the draining Formvar film. Whichever method is chosen, it will be necessary to experiment before the ideal conditions are found for the apparatus used, the room temperature and the ambient relative humidity.

For the second method, add 1–5 % water to a mixture of equal parts of 1 % Formvar in ethylene dichloride and dioxane. Shake well. Dip slides in this mixture, and allow them to drain over the solution in Coplin jars for about 30 minutes. Then dry in air. The films should be perforated with holes. If the holes do not perforate the film completely, dip the filmed grids in acetone for a few seconds. Varying the amount of water varies the size and distribution of the holes.

The holey Formvar film is then detached from the slide and floated on the surface of water. It is then picked up on grids, and a layer of carbon is vacuum evaporated on to the surface. These procedures are described in detail in Chapter 19.

After preparation, holey carbon test films can be rapidly assessed under the low power of a phase contrast light microscope, using a highly absorbing phase plate if possible. The holes should be just at the limit of resolution, but can easily be seen. Grids having an excess of torn squares should be discarded.

There are many other methods for making holey films of the required characteristics; all appear to be equally capricious. Each electron microscopist has his own particular method. It may not be necessary to prepare holey films for resolution checks on sections, because ultrathin sections frequently contain holes if they are of Araldite or similar epoxy resin, and are mounted on to bare grids, as is modern practice. A scratch line in a section can usually be irradiated with an intense electron beam to yield a few suitable holes, so that the instrument can be set up for high-resolution work on the specimen in use.

Formvar Nets It is convenient at this point to describe Formvar 'nets'. These are not used as test objects, but as supports for the most delicate specimens such as extremely thin sections. Any holey film which has an area of holes greater than film can be called a 'net'. The holes in the net should be in the 1–10 µm range of size; a micrograph of a well-prepared net is shown in Fig. 12.2(*d*). Such a network of Formvar reinforced with carbon forms an ideal support for Araldite or Epon sections thinner than about 400 Å; areas of the section can be photographed at the highest possible resolution through the holes in the net. Nets are prepared in the same way as holey films. The simplest way is to hold a drained, drying Formvar slide over the surface of boiling water for about a second, although more reproducible results are obtainable by blowing hot, saturated air over the surface of a slide held in a Coplin jar for a longer time than for making a holey film.

13.5.2 Particles

The simplest particulate test object is a deposit of carbon particles on a carbon film. A specimen is prepared simply by passing a filmed grid through the smoke from burning benzene or camphor. In a well-made preparation, chains of roughly spherical carbon particles about 200–500 Å in diameter are deposited on the grid film. A certain amount of experimentation is needed to find the best conditions to deposit well-spaced chains of suitably sized particles, a field of which is shown in Fig. 13.4. Unfortunately, the chains of particles tend to adhere to the substrate only at places spaced irregularly along their lengths. This means that individual particles will be at different focal levels, as shown in Fig. 13.4. At a median setting of the objective lens focus, some particles will be overfocused and surrounded by a fringe, some will be in focus, and some will be underfocused and show the bright line. Another disadvantage is that the chains tend to move about under the influence of the electron beam due to charge effects. However, carbon particle test objects have the advantage of being easy and quick to make. Care must be taken not to introduce grids having visible particles of soot into the microscope, or these may become detached and adhere to the objective lens aperture or polepieces where they will cause astigmatism.

The standard test object for measuring point-to-point resolution, which, as we have already seen, is the most accurate and generally acceptable way of determining resolving power, is a very thin carbon film upon which metal particles of the same order of size and spacing as the resolving power have been evaporated. Certain high melting point metals and alloys, such as platinum and platinum–iridium form discrete, highly electron-dense particles 10 Å or less in diameter when evaporated under very high vacuum. These point-resolution test specimens are made by placing the

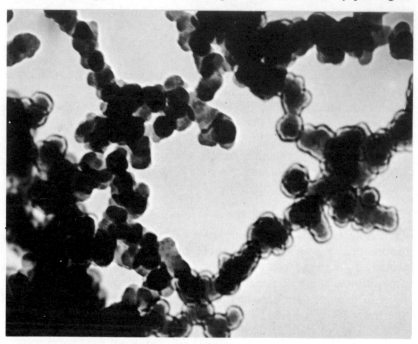

Figure 13.4   The carbon particle test object. Particles of carbon about 0·5 μm are deposited on a carbon film mounted on a grid by waving the coated grid in the smoke from burning benzene or camphor. Chains of particles attach themselves at intervals across the carbon substrate. The particles do not lie in the same plane, so a given field will contain overfocused (right), focused (centre) and under-focused (left) particles. Being unsupported, the particles tend to move randomly due to charge effects in the electron beam. This test object is very easy to prepare but is not very satisfactory. Being three-dimensional, this specimen makes an excellent test object for stereo-operation (see p. 265)

very thinnest carbon grids, selected individually in the electron microscope by inspection and measurement of electron transmission, in the bell-jar of an evaporator (see Chapter 19) at least 20 cm away from the metal source. The bell-jar is pumped down to the highest attainable vacuum, and the metal is evaporated for the shortest practicable time. Well-made speci-mens can be kept for years if handled with care.

13.5.3 Crystals

Crystal lattice planes are more easily 'resolved' than point-to-point spacings, due to the reduced effect of astigmatism on a suitable oriented specimen, and because the imaging conditions are more akin to electron diffraction than to true electron optical image formation. As we have seen (p. 271), a slight image defocus reinforces a lattice image made up of straight lines far more than a random point image, and in addition the eye can distinguish an image made up of long straight lines against a background of random electron noise far more readily than it can a point image. For these reasons, crystal lattice specimens always yield a higher apparent resolving power than point-to-point specimens, and the results claimed by instrument manufacturers using lattice spacing should always be taken with the proverbial pinch of salt. Scrutiny of these claims (see pp. 497 to 502) shows that an instrument giving a 2 Å lattice resolving power will only give a 3 Å point-to-point performance.

The specimens generally used fall into four main classes: organo-metallic crystals, minerals, metals and proteins. The last have spacings which make them more suitable for magnification standards than resolution test objects.

13.5.4 Ruled Grating Replicas

Ruled metal diffraction gratings such as are used in spectrometers can be obtained with rulings as close as 0·5 µm or less, in one direction or in two directions at right angles. Suitable test objects can be made from these gratings without damaging them by the method of surface replication (see p. 484). Briefly, a viscous solution of a plastic such as collodion is poured on to the grating surface and allowed to dry. It is then stripped off using Sellotape, and the side bearing the imprint of the rulings is lightly shadowed with platinum in a vacuum evaporator. A layer of carbon is then evaporated on top of the platinum, and the plastic strip is removed from the evaporator and floated on a suitable solvent for the plastic, which dissolves away leaving the carbon plus metal shadow floating on the solvent. This replica is then picked up on grids, and when dry is ready for examination in the electron microscope. Ready prepared grating replicas are best obtained from suppliers (see p. 504), either mounted on grids or in a colloidal suspension.

13.6 The Measurement of Resolving Power

Three basic methods are used to measure instrumental resolving power. These are:

(1) measurement of the asymmetry of a Fresnel fringe,
(2) measurement of the spacing between pairs of very small point objects,
(3) measurement of the lattice (line) spacing of crystal planes previously known from X-ray diffraction measurements.

13.6.1 Fresnel Fringe Method

This is not as objective as a measurement made on an actual object, since we are in this case examining a 'selfless object', but a very good indication of the resolving power can be obtained by photographing the fringe at very high magnifications. The actual object should be a specially prepared holey carbon film rather than a hole in a plastic section for reasons of specimen stability. The procedure is simply to focus a clean, circular hole at highest magnification, to deastigmate it as well as possible, and then to take a through-focus series of plates, starting 2 finest focus-clicks over focus (the finest steps provided are generally about 0·01 to 0·02 μm, corresponding to an image spacing of 2–3 Å) and ending 2 clicks under focus. One plate should result with the fringe almost touching at two opposite points on the hole (see Fig. 13.5). The overfocus spacing of the fringe at the opposite edges is then measured from the centre of the space between fringe and hole edge to the centre of the fringe. This distance is then converted to Ångstrom units from a knowledge of the total (electron times enlarger) magnification, and this figure gives a reasonably accurate estimate of the resolving power of the instrument.

Inspection of Fig. 13.5 will show that resolving power at around the 5 Å level cannot be estimated to within an accuracy better than about ± 20 % at best. This is why manufacturers tend to claim resolving power as, for instance, 'better than 5 Å' for their instruments, and do not claim a specific figure. All the best instruments will nowadays resolve better than 10 Å by this objective test; some can undoubtedly better 5 Å but only under the very best conditions and in the hands of very experienced and highly skilled operators.

Instrumental Stability The electron microscope itself is a far better monitor of the short-term stability of its own electronic supplies than any meter which can be devised. A quick and easy way to check the short-term stability is to watch a Fresnel fringe close to focus for some minutes. Any instability of the order of the resolving power can easily be detected from the movement of the fringe in and out from its mean position. The cause of the movement must then be found. It will generally be traced to a faulty component or soldered joint in one of the electronic supplies. Faults of

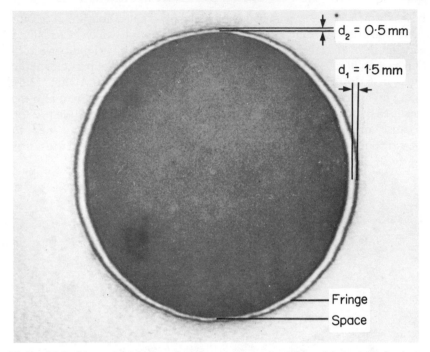

Figure 13.5 Fresnel fringe resolution measurement. This micrograph of a hole in a carbon film at × 500,000 was taken with deliberately introduced astigmatism in order to demonstrate the method. The measurements are generally made direct on the processed plate exposed at top instrumental magnification without making an intermediate print; this is a reversed print corresponding to the image on the plate. The distance between the centre of the black fringe and the centre of the white space separating it from the edge of the hole is measured with a graticule magnifier (or better a cathetometer microscope) at the point of maximum spacing d_1 and at the point of minimum spacing d_2. The fringe must be almost touching the hole at its closest point. Several measurements of each spacing are made and averaged. The distances in mm are then converted to Å from a knowledge of the magnification. Resolving power is simply the difference between the two measurements. In the example above, $d_1 = 1.5$ mm and $d_2 = 0.5$ mm. The difference of 1·0 mm at a magnification of × 500,000 is 20 Å, which is the resolving power of the instrument under these conditions of gross astigmatism applied by the objective stigmator

this nature involving very small instabilities are generally very difficult and time-consuming to trace. Fault finding will be dealt with in Chapter 15.

13.6.2 Point Spacing Method

The Fresnel fringe method just described has the advantage that it can be performed on any specimen having a sharply defined hole. Point resolution measurements demand a specially made object, which is usually a very lightly evaporated coating of platinum or platinum–iridium on a very thin carbon film. A good specimen shows large numbers of particles of the order of 5–10 Å in diameter, spaced at distances of 2–10 Å (Fig. 13.6). Being highly contrasty point-objects, they are easy to focus, and also the lens is easy to deastigmate by using the line-drawing method (see p. 282). Because of the presence of background phase granularity due to the carbon support film which may be confused with the particles, it is

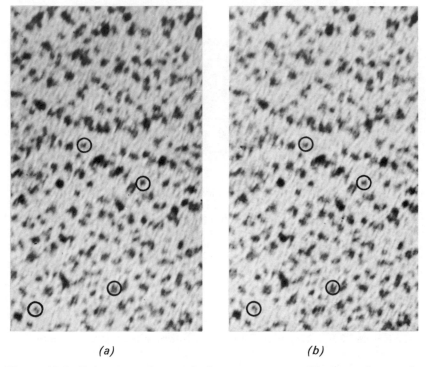

(a)   (b)

Figure 13.6 Point separation resolution measurement. The two micrographs (a) and (b) of evaporated platinum–iridium on a carbon substrate were taken in rapid succession after critical focusing. Pairs of particles (ringed) on each print show spacings of less than 1·0 mm, which at the overall magnification of approximately 2×10^6 indicates an instrumental resolving power of better than 5 Å. A trace of astigmatism in the NE–SW direction can be detected; resolution in this direction is better than resolution in the perpendicular direction. 2 plates must be taken in order to ensure that random phase granularity is not measured in mistake for actual particles. In terms of sheer resolving power, the micrograph in Fig. 12.1(d) shows slightly better resolution, but only one plate was available

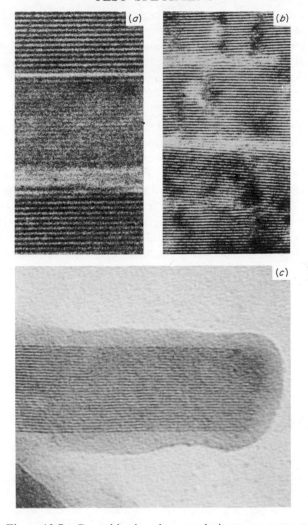

Figure 13.7   Crystal lattice plane resolution measurement

(a) Copper phthalocyanine crystal showing lattice spacings of 9·8 and 12·6 Å. Because of the presence of two readily resolvable lattices of almost the same spacing, this substance is not ideal for either magnification or resolution calibration

(b) Potassium chloroplatinate crystal showing 6·99 Å spacing. This substance is very susceptible to beam damage

(c) Platinum phthalocyanine crystal showing 12·5 Å lattice spacing

(a) and (b) courtesy Hitachi Ltd., (c) courtesy Dr. Graham Hills, John Innes Institute, Norwich

necessary to take two consecutive plates at exactly the same microscope setting. Prints from these two plates (Figs. 13.6(a) and (b)) are then searched for corresponding pairs of particles, and the minimum detectable spacing is measured. This is the practical resolving power of the microscope. The method, like the preceding one, requires a reasonably accurate knowledge of instrumental magnification.

13.6.3 Crystal Lattice Spacing Method

This method has the advantage that instrumental magnification need not be known, as the crystals are themselves an internal magnification standard, the spacings being known to a very high degree of accuracy by X-ray diffraction measurements. On the other hand, measurement of line spacing is not as satisfactory as point spacing, because lines can be sharply imaged in one direction with an astigmatic lens which will not show the same resolving power in a perpendicular direction. In practice, the most commonly used substances are crystals of copper phthalocyanine, which has lattice spacings at 12.6 and 9.8 Å (Fig. 13.7(a)), platinum phthalo-cyanine (Fig. 13.7(c)) and potassium chloroplatinate, with a spacing of 6·99 Å (Fig. 13.7(b)). These substances are very difficult to photograph in the electron miscroscope, due to their high sensitivity to beam damage.

The best crystalline high-resolution test specimens so far described are partially graphitized carbon black, and very thin gold foils. Both give sharply defined single spacings in the 2–4 Å region, and both are reason-ably immune to beam damage provided an effective anticontaminator is in use.

Partially graphitized carbon black is obtained from suppliers in the form of a powder, which is then dispersed in chloroform and spread on holey carbon grids. To be able to resolve such a closely spaced lattice, it is essential that the carbon particles be unsupported. Suitable particles are found projecting from the edge of a hole. The lattice spacing is 3·44 Å (see Figure 13.8(a)), and is best resolved at 100 kV for maximum beam penetration. It is best if a double-tilt stage can be used in order to obtain optimum orientation of the lattice to the beam; this is the case with all crystalline specimens. If a highly coherent illuminating beam is used, preferably with a pointed filament, the graphite spacing can be seen on the final image screen with the binocular magnifier at an instrumental magnification of about 250,000 ×. A print made at an overall magnifica-tion of 1,000,000 times will show the spacings clearly. An advantage of graphite is that the spacing is present in all directions, and the result is more dependable than a unidirectional lattice such as gold (see below).

Asbestos in the form of crocidolite is also fairly immune to beam damage, but is confusing in that it gives two spacings, at 9 Å and 4·5 Å. Pyrophyllite with a spacing of 4·5 Å may also be used. The highest resolving power so far reported has been made on single crystal gold foils, which yield a spacing of 1·4 Å in the 220 plane and 2·0 Å in the 200 and 020 planes (see Figure 13.8(b)).

Before making tests of resolving power, it is essential to check that the instrument is at the peak of its performance. The illuminating system and optical system must be accurately aligned about the microscope axis as described in Chapter 11, the electronic circuits must be allowed at least

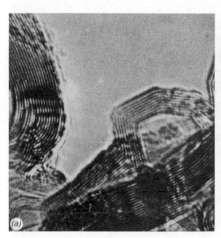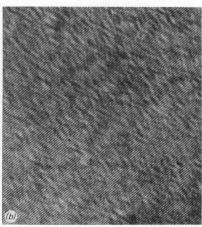

Figure 13.8 Micrographs of crystal lattice spacing in (a) graphite and (b) gold foil. In each case, the specimens have been tilted to give the optimum orientation of the lattice for maximum image contrast, using a eucentric tilt/goniometer stage. The graphite lattice has a spacing of 3·4 Å, the gold lattice has a spacing of 2·04 Å. Overall magnification as reproduced approx. 2,500,000 ×. (Courtesy Philips Eindhoven)

half an hour to warm up, and the anticontaminator if used must be allowed to reach equilibrium temperature with the specimen in place. Choose a suitable field, and do not move the stage again. Astigmatism must be compensated as accurately as possible, and a time should be chosen when there is least likelihood of disturbance to the mains power supplies, vibration due to traffic, or any other possible causes of malfunction.

13.7 Magnification Calibration

Since the measurement of resolving power is dependent on the precise calibration of magnification, it is also necessary to calibrate the magnification.

The magnification figures given by the manufacturer for various control settings may be in error, due to variations in high tension voltage, precise location of specimen plane and lens current variations due to drift in reference voltages. It is therefore essential to maintain a check on magnification, which will give a pointer to troubles arising from the above-mentioned sources.

The exact position of the specimen plane is difficult to maintain. It can alter due to bent grids, drooping of the specimen between the grid bars, bent specimen holders, defective insertion mechanism and other causes. The most accurate indication of specimen position is given by the setting of the focus control with the specimen at focus, although this also varies with the other parameters we are considering—variations in high tension and lens current. A careful record should therefore be made of the setting of the focus control at all magnification settings when the instrument is new and the high tension and lens current values and magnifications are guaranteed correct by the manufacturer. These settings should be referred to subsequently throughout the life of the instrument (Fig. 15.1). The metered value of the objective lens current at focus at each magnification setting may be recorded instead, but focus control settings are more accurate.

The most generally useful method of measuring magnification is by photographing a diffraction grating replica at focus for each given instrumental magnification setting. If a grating with mutually perpendicular rulings is used (Fig. 13.9), distortion can be measured at the same time as magnification. The spacing of the grating replica can be accurately measured by photographing it with a phase contrast light microscope. This ensures that any errors which may have arisen during the replication of the original grating (the average spacing of which can be measured with great accuracy by measuring the optical dispersion of the grating itself) can be detected and allowed for. For an accuracy of better than 2 %, at least ten lines must be present across the plate, and several plates of several fields must be counted and averaged. To avoid the hysteresis effects in the lenses which we have already mentioned in Chapter 3, magnification calibration should always begin at zero setting (the diffraction spot) and should be made consecutively in ascending order.

If a grating having a spacing of 2,000 lines/mm is used (this is typical of a generally available grating), then at a magnification of 20,000, 10 spaces (11 lines) will occupy 10 cm on the plate. If we continue to increase magnification, then the calibration will clearly become less and less accurate. This can be overcome to a certain extent by choosing two recognizable imperfections in the field, and measuring the distance between them. This

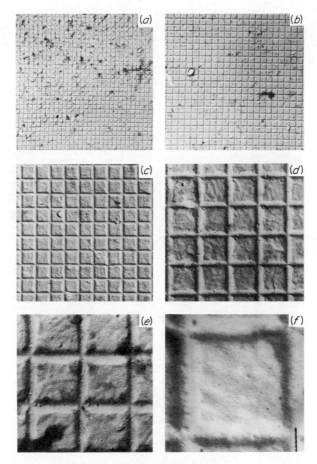

Figure 13.9 Magnification calibration with a shadowed diffraction grating replica. A square grating of 2,160 lines/mm is shown, taken at nominal instrumental magnifications of (a) 3,000, (b) 5,000, (c) 10,000, (d) 20,000, (e) 30,000 and (f) 60,000. The optical enlargement in the figure is about × 1.2. Gratings with 2 sets of perpendicular rulings are preferable to the normal single ruled grating, since distortion can easily be measured on the micrograph as well as magnification. Sigmoid distortion can be detected on (a), but total distortion is less than 2·5%; (b) shows barely measurable distortion, while micrographs taken at above this magnification are virtually distortion-free; (f) represents the upper useful limit of the method; the measurement of only one square is only valid to ± 10% or less. AEI EM–6B. (Grating replica by courtesy of GEC/AEI Ltd.)

gives reasonably accurate results up to about 50,000 times, but after this crystalline specimens must be used. Negatively stained beef catalase crystals (Fig. 13.10) are ideal for the intermediate range of magnification.

If great accuracy is required, it is necessary to add an internal magnification reference standard to the specimen. Polystyrene spheres form a useful

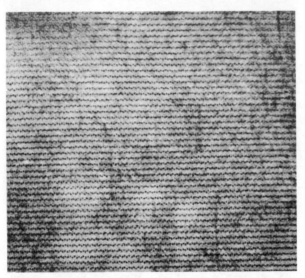

Figure 13.10 Beef liver catalase as a magnification standard. The lattice spacing can be measured to within ± 0·5 % by using the electron microscope as a low-angle diffraction camera. The actual image magnification is then determined simply by counting the lines across the image. This method is useful up to about × 500,000. The spacing of the lattice lines is 84·4 Å, and the spacing of the dots on the lines is 33·9 Å. (Courtesy Dr. Graham Hills, John Innes Institute, Norwich)

standard, for they can be suspended in water and sprayed on to specimens mounted on grids, or can be added to suspension of particles. Suitable spheres having particle diameters ranging between 880 Å and 5,570 Å are obtainable from the Dow Chemical Corporation, and are known as 'Dow latex particles'. Colloidal suspensions of diffraction grating replicas are also obtainable, and are used in the same way. For the very highest magnifications, organo-metallic crystals may be used. Other standards can be used, such as beef catalase crystals, tobacco mosaic virus particles, or phthalocyanine crystals. An internal standard of this nature has an added advantage of providing a contrasty object for accurate focusing.

13.7.1 Calculation of Magnification

A useful expression to remember for calculating the actual size of an object from a measurement on a print or plate is:

$$\text{Size of object in Å} = \frac{\text{Size of image in mm}}{\text{Overall magnification}} \times 10^7$$

13.8 Errors in Magnification Calibration

The magnification of the final image for given values of current in the microscope lenses is subject to considerable error. This is due to the 'hysteresis' effect between magnetizing flux, i.e. lens current, and the actual magnetic flux produced at the gap between the soft iron polepieces. This can be expressed graphically in the form of the familiar 'B/H' curve (Fig. 4.2) in which lens current is plotted against lens strength. If a lens starts at zero strength with zero current and is then magnetized, a certain 'remanent' flux will be retained even when the lens current is switched off. A degree of permanent magnetism will thus be retained. Therefore, the strength of the lens will depend on its previous magnetic history. Lens strength will not be the same if a given current is reached by decreasing the current from a high value as it will be if the same value is reached by increasing from a low value, as is shown in Fig. 4.2. For any given value of lens current, the lens strength can lie between two limiting values, depending on the remanence of the material from which the lens shroud and polepieces are made. The combined hysteresis effect in all the image-forming lenses can affect magnification values by as much as 10 %. Measuring the current in the intermediate lens or bringing it to a given value with the magnification control does not ensure a precise value of magnification.

13.9 Distortion Measurement

Distortion should only be noticeable at the very lowest magnifications. It is generally measured by the ratio of the magnification at the centre of the plate to that at the edge, expressed as a percentage. It usually takes the form of barrel and sigmoid distortion (see Fig. 13.7(a)), with magnification greater at the edge than at the centre. The figure quoted by the manufacturer obviously depends on the method of measurement. It should not, however, exceed about 5 % at 1,000 times, and should become negligible at 5,000 times.

13.10 Contamination Rate Measurement

Another important parameter affecting microscope performance on which a very watchful eye must be kept is contamination rate. Contamination, (see p. 159) is a slow, and, at lower magnifications, barely perceptible degradation of detail and contrast in the specimen caused by the bombardment of the surface with charged ions or radicals derived from the gases remaining in the column even at the highest vacuum. At high magnifications, detail can disappear very rapidly. Contamination (Fig. 7.6) forms a diffuse layer which can under adverse conditions measure several hundred Ångstroms in thickness over the specimen where the beam has struck it, causing electron scatter which does not contribute to the formation of the image. The resulting 'glare' reduces contrast and obscures fine detail in the image. The rate at which contamination builds up varies with the illuminating conditions and the temperature of the specimen, and on the presence and temperature of the anticontamination device. It is therefore necessary to measure contamination rate under controlled and reproducible conditions. The best conditions to choose are those which give rise to the highest rate of contamination. These are, in general, high excitation of C_1 lens leading to a focused spot of minimum area, high accelerating potential and high gun brightness. All lead to a high electron flux through the specimen. Contamination rate should therefore be measured under standard conditions of high voltage, high gun brightness, high beam current and minimum focused spot size. It should also be measured at two conditions of anti-contaminator temperature—room temperature, and the coldest attainable. A record should be kept of the results, so that cleaning operations can be undertaken when they become necessary.

Contamination is caused by backstreaming of oils from the diffusion and/or rotary pumps; dirt introduced into the column, especially in the form of methacrylate sections, which sublime in the beam; and air leaks through faulty seals. In a well-maintained instrument, the main cause should be from the inevitable backstreaming from the diffusion pump.

Comparison of contamination rate between different makes and types of microscope is not very informative unless precise measurements of electron flux in amperes per square micron through the section can be given. This is seldom possible. But it is possible to compare contamination rates on any given instrument if these are measured under standard conditions. An attempt to do this has been made by GEC/AEI for their instruments. Alderson (AEI Electron Microscope News, No. 2, November 1965) proposed that contamination rate be measured under the following conditions: maximum C_1 excitation; C_2 focused to minimum spot size; maximum (80 kV) accelerating voltage; standard conditions of filament tip position, condenser aperture diameter and beam current; and using as a specimen a holey carbon film approximately 200 Å thick photographed at an electron optical magnification of 100,000. The procedure is as follows. Set up the microscope for the standard conditions. Insert the specimen, and find a hole giving an image on the screen 3–4 cm in diameter at 100,000 times. Very slightly underfocus it, bring a plate out under the screen, and photograph the hole. Leave the plate out. Now focus C_2 to maximum intensity and minimum spot size, centring the spot accurately

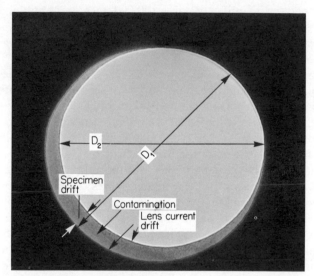

Figure 13.11   The measurement of rates of contamination, specimen drift and objective current drift. Two exposures of a hole are made successively on the same plate after a known time interval (generally 2 minutes). As the film contaminates, the hole becomes smaller, due to contamination deposited around the edge. The rate of infilling of the hole in Ångstrom units/sec is the contamination rate. If the two holes are not concentric, this is due to specimen drift. The amount of drift is the distance the centre of the second hole has moved away from the centre of the first hole, again expressed in Å/s. The photograph also shows the rate of drift of the objective lens current; the first (larger) hole is overfocused; the second is underfocused. By measuring the width of the fringes, the drift rate of the lens current can also be expressed in microns change in focal length per second. × 125,000

about the hole. Leave for 2 minutes. Do not touch anything during the test period. At the end of 2 minutes, defocus C_2 to photographic intensity, lift the screen and take a second exposure on the same plate. Remove the plate and process it. A typical result with the anticontaminator at room temperature is shown in Fig. 13.11. An image of a smaller hole is superimposed on that of a larger. This is the appearance of the larger hole after 2 minutes' irradiation; it has become smaller due to infilling by the deposition of contamination. The amount of contamination can be measured by measuring the diameter D_1 of the original hole and the diameter D_2 of the filled-in hole on the processed plate. If the time was 2 minutes and the magnification was 100,000, then the contamination rate in Ångstrom units/second is given by the expression:

$$\text{Rate} = 4 \cdot 16(D_1 - D_2) \text{ Å/second}$$

where the diameters D are measured in cm. If other magnifications or irradiation times are used, the constant 4·16 can be suitably altered.

This is a very convenient method for measuring contamination rate. The same plate also gives information on two other parameters: the rate of drift of the specimen; and the stability of the electronic supplies, obtained from the spacing of the Fresnel fringe.

The rate should be measured both with the anticontaminator cold and with it at room temperature. The actual figures obtained depend on the cleanliness of the column. After a column has been dismantled for cleaning, the rate will be very high, as water vapour and other gases are desorbed from the column. The figure may be in excess of 5 Å/second. This is completely unacceptable for high resolution work, but is hardly noticeable if the magnification is kept below 5,000. A reasonable figure for a clean column with no anticontaminator is about 0·5 to 1 Å/second. The anticontaminator should reduce this to below 0·1 Å/second. Care must be taken, however, to ensure that the opposite effect to contamination— 'stripping' (see p. 161)—does not occur.

CHAPTER 14

Photographic Techniques

14.1 Introduction

The image obtained on the fluorescent screen of an electron microscope is both impermanent and of poor resolution, due to the relatively large size of the phosphor crystals which make up the screen. If it is necessary to make a permanent record of the image, this can most conveniently and cheaply be done by taking a direct electron photograph. The high-energy electrons forming the image interact with photographic emulsions to form a latent image in much the same way as do photons of light, so an emulsion-coated film or plate can be placed beneath the projector lens and a direct recording of the image can be made by exposing the emulsion to the electrons for a suitable time. The plate or film (to which we will refer henceforth as 'the plate') can then be removed from the column vacuum, and processed in exactly the same way as a light photograph.

The photographic process can be broken down into three steps:
1. Exposure: the electrons forming the image are allowed to fall on the emulsion for a suitable length of time.
2. Development: the plate is removed from the column, the latent image is developed in a darkroom, and the resulting transparency is then fixed, washed and dried.
3. Printing: the plate bearing the transparent negative image is then placed in an optical enlarger, and a paper black-and-white positive to be viewed by reflection is exposed, developed, fixed, washed and glazed.

14.2 The Photographic Emulsion

Before an electron exposure can be made, a suitable photographic emulsion coated on to a suitable carrier must be chosen. A photographic emulsion consists of grains of silver halide held in a thin layer of solid

325

gelatin. A typical fine-grain emulsion consists of about 10 % of silver halide dispersed as discrete grains 0·1 to 0·5 μm in diameter in a gelatin layer about 25 μm thick. The latent image formed by electrons in the emulsion is identical with that formed by light, but its mode of formation is somewhat different. The reason for this is that electrons, being more massive, have very much greater energy than photons. When a single electron hits a single silver halide grain, it renders it developable to metallic silver. The 'quantum efficiency' is therefore very high—in fact, it is 100 %. A better recording medium for electrons than a photographic emulsion can hardly be devised. On the other hand, the quantum efficiency of a photographic emulsion to light is relatively low; a single halide grain must be hit by 10–100 photons before it can be developed. Therefore, the larger the grain, the more photons it will collect, and the faster will be the speed of the emulsion to light. Sensitivity to light varies as the cube of grain diameter. But because of the one-electron-one-grain relationship, all normal photographic emulsions have the same speed to electrons when fully developed. Extremely fine-grain emulsions such as high-resolution spectrographic ones are slower, because an electron may be able to pass through a very small grain without losing sufficient energy to render it developable.

This result—that all normal light emulsions when normally developed have the same speed to electrons—is a never-failing source of astonishment to those familiar with light photography. It can easily be verified by giving the same exposure successively to a slow bromide plate and to a fast pan-chromatic plate in an electron microscope. After giving each plate the correct development, it will be found that the only significant difference between the two images is that the high-speed one is very grainy. This is in spite of the fact that the panchromatic emulsion has 10^5 times the speed to light as the bromide. This is a fortunate circumstance for electron microscopists, for it means that one may work with light-slow, blue-sensitive emulsions by a fairly bright yellow safelight.

One of the most important characteristics of a developed image is its 'contrast', that is, the ratio of maximum to minimum optical density in the developed emulsion. The optical density in any given area of plate depends firstly on the total number of electrons which have hit it, and secondly on the degree of development which it is given after the exposure. Each electron striking the emulsion will give rise to at least one develop-able grain; it may rise to more than one if the path length of the electron in the emulsion is shorter than the emulsion thickness and the electron is completely absorbed. This is usually the case. If we assume that the number of photographic grains present in the emulsion greatly exceeds the number of electrons received, then the developed optical density is directly proportional to the number of hits, i.e. density is directly proportional to exposure time. This result is different from that obtained in light photography, where density is proportional to the logarithm of exposure. From this result it follows that all emulsions thick enough to absorb an electron beam completely will show the same contrast to electrons if fully developed. This is another result which the conventional light photographer finds very difficult to accept.

The most suitable emulsion for exposure to electrons therefore is one which has the finest grains to give a quantum efficiency of 100 %, and which is thick enough to absorb completely all the incident electrons. For the 20 to 100 kV electrons used in conventional transmission electron microscopy, the simple silver bromide emulsion used on transparency or 'process' plates has been found by experience to be best. This slow, blue-sensitive, very fine grain 'process' or 'line' emulsion is used commercially for making half-tone screen blocks for the type of photographic illustration used in this book, and for copying line diagrams. In spite of attempts to produce an improved electron emulsion, the 'process' or 'line' is by far the most widely used. All manufacturers produce an emulsion of this type. Various manufacturers' emulsions vary slightly in their development characteristics, so it is wise to decide on one particular type (generally the most readily available) and to stick to it. It is also advisable to order glass plates with 'no coated edges', since this avoids the uneven coating sometimes found on plates.

14.3 The Base

The bromide emulsion can be obtained coated on to glass plates, or on sheets of cellulose acetate or 'estar' film, or on rolls of acetate film. Each carrier or base has its own advantages and disadvantages. Glass does not absorb water, so plates desiccate rapidly and can at a pinch be put straight into the microscope column without prior desiccation. Glass is dimensionally stable, so the image does not change in size during processing. But glass is brittle, plates are clumsy to store, and the emulsion has a tendency to lift off glass plates on storage. Film, on the other hand, requires considerable prior desiccation, and exact measurements on the final image are not as reliable as those taken from glass plates. Film is cheaper and easier to store. Film is obtainable as 'cut film' in the same size as plates, or as 'roll film', which may be as wide as 90 mm or as narrow as 35 mm. Cut film base is usually obtainable in two thicknesses—0·005″ and 0·008″. The thicker is preferable. 35 mm film is generally placed close to the projector lens so that the whole of the image seen on the screen can be recorded. This means that a 35 mm film frame must be enlarged optically to a greater degree (generally $\times$ 8) than a plate or cut film, which is enlarged $\times$ 2 or $\times$ 3. Grain, dust particles, scatches and other imperfections if present are therefore more prominent in the final print. Most microscopists prefer the larger format for these reasons. Plates and cut film have the advantage that single exposures can be made, removed from the microscope and processed without waste of photographic material, and that each exposure can be given the optimum development time. Glass plates are now becoming very expensive and difficult to obtain, and are being replaced by cut film.

14.4 Exposure Time

Exposure time is governed by two conflicting factors. Ideally, the time should be made as short as possible in order to avoid image unsharpness due to specimen drift and instrumental instability. On the other hand, the electron beam is particulate and has a random distribution, which causes image unsharpness due to the form of graininess called 'electron noise'. The longer the exposure, the more the randomness will be ironed out, and the sharper will be the developed image. A further cause of image unsharpness is electron scatter in the emulsion. The average path length of a 50 kV electron in an emulsion is about 10 μm. It may be scattered parallel to the plate outwards from a sharp edge, leaving a trail of grains in its path. This makes the sharp edge unsharp, and image resolution is lost.

There is therefore an optimum exposure time and an optimum beam intensity which will give maximum signal-to-noise ratio and minimum instability effect. The time is not critical; it is around 0·5 to 2 seconds. It is therefore preferable to have some form of electrically operated mechanical shutter, since the minimum exposure time for an evenly-exposed plate which can be given by simple lifting of the screen is about 3 seconds.

Exposure time is also governed by the contrast of the specimen. It is here that the experience and judgment of the microscopist come in. In order to give a 'printable' negative the maximum developed optical density must not exceed about 2·0. The specimen may have a greater range of densities than this, especially when magnification is low. This excessive density range may be compensated photographically by reducing development time. But in order to record the information in the darker areas of the image, a longer exposure is required. Contrasty specimens therefore need longer exposure and shorter development. As magnification is increased, so image contrast is reduced. High magnification micrographs therefore needs the converse treatment—shorter exposure time followed by maximum development.

14.5 Exposure Measurement

It is usual to fix the time of exposure at 0·5 or 1 second and to alter exposure by varying the electron intensity of the image. Intensity can be conveniently estimated in any of three ways:

1. Measurement of total beam current from screen to earth;
2. Measurement of the total luminous output from the fluorescent screen;
3. Matching an area of the image on the screen with an illuminated optical comparator.

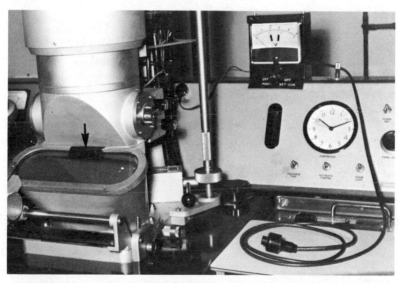

Figure 14.1   A screen luminosity meter fitted to an AEI-EM6B electron microscope. The cadmium sulphide photocell (arrowed) is fixed to the top of the viewing window with black sticky tape, and receives part of the total integrated light output from the fluorescent screen. The photocell output is measured on the meter mounted at the rear of the control desk. A reading is taken by pressing the button on the end of the extension cable; C_2 lens is then adjusted until the meter needle falls between the two black lines marked on the scale. The intensity is then correct for a preset 1 second exposure set on the shutter circuit. (For fuller details see Meek, *J. Roy. Micr. Soc.*, **88**, 419 (1968))

For the first method, it is necessary to insulate the screen from the column, and to allow the electrons striking the screen to leak to earth through a high resistance. The voltage developed across the resistance is then measured electronically with a valve voltmeter. This means that unless the instrument is fitted with an insulated screen by the makers, structural alterations have to be made, which are generally impractical. The second method is almost as reproducible, provided the luminous output from the screen remains constant. A small cadmium photocell can be stuck to the microscope window, and its output measured on a meter. Fig. 13.1 shows such an arrangement fitted to an AEI–EM6B instrument. The third method has an inherent disadvantage: which area to choose to match against the comparator? Should a highlight or a medium tone be chosen? The method clearly requires more judgment on the part of the operator than the other two methods, and consequently is not so widely used.

14.5.1 Exposure Procedure

The procedure for making an exposure in the electron microscope is described on p. 242. In the more sophisticated modern high-performance instruments, the shutter is electrically coupled to the exposure meter and the correct exposure, within limits, is given automatically. It is always necessary however to focus and check astigmatism, drift and stability.

Exposures generally have to be made with the room lights turned off. It is also necessary to have some method of numbering the exposures; it may be necessary to pencil the cassette number on a corner of the plate when loading the cassettes if no automatic plate numbering device is provided.

After exposing a plate, it is essential to record full particulars in a permanent notebook which is kept with the microscope. The notebook should be ruled in columns giving the following particulars: date; name of operator; negative number; cassette number; magnification; kilovoltage; block number; specimen; fixation; section stain; and general notes. Taking notes on loose sheets of paper is most unsatisfactory; all operators should use the microscope notebook, and abstract particulars for their own personal notes at a later time.

14.6 Development

After exposure, the photographic material is removed from the column and processed. It is not good practice to develop plates in the microscope room, because of the possibility of column contamination with dust from dried-up spilled solutions. It is advisable to leave the specimen in the microscope while the plates are being developed, so that any exposures which turn out badly may be rephotographed. The time taken to pump down to working vacuum after a plate change is generally about the same as the time taken to process the plates.

The development process is essentially one of chemical reduction. The developer is an organic reducing agent in an alkaline solution of the correct pH (generally about 10–11) for optimum activity. The solution also contains other components to prevent oxidation by the air, and to restrain the formation of 'fog' (see below). When the emulsion is placed in this solution, each grain which has been activated by a hit from an electron will begin to be reduced to metallic silver. The process starts from a nucleus in a grain, and spreads through the grain; it requires several minutes for completion. In addition, some grains which have not suffered hits will develop; this random silver grain formation gives rise to a slight uniform density over all the plate which is called 'background fog'. The effect can be reduced by suitable 'restrainers' in the solution. If development is carried beyond the point where each grain is fully developed, not only will fog increase, but the developed grains will begin to activate their neighbours, causing grain clumping. This will increase grain size in the processed emulsion, and resolution will be lost. There is therefore an optimum development time, which varies according to the emulsion and the type, concentration and temperature of the developer. With a very contrasty image, it is better not to carry development to completion, or the dark parts will be excessively dense. An image lacking in contrast, on the other hand, needs as full development as possible. For best results, therefore, development time cannot be standardized. In general, however, a set of plates taken of the same specimen at the same magnification can be processed together. It is better however to process plates taken at very high or very low magnification separately.

The object of development is to produce a transparency with a range of densities (contrast) which is suitable for transfer to a reflective paper to make a positive print, with the minimum loss of information. The print has an inherent disadvantage over the plate, because the ratio of reflectivity between the blackest possible black and pure white has a maximum of about 50:1, whereas the transparency range of a plate can easily be made as great as 1,000:1 or more, i.e. a maximum optical density of 3. All the information on such a plate can clearly not be transferred to paper, for a print made from it will either lose detail in the highlights (dark areas) or in the shadows (light areas). Two prints, one for highlight detail and one for shadow detail, would have to be made. Development should only be carried to the point where the transmission of the highlights is about 1/50 that of the shadows, i.e. a maximum optical density of about 1·5–2·0. Such a plate may look a little 'thin' to a light photographer.

Correct development clearly calls for a degree of judgment and experience on the part of the microscopist. It is the most important single stage in the photographic process, for it can make or mar an electron micrograph. A perfect print can only be made from a perfect negative. No amount of doctoring by so-called processes of 'intensification' or 'reduction' can help a poor plate. It is best to throw it away immediately, and return to the microscope and retake the plate. The choice of developer is in the hands of the microscopist; almost any standard developer is capable of giving excellent results when used correctly. If a developer is chosen which contains two or more reducing agents (e.g. metol plus hydroquinone), care must be taken to standardize development temperature accurately. This is because each component does not necessarily have the same temperature coefficient of activity, and different results may be obtained if the developer is used cold, and an attempt is made to correct for this by increasing development time, as is frequently advised. This is most particularly the case if quantitative results are sought. Developer is most conveniently obtained as a concentrated solution which only needs dilution with water at the correct temperature. A fresh batch of developer should always be made up for each batch of plates, and dissolving powders can be very tedious. It is very poor economy to try to develop a batch of plates in used developer; the cost of the plates may easily be one hundred times as great as the cost of the developer. Plate development is best carried out in stainless steel racks placed in plastic tanks of diluted developer. The degree of agitation given during development is as important as the temperature of the developer. If the solution is not frequently moved over the emulsion surface during development, patchily developed plates result. The easiest way to agitate a rack of plates is simply to lift them out of the bath and replace them. This should be done at half-minute intervals. Developer dilution should be such that full development takes about 5–10 minutes, so that contrast can be controlled by inspection of the plates as they are lifted out for agitation.

14.7 Fixing

After development has proceeded to the required degree, the unactivated, undeveloped silver halide grains must be removed from the emulsion in order to render the unexposed areas transparent and to prevent further image darkening by the action of light. This process is called 'fixing'; it must not be confused with the 'fixation' of tissue, which is a stage in the preparation of biological material for electron microscopy. Fixing converts insoluble silver halide into soluble silver thiocyanate by immersing the developed plate into a bath of sodium or ammonium thiocyanate solution, the latter being preferred due to its more rapid action. The fixer must contain an acid to neutralize any alkaline developing agent carried over from the developing bath, which might oxidize to colloidal sulphur in the fixing bath and give rise to stains on the fixed plate called 'dichroic fog'. It is good practice therefore to transfer developed plates firstly from the developer to a bath of weak acid; a 5 % solution of acetic acid is generally used. This stops the action of the developer immediately; hence the term 'stop bath'. The use of a stop bath prolongs the life of the acid fixer, which would otherwise rapidly become alkaline and cause staining.

After neutralization in the stop bath (a rinse of a few seconds suffices), the plates are transferred to the fixer. A fresh solution of ammonium thiocyanate will dissolve the silver halide and render the plates 'clear' in about 30 seconds. It is good practice, however, to add a gelatin tanning agent or 'hardener' such as 2 % potassium alum to the fixer bath and allow the emulsion to stay in the fixer for about 3 minutes in order to harden the gelatin before the final washing and drying. This protects it to a certain extent against scratching. The time for fixing is not in the least critical, once the emulsion has cleared. If, however, the plates are left in the fixer for a period of many hours, the contrast will drop slightly due to solution of the silver image. The fixer bath should be replaced with fresh solution when the clearing time has doubled.

After fixing and hardening in the acid-hardener-fixer bath, the emulsion gelatin must be washed free from trapped salts, or the silver image will fade with time. The rack of plates is transferred to a bath of gently running water and left for 10–20 minutes. It is important to see that the temperature of the washing water does not differ greatly from the temperature of the processing solutions, otherwise the gelatin may shrink in cold or swell in hot water, causing 'reticulation' or 'frilling' respectively of the emulsion. Since tap water usually contains dissolved substances, it is wise after washing to immerse the plates for 5 minutes in a bath of distilled water. The plates are then placed on edge on wooden racks, drained and dried in a warm, dust-free place. A note of the number of the rack should immediately be made in the microscope plate book to avoid confusion.

Roll and cut film is processed in exactly the same manner, except for the mechanics of supporting the non-self-supporting carrier. Roll film is best wound on to a plastic spiral and processed in circular tanks. Cut film can be hung in racks from stainless steel clips, or can be placed in special stainless steel or plastic holders.

Before drying, the plates should be laid with one edge supported by a strip of glass on a waterproof cold light box, and scrutinized minutely with a × 5 hand lens. Any defective plates, e.g. those showing astigmatism, out-of-focus, drift, uneven illumination, incorrect contrast, etc., should be thrown away immediately and taken again. Everyone makes mistakes sometimes, and only in this way can the microscopist preserve his reputation.

After the plates have dried, each must be given a serial number with Indian ink in one corner of the emulsion, unless the microscope has an automatic serial numbering device. The number must be recorded in the plate book at the same time. Each plate, cut film or strip of film should then be placed in a transparent envelope for storage, marked with the same serial number. This will protect the emulsion from damage. The bagged plates are then stored in boxes or more sophisticated filing systems; the manufacturers' boxes make excellent storage files.

14.8 The Enlarger

A positive print is next made from each plate by placing it in an optical enlarger and enlarging it to a convenient size, which may be whole plate, 6″ × 8″ or 8″ × 10″. The larger size of print is easier to examine and write on with wax pencil. This stage is essential to the amassing and codifying of information from the electron microscope; it is indeed the whole point of the rather lengthy and complex exercise.

The first essential is a suitable high-grade enlarger. It is astonishing how some purchasers will spend a vast sum of money on the finest electron microscope obtainable, and then 'economize' with a cheap enlarger, the lens of which cannot possibly image sharply the resolution present on the electron micrograph! It is essential to have an enlarger equipped with the finest lens which can be obtained, and fitted with a high-contrast point-source condenser system. The mechanics of the enlarger are immaterial, although the vertical wall-mounted type is preferable to the free-standing type for rigidity and lack of vibration. It is on the whole a waste of money to have an automatic focusing attachment on a plate enlarger, since it is generally used routinely at a fixed magnification of around × 2·5. An automatic focusing 35 mm enlarger, however, is well worth the money if 35 mm film is used routinely, since the focus setting is more critical. The enlarger lamp must on no account be a large area fluorescent type; this reduces image contrast by light scatter in the emulsion (this effect is desirable in portraiture but highly undesirable in electron microscopy). A clear tungsten bulb or a low-voltage point source lamp, preferably of the iodine–quartz type, should be fitted to the lamphouse, which must have a double-lens condenser without a diffuser. The light source and condenser must be set up correctly by focusing an image of the lamp filament on the plane of the aperture of the enlarger lens. Even illumination of the plate is essential; the lamphouse condenser must have a diameter significantly greater than the diagonal of the plate. If $3\frac{1}{4}'' \times 4''$ or quarter-plates are to be used, a $5'' \times 4''$ enlarger is necessary. For full plate coverage in this size, an enlarger lens of $6''$ focal length is necessary, which must be of the very highest definition obtainable. High quality $4''$ and $2\frac{1}{2}''$ focal length lenses are also desirable for higher magnification of selected areas of the plate. A $2\frac{1}{2}''$ lens will give a maximum magnification of $\times 25$ with the normal set-up. The enlarger magnification with each lens must be calibrated accurately on a scale alongside the plate carrier. The bromide paper must be held securely with a masking frame or easel on the table or enlarger base, and suitable processing trays for the solutions (at least 25 % bigger than the paper size to be used) must be provided.

14.9 Enlarger Magnification

The image on the photographic plate is made up of discrete grains which vary in size according to the density of the image. In the very dark parts of the negative, grain clumping inevitably takes place, giving grains 25 µm or more in diameter. In overdeveloped plates, grains may be more than twice this size. In the light areas of a negative, the grains may be 2 µm or less. The maximum magnification which can be tolerated without giving an impression of graininess in the print is simply the ratio of the size of the grain to the resolving power of the eye. Lightly developed areas will stand $\times 25$ or even more; heavily developed areas will stand up to $\times 10$. The enlargement required to bring a $3'' \times 4''$ plate up to $8'' \times 10''$ is 2·5 times; a 6×9 cm plate requires 3 times. Such a print can be scrutinized with a $\times 5$ hand-lens before grain can be made out. The grain which will be visible will in any case be more likely to be the defocus granularity. In the case of 35 mm film negatives, the magnification must be $\times 8$ for full $8'' \times 10''$ prints. Since the negatives have to be developed to higher contrast, grain may become apparent. Dust and other imperfections on the negative also become more obvious.

14.10 Printing

The plate or film is first placed in the negative carrier of the enlarger, emulsion side down. A glassless film carrier is to be preferred, to avoid the formation of Newton's rings. The lamp is turned on, the lens opened to full aperture, and the image focused accurately on to the surface of the easel using a reflecting focusing magnifier. The negative grain is focused on. The lens is now stopped down to an aperture previously known to give a paper exposure of 5–15 seconds.

The choice of paper grade is most important. Paper emulsion is made in various grades, which vary in the amount of exposure needed to give maximum black together with the barest discernible shade of grey. This is so that the contrast ratio of the negative can be matched to that of the paper. For example, if a negative is very 'thin', with the lightest part transmitting only 5 times as much light as the darkest part, then a paper must be used which gives maximum developed black with an exposure five times that necessary to give the first shade of grey. Such a paper is called 'very hard'; it matches the inverse 'very soft' negative. On the other hand, if the lightest part of a negative transmits 50 times as much light as the darkest part (corresponding to an optical density ratio of 1·7), then a paper must be used which gives maximum developed black with an exposure 50 times that necessary to give the first shade of grey. Such a paper is called 'soft', and matches a 'hard' negative. Paper can be obtained in the following grades: No. 0 (very soft); No. 1 (soft); No. 2 (normal); No. 3 (hard) and No. 4 (very hard). Stocks should be kept of Nos. 1, 2, 3 and 4 papers. No. 2 paper is suitable for negatives with a contrast ratio of 1·3–1·5, which should always be aimed at. Negative contrast can always be reduced by shorter or less active development; specimens of very low inherent contrast must have their contrast enhanced by maximum plate development followed by printing on hard paper.

The first stage in printing, therefore, is to examine the image on the easel (which should have a base made of white plastic) and to decide which paper grade to use. This comes only with practice and experience. Electronic devices can be obtained which measure image contrast and thus indicate the paper grade required. They do this by measuring the light intensity at the enlarger easel of the lightest and darkest part of the image, using a photocell of small area. Such devices can be of the greatest help to the beginner, but the experienced printer can choose the correct paper grade at once by inspection. The paper grade required depends also on the degree of enlargement of the negative. As magnification is increased, so contrast falls, and a harder paper is required.

The aim of printing is to transfer all the information contained in the negative to the paper. As we have seen, if the contrast ratio of the negative exceeds about 1·7–2·0, this cannot be done on a single print. It is essential that the tones in the print should range from the barest detectable shade of grey to full black, otherwise information will be lost. It is therefore essential to expose the paper so that maximum black is fully developed at the same time as the first grey tint is appearing in the lightest part. The paper must be developed fully; it is a good idea to give at least twice the manufacturer's recommended development time. There is therefore virtually no latitude in paper exposure·time; it must be correct to within 10 %. The harder the paper grade, the slower the emulsion and the more critical the exposure time. To obtain a perfect print for publication, it is wise to make several trial exposures of the whole field and examine each one critically when dry.

A print may sometimes be improved by giving differential exposure to selected areas in order to 'burn in' detail. If the plate was correctly exposed and developed, and the microscope was set up with correctly centred illumination, this trick, which is known as 'dodging' or 'shading', should be unnecessary; to use it is partly a confession of failure. The method is to cover selected areas of the image with an out-of-focus mask, which may be cut out of card or may be the bare hand. The mask is held about halfway between the enlarger lens and the easel, and is moved about slightly during exposure. In this way, uneven illumination can be compensated for. However, uneven plates should be recognized at the development stage and should not be allowed to enter the printing darkroom.

Prints are developed in a very active, contrasty, 'hard' developer. An overexposed print must never be 'pulled', or removed from the developer before it has been developed fully. This leads to poor, 'muddy' blacks; the resulting print lacks crispness. After full development (the image will look far too dark by the orange safelight) the print is placed in a dilute acetic acid stop bath, and is then fixed in the same ammonium thiocyanate hardener-fixer solution used for plates. After 2–5 minutes fixing, the print is washed very thoroughly in running water for at least 30 minutes (the paper base retains the processing reagents fairly tenaciously), and is then drained and glazed by pressing it on a hot stainless steel plate or drum. The print should be permanent, and of a quality suitable for reproduction in a journal if necessary. The negative number should be written on the back of the print; this is preferably done in soft pencil before the paper is exposed. It is convenient to have a rubber stamp which can be impressed on the back of the finished, glazed print, on which details such as specimen, block number, magnification, etc., can be recorded in a standard form.

Glossy prints can be written on with a soft-wax pencil of the type used for writing on glass. The marks can easily be removed if necessary with a soft cloth without damaging the print. The microscopist should examine each print with care, using a hand-lens and noting the salient features with the wax pencil. After noting details in a notebook, the prints can then be filed away on edge in folders in a filing cabinet. If kept on the bench, prints have a tendency to curl. Many thousands of prints can be amassed in a relatively short time from a well-used electron microscope, and an efficient filing system for prints and plates is essential. Plates must be filed away in a place where the relative humidity is moderate; if kept in a damp cellar, they will be attacked by fungus; if kept in a very dry place the emulsion tends to lift off the glass backing.

14.11 Automatic Processers

These are paper processing machines which will produce an almost dry print in a few seconds after the paper has been removed from the enlarger easel. A special paper is used which is available in 3 grades and has roughly the same speed as bromide paper. The paper emulsion contains the developing agent. It is developed by moistening it with an alkaline 'activator'. The paper is exposed in the enlarger in the normal way, and is then fed, emulsion side down, into a machine which consists essentially of a series of rubber rollers which dip into the processing solutions. The paper is rolled along over the upper part of the rollers, and the emulsion is moistened first by the activator, then by the fixer or 'stabilizer' and then by a water rinse. It is then squeezed almost dry between a final pair of rollers, and emerges in 8–10 seconds as a finished print. It requires only a short final drying on the bench, since the paper backing is hardly wetted.

This type of machine, of which the Kodak 'Autoprocessor' is an example, greatly lightens the work of printing, since all the operations of developing, fixing, washing and glazing are eliminated. The paper has a semiglossy surface, and the total output from a day's heavy use of an electron microscope can be printed in an hour or so. The snag is that there is no margin for error in exposure time, since processing is fixed and invariable. Dodging is of course possible, but really good prints can as always only be obtained from really good negatives. The prints tend to fade with time, and as the emulsion is very thick, they tend to curl heavily. They are quite adequate for the first assessment of electron micrographs; really good bromide prints can later be made of the best plates if required. The prints can be made permanent by fixing, washing and glazing, but this naturally removes most of the advantages of the system. The roller machine must be kept scrupulously clean, and the solutions must be changed frequently.

14.12 The Darkroom

The ideal arrangement is to have two separate darkrooms. A small plate darkroom, which need be little larger than one person can get into comfortably, is used for processing and reloading plates from the electron microscope. Processed plates can then be dried in a drying cabinet which can be placed in any suitable position in a corridor. The main darkroom must have a dry bench for the enlarger and paper, and a separate wet bench for paper processing, which should have two sinks, or one sink plus a washing tank. The glazing machine is best kept out of the darkroom, otherwise the heat from the drum makes the darkroom unbearably hot. The walls and ceiling of the darkroom should be painted white for maximum reflectivity from the safelights, which should be of the ceiling reflection type. It is also a great convenience to have in the main laboratory a large 'clean bench' where micrographs can be examined without the risk of having reagents spilled on them.

14.13 The Ideal Negative

Fig. 14.2 is a print from an electron micrograph of an unsupported thin section (c. 60 nm thick) of part of the cytoplasm of a glutaraldehyde–osmium fixed cell, embedded in Araldite and stained with lead hydroxide solution.

A microdensitometer tracing (Fig. 14.3) was made on the negative along the line marked A–B. This scan line covers the full range of densities on the negative, from the highest (a hole in the section) through mid-tones (bare section and lightly staining elements) to heavily electron scattering areas (glycogen and membranes), and includes the basic fog density. It will be seen from the tracing that the lightest (negative) cytological detail (heavily staining glycogen) has an optical density of about 0·6—well above fog level, which has a density of about 0·1. All the cytological information in the negative falls between the density values of 0·6 and 1·8. The density of the image of the hole (the unimpeded beam) is 2·2. The total density range between hole and glycogen is therefore 1·6, which is a transmission ratio of about 1:40. This is equal to the reflectivity ratio of the glossy paper. The whole of the information can therefore be accommodated on a single print on Grade 2 paper.

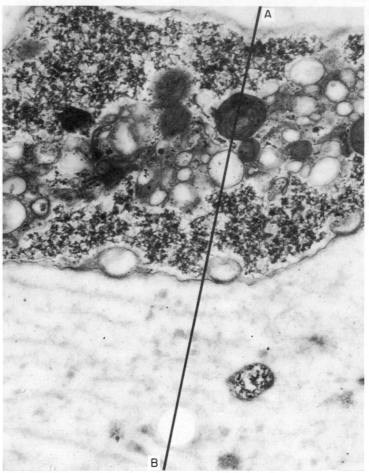

Figure 14.2   A print showing the full range of tones which should be obtainable from a correctly exposed negative correctly enlarged. The negative was exposed for 1 second in a Philips EM 200 electron microscope at an electron optical magnification of × 23,000 and at an average electron beam intensity at the screen of about 5×10^{-11} amp/cm^2. The specimen was an ultrathin (about 60 nm thick) section of part of the cytoplasm of a cell fixed in glutaraldehyde–osmium, embedded in Araldite and stained with lead solution. The negative material was Kodak Process Sheet Film, developed for 8 minutes in Ilford P–Q Universal developer diluted 1 + 24 at 20 °C. The print was made on Grade 2 Kodak WSG Bromide paper using a de Vere 5″ × 4″ 'Jualite' wall mounted enlarger equipped with a 150 W clear tungsten bulb and 2-lens condenser. The enlarger lens was a 6″ focal length Taylor–Hobson 'Ental II'. The paper was developed for 5 minutes in Kodak D–163 developer diluted 1 + 3 at 20 °C, fixed in Ilford Hypam with hardener for 2 minutes diluted 1 + 4, and glazed on a Priox drum glazer. × 31,000

If the original negative is underexposed, the contrast range will be diminished. In this case, one quarter of the exposure gave a negative with glycogen density of 0·2 (only twice that of the fog) and a hole density of 0·8—a total range of only 0·6. The negative required a Grade 4 paper, and

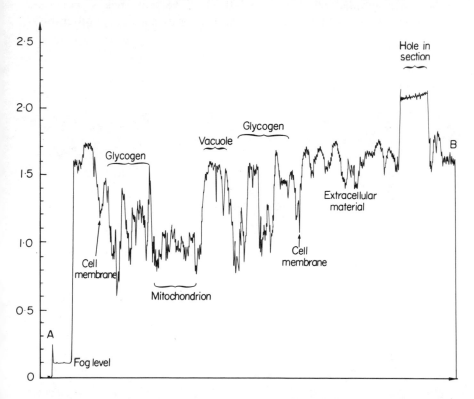

Figure 14.3   A microdensitometer tracing along the scan line marked A–B in Fig. 14.2. The whole of the detail falls within the optical density values of 0·6 and 1·8, a range of 1·6. This is a transmission ratio of about 1:40, and is about the maximum which can be transmitted to a paper print without loss of information

gave a 'muddy' and unsatisfactory print. Halving the exposure gave a negative with a density range of 0·95, which gave an almost perfect print on a Grade 3 paper.

If the original negative is overexposed, the contrast range will be increased, and may not be printable on a Grade 1 paper. Doubling the exposure in this case gave a density range of 2·0, and although the print on Grade 1 paper was almost perfect, some information was beginning to be lost in the blacks. The exposure time in the enlarger was also increased by a factor of 5 to 40 seconds, which is a considerable disadvantage.

It will be seen therefore that very careful control of exposure time and negative development procedure is essential for the production of first-class micrographs. The exposure latitude is very much less than that commonly found with the high-speed emulsions used in light photography.

CHAPTER 15

Maintenance and Fault-finding

15.1 Introduction

Considering its complexity, the modern electron microscope is remarkably reliable. It is in fact so reliable that care must be taken not to give it excessive maintenance. The most important precept to bear in mind is:

LEAVE WELL ALONE!

This should be printed large and hung up in every electron microscope room. It is especially necessary when the instrument is new and the person in charge is eager and zealous. It must always be remembered that when a column is opened and a part is removed for cleaning, this disturbs the instrument, and may at best cause a poor vacuum for several days, or at worst may cause a leak or a misalignment due to unskilful reassembly. It may take up to a week for a column to pump down to its ultimate vacuum, due to outgassing of cleaned components. Therefore, avoid the temptation to dismantle a column to 'clean' a suspected component until all other avenues have been explored, and it is absolutely certain that the column must be opened.

If you intend to do the major part of the maintenance of your own instrument, it is a very wise plan to attend the manufacturer's training course before the microscope is delivered. Then, while the microscope is being erected and commissioned, assist the engineers. This will enable you to find out if you have an aptitude for this kind of work. You must have a 'feeling' for fine instrument making, together with some basic engineering and electronic knowledge and a basic ability to use hand tools.

It is generally necessary to invest in a kit of first-quality tools. The manufacturer may provide everything that is needed, or he may only provide the special tools such as extractors and C-spanners. It is necessary to have a set of open-ended and ring spanners (of metric sizes for imported

343

instruments); Allen keys; a rubber hammer; properly ground screwdrivers of various widths and lengths; together with forceps, snipe-nosed pliers and other tools for fine work. The tools should be kept in a box in the microscope room and used for nothing else. A good engineer is never seen with an adjustable spanner or 'monkey wrench'; this is the trade mark of the bodger. Nothing looks worse than to see an instrument with nuts and screwdriver slots burred and scratched from the unskilful use of unsuitable tools. For the electronic circuits, a high-class, sensitive multimeter for measuring current, voltage and resistance is essential. An oscilloscope is so rarely needed that it can be borrowed when required. A fully equipped workshop is quite unnecessary.

When servicing an instrument, remember that far more damage has been done by unskilful servicing than was ever done by fair wear and tear.

AEI EM6B Record Sheet

Date_____

Record every 4th Monday morning.
Pump for at least $\frac{1}{2}$ hour before recording

1. Console
SET: H.T. OFF, C_1 AT 3, C_2 FULL, OBJ. FULL

VACUUM (PENNING) _____ SCALE UNITS.
FIL. METER _____ HOURS.
TIME RUN _____ HOURS.
LENS CURRENTS

kV	C_1	C_2	OBJ.	P_1 30	250	P_2 1·5	20	BEAM CURRENT H.T. ON, FIL OFF
30								
40								
50								
60								
80								

2. Power Unit Monitor Meter
SET: H.T. ON, FIL. ON (SAT.), BEAM CURRENT 200μA

kV	H.T.+ OSC.	FIL.* OSC.	1 (14)	2 (13)	3 (25)	4 (23)	5 (21)	6 (17)	7 (12)	8 (11)	9+ (49)	10* (19)	11 (-)
30													
60													
80													

+ Depends on beam current setting
* Depends on age of filament and setting of filament knob

Figure 15.1 An example of an electron microscope monthly record sheet

Think twice before dismantling a column or opening an oil-filled high tension generator tank. It is generally cheaper in the long run to call the service engineer.

15.2 Keeping Records

The key to proper maintenance is the continuous assessment of performance, and comparison with the performance of the instrument when it was new and perfect. This can only be done by keeping proper records. It should be the task of the person in charge of the instrument to monitor lens currents and compare the settings of the controls at several magnification settings and at all available kilovoltages. A specimen record sheet for an AEI–EM6B is shown in Fig. 15.1, and should be filled in and filed once per month. The specimen-change time must be monitored continually by the operators, and the fact reported when it becomes unduly long. This gives information on the state of cleanliness of the pumps. A log book must be kept for each instrument in which all the incidents in the life of the microscope are recorded. All filament changes are recorded in this book, together with filament life and full details of any breakdown which occurs. Symptoms, action taken, cause of breakdown and precautions to be taken to avoid a repetition of the breakdown should all be noted down. Each month, as records are being taken, the performance must be checked. Contamination rate must be measured, together with ultimate resolution. This rigorous routine checking takes about one morning per month, and is essential if unnecessary shut-downs due to neglect are to be avoided.

15.3 Filament Life

The filament is a hairpin of drawn tungsten wire about $0.005''$ (0.12 mm) in diameter. The thinner the wire, the smaller and more intense the cross-over source can be made, but the shorter the life of the filament. The life expectation is usually about 10–20 hours if the gun is used at high brightness at 100 kV at magnifications of 100,000 and over; it is 50 hours or more if the gun is used at low beam currents and moderate magnifications (up to 50,000). As the filament ages, the wire becomes thinner and its resistance increases, due to the evaporation of metal from its surface. This causes the voltage required to drive it to change. The actual voltage variation between a new filament and one about to 'blow' is not great; about 10 % is usual. The voltage can be monitored continuously by means of a meter, or more accurately by means of the position of the filament heating control. The most accurate method of estimating the life remaining in the filament is to measure the actual number of hours it has run; a time recording clock actuated by the filament on-off switch should be fitted. The clock reading is noted each time the filament is changed, and the average·filament life is noted. Short life is generally due to overheating by the operator.

15.4 Filament Changing

To change the filament it is generally necessary to fill the column with air so that the gun chamber can be opened. Fig. 15.2 shows the gun chamber of a modern instrument in the opened position. A few instruments have an airlock between the gun chamber and the rest of the column, so that only the gun chamber need be opened to atmosphere when changing the filament. This is a rather unnecessary luxury, as the filament in a properly used instrument should last a month or more. When the column is filled with air for a filament change, the opportunity should be taken to change the condenser and objective apertures.

In practically all modern instruments, the filament, filament holder and bias shield are assembled as a unit which plugs, screws or is bayonetted

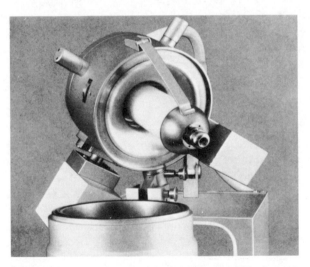

Figure 15.2   The gun head of a Siemens 101 electron microscope in the opened position. The gravity-operated earthing strip can be seen in contact with the shield. The filament plugs into a 2-pin socket and is factory pre-aligned. (Courtesy Siemens Ltd.)

on to the lower end of the gun insulator. Three contacts must be made: two for the filament and one for the bias shield. The manufacturer generally provides two complete filament-shield assemblies, so that a clean assembly can be kept fitted with a new filament correctly centred and ready for immediate use. Changing a blown filament is then simply a matter of filling the column with air and opening the gun chamber; unplugging the blown filament-shield assembly and plugging in the new one; then closing the gun chamber and pumping down again. The whole procedure should not take more than 15 minutes.

The blown assembly must be cleaned and repaired immediately. First remove the bias shield, taking care not to break the brittle remains of the old filament. Examine the old filament with a low power stereo-microscope. This will show if the filament was blown by misuse or if it expired of old age. Fig. 15.3(a) shows the appearance of a filament which has given good service before blowing. It is thinned regularly towards the tip, and has broken at the thinnest part well away from the tip, leaving a pair of almost imperceptible rounded droplets. Fig. 15.3(b) shows a filament

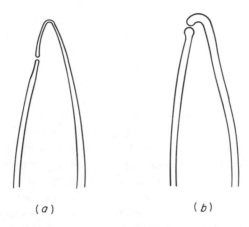

(a)   (b)

Figure 15.3(a)   A filament which has blown after a long period of use. Evaporation and gas' attack have thinned the point evenly away, and the break has occurred in one limb

(b) A filament which has been blown by accidental overheating. The wire has not thinned, and the break has occurred close to the tip. Approximately × 25

which has been blown by overheating. It has melted close to the tip, leaving a large pair of rounded droplets at the break. The wire shows no signs of thinning.

The inside of the bias shield will be covered with a bluish coating of evaporated tungsten plus contamination. This coating is most easily removed by soaking the shield in 5% aqueous ammonia solution for 5 minutes followed by polishing with a non-abrasive polishing medium such as powdered chalk or fine alumina. Use a pad of cotton wool rolled on to the end of a chisel-pointed throat swab stick. The swabs sold for cleaning babies' ears make excellent polishing pads. The abrasive must be softer than the stainless steel bias shield or it will become scratched. Be very sure that the aperture in the bias shield is absolutely clean; this is the most important part. Rinse all the polishing medium away with fresh 5% ammonia made up in glass distilled water and allow it to dry in a dust-free place. Handle it from now on with nylon gloves or lint-free paper ('Velin'). The shield must be free from grease, fingermarks or dust specks; these act as nuclei for electrical discharges between shield and gun chamber, causing H.T. instability. Dirt in the shield aperture will similarly cause instability in beam current, giving rise to illumination flicker.

A new filament must now be inserted into the filament holder and clamped firmly. Some instruments use factory precentred filaments; on others the filament must be set for height and centration by means of screws which move the filament axially and laterally with respect to the bias shield aperture. The exact method of centration varies with the design, and the maker's instructions must be followed very carefully. The bias shield must now be fitted to the filament holder, which, being in a field-free place, can be handled with the bare fingers. The outer surface of the bias shield must never be handled with anything which may contaminate it. While screwing the shield down, watch the filament tip very carefully through the bias shield aperture, making sure that the delicate filament wire does not come into contact with the shield. The tip must now be positioned correctly in the exact centre of the shield aperture, using the centration and height adjustments, or the illumination will be uneven. The height must be correct, or the beam current and hence the brightness will be incorrect. The tip position in the shield aperture is critical to within 100 μm, and should be set using a low-power stereo microscope. When the filament change has been completed, wrap the bias shield assembly in lint-free paper, place it on a glass Petri dish and cover it with an inverted beaker. Store in a dust-free place ready for immediate use when required. It is wise to clean the anode at the same time as the filament is changed.

15.5 'Preflashing' Filaments

The filament itself (see Fig. 6.6) is spot welded to two thick tungsten support wires which are in turn firmly held in a sintered glass or ceramic base. The spot welding process gives rise to stresses in the thin filament wire which are relieved when the filament is first heated to operating temperature. The stress relief usually causes movement of the filament tip away from its carefully preset position in the shield aperture. If the movement is more than a few micrometres, the illumination spot will become asymmetrical and the gun chamber will have to be opened again in order to recentre it. This can be avoided if the filament is heated to operating temperature in a vacuum to relieve thermal stresses before it is mounted in its holder. This procedure is called 'preflashing'. Some manufacturers treat all replacement filaments in this way; others do not. Enquiries should be made. If trouble is found with filament movement directly after a filament change, the operator should preflash them. It is a simple matter to make a dummy filament holder to fit into the bell jar of a vacuum evaporator. An externally variable source of about 5 V maximum must be available within the bell jar. Connect the filament to the source and pump the bell jar down to a pressure of 10^{-4} torr or better. Bring the filament to red heat (about 1–1·5 V) and leave it for 5 minutes. Then increase the voltage to bring the filament up to white heat (about 2·5 V) momentarily. Be careful not to blow it. Reduce the temperature slowly. This procedure should prevent further thermal movement when the filament is brought into use in the electron microscope.

After a new filament has been fitted into the microscope, always run it at a low voltage (red heat) for a few minutes before bringing it up to operating temperature. This procedure has been found to increase the life of the filament materially.

15.6 Pointed Filaments

The smaller the area of the filament actually emitting electrons, the more coherent and intense the source will be. The area at the tip of a bent hairpin can be considerably reduced by 'pointing' the filament (Fig. 15.4). The stages in accomplishing this very delicate operation are shown in Fig. 15.5. The circular section filament tip (a) is first squeezed very tightly between polished hardened steel cheeks in a vice or press to flatten the metal and cause it to flow into the shape shown in (b). Excess metal is now ground away using a stone or very fine revolving emery wheel to give the chisel-pointed shape shown in (c), side view at (d). This type of chisel or lancet pointed filament when used in conjunction with a small (500 μm diameter) shield aperture gives a very intense and coherent beam, but filament life is necessarily rather short (10 hours or less). Pointed filaments are a great advantage for very high magnification work where ultimate resolution is sought, but are of little or no advantage in routine medium-magnification work.

Figure 15.4   A lancet-pointed filament used in the Siemens 101. (Courtesy Siemens Ltd.)

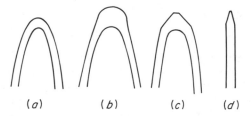

Figure 15.5   The stages in making a pointed filament. The tip is first squeezed in a vice to squash it flat (b). The flattened point is then ground to a chisel shape (c), side view (d)

15.7 Column Cleaning

The parts of the column which are exposed to the electron beam slowly become coated with contamination—the brown, tarry substance derived mainly from the hydrocarbon pump oils. The greater the electron bombardment, the greater the rate of build-up. The parts which contaminate most rapidly are the lens apertures, especially the condenser; the gun components, especially the anode and the inner surface of the bias shield aperture; the condenser lens polepieces; the gun chamber and insulator; the anticontaminator device and the specimen holders. Contamination in the parts of the column below the objective lens should be negligible.

It is a good plan to change the anode and the condenser apertures each time the filament is changed. Dirty condenser apertures make the illuminating beam astigmatic and reduce the brightness of the focused beam, although they do not degrade the final image. Dirty objective apertures and anticontaminators show up at once by the astigmatism they cause in the final image. They must be changed immediately. Dirty gun components cause random illumination flicker. The frequency with which cleaning of column components is necessary varies with the length and kind of use of the instrument (high or low beam, high or low resolution) and on the design of the microscope. Some gun designs are far more susceptible to contamination on the insulator and bias shield than others. Objective apertures, anticontaminators and specimen holders need frequent changing or cleaning (weekly, possibly for high resolution work even daily). The gun and bias shield need monthly cleaning, but the inner column components such as lens polepieces should not need more than annual cleaning.

Contamination is best removed from large surfaces by dissolving it in 5 % ammonia made up in glass distilled water. Very heavy deposits (which should never have been allowed to accumulate in the first place) are removed by abrading them with a very soft, fine powder such as chalk or alumina. Never under any circumstances use any abrasive which might scratch the surface of the component. Cleaned components must never be handled with the bare fingers; they must always be picked up with nylon gloves or, better, with lint-free paper.

15.8 The Insulator

It is essential that there should be no conducting path between the electron gun and the earthed gun chamber. The gun itself is therefore mounted on a large vacuum-tight ceramic holder, the insulator. This may have a glazed or unglazed surface, which gradually acquires a coating of contamination. Sooner or later this surface coat becomes conducting, giving rise to a barely perceptible leakage of high tension to earth. This causes H.T. instability with resulting image degradation, finally leading to catastrophic 'flash-overs' when the H.T. is applied, rendering the instrument unusable.

When H.T. instability ('gun flicker') occurs, the surface of the insulator should be carefully examined. Conductive paths can frequently be traced on its surface as fine, branching brown streaks. The insulator must then be cleaned, or the contamination may become burnt into the surface. It may be possible to clean it *in situ*, or it may be necessary to dismantle the gun head completely and remove the insulator. Because of the presence of cable insulating oil and special seals, the latter procedure is generally beyond the capabilities of most operators, and should be left to the service engineer. If it can be cleaned *in situ*, first cover the open gun chamber with a blanking plate and a clean towel to prevent the ingress of cleaning fluids into the column. Remove the filament-shield assembly and clean the insulator by judicious rubbing with cotton wool and 5 % ammonia. Do not use an abrasive unless absolutely necessary, and then only powdered chalk or Vim. After cleaning, rinse very thoroughly with glass distilled water and dry with hot air from a hair dryer. Replace the filament-shield assembly, close the gun and pump down. Make absolutely sure no water has entered the column before closing the gun. In general, it is inadvisable to attempt to clean unglazed insulators *in situ*, although this may be done successfully with glazed ones.

15.9 Objective and Condenser Lens Apertures

These are made by drilling minute holes in discs or strips of molybdenum or platinum; the latter metal is preferable. The commonest method of cleaning the apertures is to heat them to yellow heat in a vacuum. This method is unsatisfactory, because some of the components of contamination are not volatile, and burn into the metal. Also the method encourages crystal growth of the metal, making the circular apertures out-of-round. Molybdenum especially grows pointed crystals into the holes, which, being only 50 μm in diameter in the case of objective apertures, are useless after a few cleanings. A better method is to use hydrofluoric acid. The procedure must be done in a fume cupboard while wearing rubber gloves. First place about 1 ml of concentrated hydrofluoric acid in the bottom of a small polythene beaker. Place the dirty apertures on a strip of molybdenum and heat them to redness for a minute in an oxidizing bunsen flame. Drop the red-hot apertures into the acid, which will penetrate into the tiny aperture bores and attack the contamination. Leave for one minute, then dilute the acid with 5 ml of glass distilled water. Leave 5 minutes, pour off the acid and replace with 5 % ammonia solution. Transfer to a small glass beaker and boil for 5 minutes. Then boil in glass distilled water, remove the apertures and allow them to dry in a covered Petri dish on lint-free paper. Examine the dry apertures (handle on clean paper with clean No. 4 forceps) with a $\times 50$ stereo microscope, and reject any with dirty or misshapen bores. Finally, try each aperture out in the electron microscope and test the astigmatism before storing them, preferably by sealing each individual aperture in a thin glass tube. Do not store cleaned apertures in gelatin capsules. Condenser apertures do not require such rigorous treatment. The hydrofluoric acid method is best used with platinum apertures; molybdenum is slightly attacked. Apertures which still contain dirt after this treatment cannot generally be reclaimed, although it is worth trying further cleaning by heating in vacuum. Platinum apertures respond well to the hydrofluoric acid cleaning treatment, which can be applied indefinitely.

15.10 Lens Polepieces

The objective lens polepieces should under no circumstances be dismantled by the operator unless he is confident that he knows exactly what he is doing. They are mated together in pairs, and are aligned to each other for minimum astigmatism. Cleaning them can make astigmatism worse. Condenser polepieces are far less delicate, and require cleaning more often. It should not be necessary to dismantle the objective lens for cleaning more often than once in every two years, provided that volatile specimens which give rise to contamination such as methacrylate sections are kept out of the column. Condenser polepieces are protected in some instruments by a thin bronze tube passed down the centre, which can easily be removed for cleaning. The actual polepieces in such a design should not need to be touched.

Polepieces are cleaned after removal from the lens by treatment with 5 % ammonia and powdered chalk, as was described for the bias shield. Great care must be taken to ensure that they do not fall on the floor, or excessive astigmatism may result from slight distortion caused by the blow. Since they are made of soft iron, great care must be taken to prevent rust. If polepieces are to be stored, they should be liberally coated with diffusion pump oil and stored in sealed polythene bags.

15.11 Viewing Screens

These are generally made of thin aluminium sheet coated with a very thin layer of an activated zinc–cadmium sulphide fluorescent powder which is held on the surface by a trace of an adhesive such as collodion. Many different powders are available, having colours ranging from a dull olive-yellow to a bright apple-green. All have similar light outputs for a given intensity of electron bombardment; the colour is a matter of personal preference. Screens must be handled with the greatest care; the lightest touch on the surface will affect light output from the disturbed area. Specks of dust may only be removed by blowing them off with a bulb blower, never by brushing with either a camel-hair brush or a feather.

After a screen has been in use for some time, the centre will become discoloured to a dirty brown, and the light output from this area will fall off. A spare screen should always be kept as a replacement for immediate fitting, but it is a fairly simple matter to resettle the old screen. First, soak it in acetone until the old coating is soft enough to be rubbed off with soft tissue or cloth. Do not scrape it; this will scratch the surface. Lay the cleaned screen face up in a large Petri dish with a lid. Next, make up a suspension of about 5 % phosphor powder in a 1 % solution of collodion in redistilled acetone. Shake the suspension vigorously, allow the larger particles to settle, and then pour it quickly into the dish at the side of the screen. Rock the dish to ensure even distribution, then cover the dish and leave until all the powder has completely settled. Next, remove the liquid by pipetting it off, taking the greatest care not to disturb the settled particles on the screen. Allow the screen to dry without removing it from the dish. Store it in a clean Petri dish sealed with adhesive tape, or in a special box.

Lines indicating the photographic area may be scribed on to the surface using a mounted needle, but this is not to be recommended unless a special jig is available. The centre point is best marked by burning-in a very tiny spot with the microscope beam. First, align the column accurately. Then focus the diffraction spot at maximum beam current, and withdraw the condenser aperture momentarily. This should burn a spot about 0·5 mm diameter at the exact optical centre of the microscope for convenience in alignment.

15.12 The Camera

Cameras are frequently removed from the column and handled, and contain shaft seals on controls which are frequently moved. They are therefore potential sources of leaks and contamination, and the mechanism may jam if neglected.

The camera body should be inspected frequently for any chippings from glass plates, paint from cassettes or flakes from the fluorescent screen. These must be carefully removed, and the camera blown out with dust-free low-pressure dry air. Cassettes must be inspected frequently for flatness, bent lugs, damaged paint, etc., and replaced if necessary. The camera mechanism may or may not require lubrication; the maker's instructions must be scrupulously followed. If a lubricant must be used, only the barest trace of high vacuum grease or diffusion pump oil can be tolerated. The main sealing ring must be inspected frequently and any adherent dust particles carefully removed. Great care must be taken not to allow used grids to get into the camera mechanism or to stick to the sealing rings.

The camera mechanism may jam during operation. This is almost invariably due to incorrect loading of plates in cassettes, or cassettes in the camera body, but may be due to clumsy, hesitant or too rapid manipulation of the cassette changing control. It may be due to glass chippings fouling the mechanism, or to bent cassettes or lugs. Never use force on the camera mechanism, or expensive damage will ensue. Always exercise patience and restraint, remove the camera and inspect it for the cause of jamming. In general, the fewer the moving parts in a camera, the better. Roll film cameras are generally less prone to jamming than plate cameras.

15.13 The Vacuum System

The vacuum system needs attention if (*a*) the column will not pump down to its ultimate vacuum, or (*b*) if the time taken to pump down after changing the specimen or the plates becomes excessive. Maintenance consists of (*a*) tracking down leaks and replacing the faulty seals; (*b*) dismantling and cleaning the diffusion pumps; and (*c*) changing the oil in the rotary pumps and replacing the driving belts. The pumps should not need cleaning more than once a year.

15.14 Leaks

Excessive column pressure is almost always caused by a leak. However dirty the pumps may be, they will pump a leak-free system down to ultimate vacuum if given sufficient time. Leaks occur almost invariably at movable vacuum seals. Fixed seals, such as the O-rings between lenses, hardly ever leak spontaneously. The exception to this is in the condenser lenses, where the seals are constantly bombarded by X-rays and in time crack owing to hardening of the rubber. When the microscope is new and known to be leak-free, it is a good idea to make a note of the rate of rise in pressure when the pumps are shut off from the column with all photographic material removed. If the Penning gauge is connected directly to the column, it is a simple matter to close the main vacuum valve and take the pressure gauge reading at ten-second intervals. If the gauge is connected to the pumps, a less accurate reading can be obtained by closing the valve, waiting 1 minute and then opening the valve quickly. The amount of pressure rise before the pumps start to reduce the pressure gives a good indication of whether a leak is present in the column or not.

Leaks are hunted by manipulating suspected seals such as aperture drives, fluorescent screen spindle, camera drive, specimen airlock etc. while watching the pressure gauge closely. A faulty seal will generally show up by corresponding movements of the gauge needle. Small leaks are infuriatingly difficult to find, and do not always show up in this way. It may be necessary in the last resort to dismantle the column lens by lens, closing off with blanking plates and new seals and testing at each stage. In this way, a leak can be traced to a particular unit. It is then best to dismantle the suspected unit completely, replacing every seal. Remove O-rings only with the thumbnail or a chisel-pointed wooden stick. New O-rings must always be examined carefully with a low-power microscope before fitting them; flaws are not uncommon. The grooves in which the O-rings seat must be carefully cleaned with a grease solvent such as chloroform, using a moistened pad of cotton wool on a wooden stick. Fixed seals should be assembled dry, without grease. Movable seals require a trace of grease to reduce friction. Only the barest amount to moisten the O-ring may be used, and the grease must be vacuum grease and not lubricating grease, which contains volatile components. Special hard rubber O-rings may be needed in some situations. Experience and intuition play a great part in tracking down leaks. If a leak cannot be found quickly by the operator, the service engineer should be called. Much damage can be caused by unskilled 'leak detection'. It may be necessary for the engineer to use a small portable mass spectrometer which detects hydrogen. This is connected to the vacuum system of the electron microscope, and a fine jet of hydrogen is played about each seal in turn. This very sensitive method requires elaborate equipment which is normally only available to the manufacturer.

15.15 Pump Cleaning

If the general pumping speed becomes slow, as indicated by a doubling of the time needed to return to normal working vacuum after a specimen change, the pumps will probably have become dirty. The oil used as pumping fluid in the main diffusion pump breaks down with time into a hard, tarry substance that is deposited on the cooled walls and umbrellas of the pump. The annular space between umbrella and wall becomes choked, and the pumping speed falls. At this stage, the pumps will still attain their ultimate vacuum eventually, but if the tar is not cleaned out at this stage, the ultimate vacuum may also deteriorate. To clean the pumps, they must first be allowed to cool to room temperature, after which the whole vacuum system is filled with air. This generally entails opening the high vacuum valve and then letting air into the column. Since there is generally an interlock to prevent this from happening accidentally, some means, which the manufacturer may or may not specify in the handbook, must be used to overcome the interlock. It is unwise to admit air by breaking an O-ring seal; the uncontrolled rush of air may carry the O-ring into the system. Having let air into the system, the cooling water supply is disconnected, the diffusion pump is disconnected at the high and low vacuum joints and electrical connections and is then dismantled on the bench. It may be necessary to scrape the hard tar off the umbrellas with a blunt knife. Great care must be taken not to scratch or distort them. If scraping is necessary, maintenance has been delayed too long. All the inner surfaces of the pump are then polished with a scouring powder such as Vim, after which all parts are rinsed and dried in a cool oven. Coat all parts with fresh diffusion pump oil before assembly. Fill the pump by pouring a fresh charge of oil in from the top, and replace all the O-ring seals when reassembling it into the microscope.

Mercury pumps generally have small venturi jets which tend to become blocked with a solid dross. This is very easily cleaned out with a pipe cleaner. Nothing should be done to a steel-bodied mercury pump other than to brush away the dross lightly and refill with fresh, redistilled mercury. Do not forget to clean the trap between the oil and mercury pumps; pumping speed is reduced by the condensation of mercury on its surfaces.

If the pressure in the backing line becomes poor, this is generally due to the presence of volatile substances which have condensed in the rotary pump oil and which the pump is unable to remove. If a gas ballast valve is fitted to the rotary pump, it is a very good plan to run the pump with the ballast valve open and the high vacuum valve shut for an hour or so each week in order to sweep the accumulated volatiles (mainly water) out of the oil. When the oil eventually needs replacing, this is simply done by placing a tray beneath the pump to catch the old oil, removing the drain plug and draining it out. Turn the pump over by hand several times to expel the old oil from the rotor. Place about a quarter of a fresh charge in the pump, turn it over by hand to flush out all the old oil, and drain out the flushing charge. Refill with fresh oil. Do not under any circumstances use motor car engine oil; this has not been vacuum distilled and contains volatile fractions. When replacing the rotary pump oil, examine the vee-belt which drives the rotor, and replace it if it shows signs of fraying. The capacitor–start A.C. electric motor generally has factory sealed bearings and needs no maintenance. After a rotary pump has been in continual use for many years, the clearances may wear to an excessive degree and the pump can no longer attain its ultimate vacuum. It is then necessary to replace the complete pump for a new or reconditioned one.

Pump cleaning and oil changing should be done annually, whether slow pumping symptoms arise or not. It is better not to allow the oil diffusion pump to become excessively dirty. Pump cleaning is a messy job, and is best done by a service engineer during the annual visit.

15.16 Water Cooling Circuits

Water filters should be installed with a pressure gauge on both inlet and outlet side (Fig. 9.1). When the pressure drop across the filter exceeds 10–20 p.s.i., the filter should be dismantled and cleaned. The frequency of cleaning depends on the cleanliness of the mains water supply. It should not be necessary more than twice per month.

Water passages inside lenses should be checked from time to time to ensure that water can flow freely through them. In some microscopes, the water passages are connected in series, in others in parallel. With the latter system, one lens can easily become blocked without being noticed. Blocked water passages can generally be cleaned by forcing a weak solution of a proprietary motor car radiator cleaning fluid through them with a hand pump. Water passages are frequently fabricated by soldering a metal plate on to the lens casing. Care must be taken to avoid using too much pressure or the lens may be distorted or, worse, the casing split. Care must be taken to ensure that the water passages do not become immovably blocked, or the lens will have to be returned to the manufacturer for repair.

It is far better to instal a closed-circuit refrigerated cooling system, which obviates blocking of lens water passages.

15.17 The Electronic System

Modern electronic circuits, especially transistorized ones, are almost incredibly reliable. They should, and generally will, go on working for ten hours a day, year after year, without major breakdown. In general, the only components which need replacing are the reference batteries, valves and relay contacts. Reference batteries have a predictable life, and should be replaced at the intervals stated by the manufacturer, even if the voltage appears to be correct. A very careful check must be kept on the setting of the focus control. This is the most sensitive indication both of high tension voltage and of lens currents. These are tied to reference battery voltage, and experience will soon tell which batteries are causing the drift. Valves, on the other hand, are unpredictable and fail suddenly. The lives of some valves are shorter than others; this is where the service engineer's knowledge and experience comes in. Some heavily-used valves need annual or even more frequent replacement; others last for many years. A complete set of spare valves must be carried; it is generally only necessary to have one of each type. If a valve breakdown is suspected in a circuit, first examine it when running to see if any of the heaters is not glowing. Next, switch the circuit off and feel each valve in turn. A valve which is not emitting will be colder than the others. Finally, replace the valves one by one with known good spares. If the fault is not cured, replace the original before substituting the next. If this procedure does not cure the fault, then the valves are not to blame, and it is better to call the engineer. If a fault is known to be in a particular circuit, the engineer may be able to bring a complete, tested circuit along and substitute it. This is becoming an increasing tendency in electronic circuit maintenance, rather than the engineer attempting to repair the circuit on site. A factory repair is much more likely to be lastingly satisfactory than a site repair, since more rigorous testing can be applied at the factory.

The key to proper electronic circuit maintenance is the keeping of proper records, and observing trends. There is no substitute for this.

15.18 Fault Finding

It would not be practicable to give a universally applicable fault-finding chart, since electron microscopes vary so much in detailed design. Faults however fall into a basic system of classification, and pointers to the basic causes can be suggested from which the actual fault can be determined by deduction from the information given in the instruction manual. It is always a good idea to pinpoint a fault to a basic cause even if the service engineer is to be called. It may save him a visit, as he may be able to diagnose the fault over the telephone and tell the customer exactly what to do. If he has to come, it will enable him to bring any requisite spare parts or reconditioned circuits with him. In either case, the serviceman's time and the customer's money will be saved.

15.19 Classification of Faults

The commonest faults can be classified into the following categories:

1. There is no image when the H.T. is switched on.
2. The image is distorted in some way.
3. The image or illumination is unstable.
4. The image suddenly disappears when operating.
5. The image drifts sideways.
6. The image will not focus.
7. The illuminating spot will not focus.

The causes of these faults will now be examined in greater detail.

1 No image when H.T. is switched on

(*a*) Is the main electricity on? Check meter readings.
(*b*) Has the water supply failed? Check pressure and flow.
(*c*) Is the vacuum sufficient? Check pressure gauge readings.
(*d*) Is a safety device, e.g. a switch on a cabinet door, operating?

The basic precept to follow here is to ask: 'What did I do last?' If the instrument was operating satisfactorily before, e.g. the specimen was changed, then it is 99 % likely that the fault is due to some procedure in the specimen change having been forgotten or incorrectly performed.

If all appears to be in order, check the following:

(*a*) Is the screen down?
(*b*) Is the auxiliary screen stopping the beam?
(*c*) Is the intermediate screen obscuring the beam?
(*d*) Is the shutter closed?
(*e*) Are the stage controls centred?
(*f*) Is a grid bar in the way?
(*g*) Is low magnification selected?
(*h*) Is C_2 focused?
(*i*) Is there beam current?
(*j*) Is the filament voltage correct?
(*k*) Has the filament blown?

If these all check, then there must be something in the path of the beam. Check the following:

(a) Remove the specimen and try again.

(b) Remove in turn: objective, diffraction, condenser apertures and anticontaminator and try again.

If there is still no screen glow and it is certain that the gun is producing a beam, some part of the column, probably the gun, is grossly misaligned. Switch off all lenses at minimum beam current and look for the central spot. Realign the column as described on p. 249.

2 The image is distorted

§ 2.1. The image is out of focus in one direction. The basic cause is astigmatism, due almost certainly to the presence of non-conducting material in or close to the beam in the neighbourhood of the plane of the specimen. Examine in turn for dirt:

(a) Anticontaminator blades—ice formation?

(b) Objective aperture—does the out-of-focus axis move as it is moved?

(c) The specimen.

(d) The specimen holder.

(e) The objective polepieces.

§ 2.2. The image changes in magnification radially outwards. This is true distortion—see p. 84. The cause may be:

(a) The intermediate lens is being used too close to crossover;

(b) The currents in the intermediate and projector lenses are not balanced for minimum distortion;

(c) The wrong projector polepiece is in use.

§ 2.3. The image changes in magnification and focus from the centre outwards—this is chromatic change in magnification. The causes are:

(a) The specimen is too thick;

(b) The H.T. voltage is too low.

3 The image is unstable

§ 3.1. The illumination flickers with little or no change in focus or magnification.

(a) The bias shield aperture is dirty.

(b) The filament is old, thin and about to blow.

(c) The gun insulator is dirty.

(d) The gun chamber is dirty.

(*e*) The condenser apertures are dirty.

(*f*) The condenser polepieces or screening tube are dirty.

§ 3.2. The image changes in magnification and focus without rotating: The H.T. voltage is varying at random. Examine reference batteries and any valves in the H.T. generator circuits.

§ 3.3. The image changes in magnification and focus and also rotates about the screen centre: the current in one or more of the lenses (most likely the objective lens) is varying at random. Examine reference batteries and valves in the lens generator and reference circuits.

4 *The image suddenly disappears when operating*

(*a*) The filament has blown.

(*b*) The water pressure has dropped.

(*c*) The water filter is choked.

(*d*) A fuse has blown.

(*e*) The mains electricity has failed.

(*f*) The vacuum has failed.

(*g*) A safety device has operated.

5 *The image drifts sideways*

(*a*) The specimen support film is broken.

(*b*) There is a large particle of dirt on the specimen.

(*c*) The specimen support film is not adhering to the grid.

(*d*) The grid is not properly clamped in the holder.

(*e*) The holder is not correctly placed in the stage.

(*f*) The stage controls have been strained beyond their stops.

(*g*) The bearing surfaces of the stage are dirty.

6 *The image will not focus*

(*a*) The specimen is in the wrong plane—bent, incorrectly clamped; holder incorrectly set in the stage, inserted upside-down, or bent.

(*b*) The imaging lens currents are incorrect—suspect reference batteries, valves.

(*c*) The value of H.T. is incorrect—suspect reference batteries, valves.

7 *The illumination spot will not focus*

(*a*) The condenser lens currents are incorrect—reference batteries or valves.

(*b*) The value of H.T. is incorrect—again, reference batteries or valves.

15.20 Spare Parts

It is essential to carry a reasonable stock of commonly required spare parts to avoid possible lengthy waits for spares which may be temporarily out of stock at the agent or maker. The following spares are the minimum which should be carried:

1 full set of tools;
1 amps–volts–ohms multimeter;
12 filaments;
1 bias shield assembly;
6 of each size of aperture used;
1 full set of sealing rings and gaskets;
6 of each of the commonly used O-rings (specimen airlock, camera, gun, aperture drives);
1 heater for each diffusion pump;
1 charge of fluid for each diffusion pump;
1 of each type of electronic valve;
12 of each type of fuse;
1 of each type of indicator lamp;
1 gallon of rotary pump oil;
1 gallon of transformer oil for H.T. oil tank;
1 tube of high vacuum grease (NOT silicone);
1 vacuum gauge head;
1 set of column blanking plates.

15.21 Maintenance Contracts

Most manufacturers prefer to enter into a maintenance contract with the user. There is generally a guarantee period of one year during which time the instrument is serviced free of charge. This enables the user to see just how much maintenance is required, and whether he or his staff can cope with it unaided. Maintenance contracts range from one annual visit lasting about a week to clean the column and pumps and to check performance, to full contracts where the engineer can be called an unlimited number of times. The services of an engineer are charged at £40 per day or more plus expenses, so an unlimited contract may cost £1,000 per annum. A contract for a single annual clean and check would cost around £100–200, depending on the complexity of the instrument.

Whether or not a maintenance contract is entered into depends on the skill and experience of the technical staff responsible for the instrument. Each microscope has its own peculiar common faults which can be recognized instantly by a competent serviceman, and he can put them right in the minimum of time. Such faults can be very obscure, and might take a skilled electronic engineer unfamiliar with electron microscopes a week to diagnose and ten minutes to repair. The cost of such lengthy shut-downs must be weighed against the cost of a service contract. In general, most users would be wise to contract for an annual clean and check. Any fault which the technical staff cannot deal with must then be repaired by calling an engineer. The disadvantage here is that this may entail a wait of several days if all the available engineers are out on contract work, which is generally given preference. The quality of service varies from manufacturer to manufacturer, and one of the points to bear in mind when choosing an electron microscope is the quality of the service available. This can only be discovered by personal enquiries from a number of users.

CHAPTER 16

High-voltage Electron Microscopy

16.1 Introduction

The term 'high-voltage electron microscope' (subsequently referred to as HVEM) is generally applied to a conventional flood-beam TEM operating at an accelerating voltage greater than 100 kV. The 'megavolt' HVEMs currently obtainable commercially are designed to operate between 200 kV and 1,000 kV (1 MV) in steps of 100 kV, although there are two or three research HVEMSs in existence which are designed for 3–10 MV.

Three major advantages are to be gained by raising the electron accelerating voltage. Firstly, specimen penetration is increased, enabling thicker specimens to be examined. Secondly (and somewhat paradoxically) specimen damage by the beam is reduced. This is because energetic electrons spend less time in the vicinity of specimen atoms, and thus interact less with them. Thirdly, resolving power is increased, due mainly to the reduction in chromatic aberration at the image arising from the reduced probe–specimen interaction. A major disadvantage inseparable from the reduced specimen interaction is a considerable diminution in image contrast, which has so far limited the major applications of the HVEM to the field of metallurgy and materials technology.

16.2 Increased Penetration

It was this aspect which led to the initial interest in the HVEM. Before it was possible to cut thin sections, very few transmission specimens could be examined. As early as 1947, van Dorsten and le Poole were experimenting with higher accelerating voltages in attempts to penetrate bacteria and metal foils. The development of satisfactory thin-sectioning techniques in the early '50s together with improved thinning techniques in metallurgy led to a waning in interest in the HVEM. The present-day requirements of metallurgists and materials scientists in particular, however, have led to a great upsurge of interest in the development of the HVEM.

The thickness of a crystalline metal sample which can be examined at 1 MV is increased by a factor of about 5 over the usable thickness at 100 kV, depending on the orientation of the crystal lattice. It is possible to penetrate a thickness of 10 μm of aluminium and still obtain a usable image. This is a greater increase than predicted by theory. It makes specimen preparation much simpler and quicker, and makes it easier to find such things as lattice defects in a specimen. The benefits of increased penetration are stimulating the construction of 10 MV TEMs. The usable thickness of an amorphous biological plastic-embedded section is increased even further: from 1000 Å at 100 kV to 5 μm or more at 1 MV: an increase of 50 times. However, this greatly increased specimen thickness brings great difficulties in image interpretation in its train. Because of the enormous depth of field of the TEM (see p. 90), the whole thickness of the specimen is in simultaneous focus at the image plane. The morphological information in the thick section is therefore presented as a superimposed clutter which is difficult or even impossible to sort out and interpret. However, the possibility of using thick sections has led to a revival of interest in the three-dimensional techniques of high-angle specimen tilting or stereoscopy (see p. 265). The thicker the section, the greater is the 3-D effect obtainable from a stereo pair. Provided the specimen embedded in the plastic is of high contrast and is of an openwork nature, e.g. chromosomes, spindle fibres, isolated membranes etc., excellent stereo pairs can be obtained from sections 2–5 μm thick. The three-dimensional distribution of the stained objects in the unstained plastic matrix can be visualized directly without the tedium of serial sectioning. Useful results are currently being obtained on the Golgi apparatus (stained by the original metal impregnation techniques), chromosomes and spindles, and plant cell walls.

A disadvantage of increased penetration is the reduced interaction of the high-energy electrons with both the fluorescent screen and the photographic emulsion. This leads to marked loss in screen brightness at 1 MV, and the exposure time for the usual type of photographic recording emulsion has to be increased by a factor of 5 or more. Both effects can be reduced by increasing the thickness of material struck by the imaging electrons by a corresponding factor of 5 or so. This unfortunately tends to lead to loss of micrograph resolution and difficulty in focusing precisely.

16.3 Specimen Damage

When an energetic electron beam interacts with a specimen, irreversible structural changes take place which are of two main types. Firstly, specimen atoms are ejected from their original positions and come to rest randomly. This is known as 'displacement damage' and is the principal form of damage suffered by a crystalline specimen. It is of great importance in metallurgical samples. Secondly, electrons are ejected from specimen atoms, giving rise to ionization. This in turn affects the electron distribution in the covalent bonds which occur predominantly in non-conducting biological specimens, and leads to rapid degradation of molecular structure. This effect has been measured on crystalline non-conducting polymers by recording the length of time a specimen may be irradiated before the electron diffraction pattern of the polymer finally disappears. Using polyethylene crystals, the disappearance time was found to be 50 s at 100 kV and 250 s at 1 MV—an increase of a factor of 5. An increase of a factor of 3 was found for lead acetate crystals. The electron intensity in μA per unit area was the same in each case. Protein and enzyme crystals show a similar effect.

16.4 Resolving Power

A significant improvement in instrumental resolving power may be obtained by increasing the electron accelerating voltage. This is due to the reduction in wavelength of the electron 'matter waves' due to the increase in the 'relativistic mass' of the electrons as the velocity approaches the velocity of light. The equivalent wavelength at 100 kV is 0·037 Å, but at 1 MV it is 0·0087 Å and at 3 MV is 0·0036 Å. As we have already seen, the instrumental resolving power of the TEM is determined theoretically mainly by two factors: the spherical aberration coefficient and the optimum aperture angle of the objective lens. Spherical aberration unfortunately increases as voltage is increased, due to the increased focal length of the larger lenses necessary; the increase between 100 kV and 1 MV is a factor of about 2. But the optimum aperture angle can be shown to diminish with increasing voltage, and since this enters into the relationship as the cube of its numerical value, the final result is an increase in theoretical resolving power of about 2·5. Calculated r.p. is about 2·7 Å at 100 kV and 1·1 Å at 1 MV.

However, as has already been pointed out (see p. 275), the actual image resolution obtained in practice from biological ultra-thin (approx. 600 Å thick) sections is at least a factor of 10 worse than the instrumental resolving power. It is seldom possible to achieve 20–25 Å using 60 kV. The chief reason for this is chromatic aberration introduced by the specimen (see p. 87). A proportion of the electrons scattered by the specimen is bound to enter the aperture of the objective lens, since only fairly widely scattered electrons are in practice removed from the transmitted beam by the physical aperture disc. These scattered electrons have lower velocities than those of unscattered beam, and therefore suffer chromatic aberration as they pass through each imaging lens in turn. A point scattering source in the specimen will therefore be imaged as a disc of finite diameter. It is possible to calculate the diameter of this disc for differing specimen mass-thicknesses and for differing accelerating voltages. At 100 kV, a thin film specimen of graphite 100 Å thick gives rise to a disc 8 Å in diameter. A 1,000 Å thick specimen increases this by a factor of 10 to 80 Å. This observation is frequently known as 'Cosslett's Rule', and states that the resolution obtainable from a specimen of density = 1 is about one-tenth of the specimen thickness.

As the accelerating potential is raised, the energy loss in the specimen decreases very rapidly up to about 200 kV, and then falls more slowly to a constant minimum at around 3 MV. A 1,000 Å thick graphite specimen has a chromatic aberration disc of approximately 80 Å dia. at 100 kV, 8 Å at 500 kV and 3·6 Å at 1 MV. The resolution obtainable in practice from a 'thick' specimen therefore increases by about a factor of 20 when the operating voltage is increased from 100 kV to 1 MV.

16.5 Image Contrast

As we have seen in Chapter 5, the amplitude contrast which gives rise to the final TEM image is produced by electron scatter alone. As the accelerating voltage is increased, specimen interaction is reduced, and therefore both scatter and final image contrast are reduced. A conventional 600 Å thick Araldite section of biological tissue stained with osmium, uranium and lead and which gives excellent images at 60 kV, disappears completely at 500 kV. It is of very little practical value to increase resolving power if during the process the image can no longer be perceived. Two methods can be used to increase contrast in the HVEM, both of which unfortunately reduce resolution. The first is simply to increase specimen thickness, in which case one is more or less back to square one, but with the advantage of increased penetration and the possibility of obtaining stereo pairs. The second is to use the dark-field technique (see p. 257) in which the direct transmitted beam is blocked out and only the scattered electrons are used to form the final image.

The result is a marginal increase in resolution at 1 MV over that obtainable by conventional bright-field techniques at 100 kV.

16.6 Environmental Specimen Microchambers ('Wet Cells')

One advantage of the greatly increased penetrating power of high-voltage electrons is that the specimen may be sandwiched between two thin windows of evaporated carbon and may in this way be removed from the colum vacuum. It may then be surrounded by a liquid or gaseous atmosphere at atmospheric pressure, and dynamic studies of processes such as oxidation, reduction, nitriding, hydration, corrosion etc. may be made. An improvement for use with gaseous atmospheres is the 'window-less' chamber, in which the carbon windows are replaced by two objective aperture discs drilled with 50 μm holes, and spaced 1–2 mm apart. The gas leak from the holes can be dealt with by an auxiliary pumping system on the specimen stage chamber, and the column vacuum is not significantly degraded. These chambers are further discussed on p. 403.

These 'wet cells' hold out the tantalizing hope to biologists that it might be possible to study living cells in their natural aqueous environment, but the results obtained so far have been very disappointing. The lifetime of an enzyme molecule exposed to 1 MV electron bombardment is only of the order of seconds, and in any case the problem of obtaining sufficient contrast from unstained living cells seems to be insuperable. This limits the maximum usable magnification to about × 1000. Radiation-resistant bacteria have been placed in such cells, and have been removed after imaging at 1–3 MV. They have then been cultured and some were found still to be viable and continued to divide. The interpretation of such experiments is, however, high controversial, and no significance has yet been attached to the results.

16.7 Instrumentation

High-voltage electron beams require larger and more powerful magnetic fields to focus them. Also, the X-rays given off by the interaction of the probe beam with the apertures, the specimen and the fluorescent screen are extremely penetrating, and require very heavy shielding. The lead-glass window through which the screen is viewed is generally some 25 cm thick. The lenses for the present generation of HVEMs are designed on conventional principles, but are three to four times the diameter of lenses designed for 100 kV. A typical objective lens with stage is about 80 cm in diameter, 50 cm high and weighs about half a tonne. The column,

Figure 16.1   The column of the high-voltage electron microscope at the C.N.R.S. Laboratories, Toulouse, France. The
massive size of the instrument, designed by Professor Dupouy
and Dr. Perrier to run at 1·5 million volts accelerating potential,
can be gauged from the man standing beside it. This rear view
shows the diffusion pump and vacuum manifold, and the pile of
lead bricks shielding the viewing chamber. (Courtesy Prof.
G. Dupouy and *J. Microscopie*)

comprising two condenser lenses, three or four imaging lenses, the viewing chamber and camera system, is about 3 m high. The high tension is generated and stabilized in a linear accelerator tank filled with an insulating gas such as sulphur hexafluoride under pressure. This is mounted above the gun and is supported on a sub-floor which must be built into the building housing the HVEM (see Figs. 16.1, 2 and 3). The complete installation is around 10 m in height and weighs 20–30 tonnes. It costs around £300,000 or 'a dollar a volt', and requires a vibration-proof concrete building costing as much again as the microscope itself.

Figure 16.1 is a photograph of the Toulouse 1 MV instrument, the first to operate continuously at this voltage. The size of the microscope and its ancillary pumping equipment can be judged from the man standing

Figure 16.2   A photograph of a model of the GEC-AEI million volt electron microscope, the EM–7. The two water-cooled cylinders above are the Haefely million volt generator and stabilizer. To the left of the column and main control desk stands a fork-lift truck for lens removal; each lens weighs of the order of half a tonne. A television remote control operating desk is on the extreme right, beyond the lens current generating cabinets. (Courtesy GEC–AEI Ltd.)

beside it. The column alone weighs over 4 tonnes; the objective lens has a focal length of 4 mm and weighs over a tonne. The high-voltage generator and stabilizer are mounted on a floor above the column, and the whole instrument is housed in a spherical building about 20 m in diameter.

Figure 16.2 illustrates a model of the GEC–AEI EM–7 instrument. The overall height exceeds 10 m, over half of which is taken up by the Haefely 1 MV generator and stabilizer. The column is 90 cm in diameter, and each lens weighs half a tonne. The fork-lift truck shown on the left is used to transport the lenses when the column is being worked on. The operator can sit at the console directly, or can operate it remotely from a TV-equipped console on the right. This is in order to remove him from the hazard of secondary X-ray emission if this is thought to be necessary. The viewing window is of lead glass 30 cm thick. Six filaments are carried in a rotating turret in the gun, so that is unnecessary to dismantle the column to change a blown filament. The filament is powered by nickel–cadmium batteries contained in the accelerator tank, and the turret can be changed through an airlock without depressurizing the gas (SiF_6 or Freon) in the tanks. The accelerator is a 24-stage Cockroft–Walton type with claimed stability of ± 3 parts per million per minute. A continuous voltage range between 100 and 1,250 kV is available. The quoted resolving power at 1 MV is 5 Å; the magnification range is from $\times 63$ to $\times 1,250,000$ with rotation-free imaging from $\times 4000$ upwards. At the time of writing, a total of 10 of these instruments have been sold in the U.K. and the U.S.

Figure 16.3 shows very effectively the size of the JEOL JEM–1000D. The 14-stage Cockroft–Walton generator and stabilizer are contained in the upper pressure chamber filled with Freon at 3 kg/cm². 500–750–1,000 kV are available with a stability of 2×10^5 per minute. The magnification range is from $\times 1000$ to $\times 150,000$ and the claimed resolving power is 7 Å.

Figure 16.4 shows the column of the experimental 3 MV HVEM in Prof. Dupouy's CNRS Laboratory, Toulouse. The outside diameter of the lenses has now increased to about 1 m; the objective lens of 10·5 mm focal length weighs 2,240 kg. The stainless steel tank of the Cockroft–Walton generator is 3·5 m in diameter and 8 m high and is filled with SiF_6 at a pressure of 4·8 atm. The current stability is better than 1 part in 10^6 over a period of 10 minutes. The advantages to be gained by increasing accelerating voltage from 1 MV to 3 MV are not very great. Penetrating power is increased still further; the usable thickness of a silicon specimen can be increased to about 15 μm. Chromatic aberration at the specimen is even further reduced; it is about 65 times less important at 3 MV than at

Figure 16·3 The JEOL-JEM-1000D million-volt electron microscope. The platform supports the pressurized Freon-filled high-voltage generator-stabilizer-accelerator tank mounted above the column. The overall size can be gauged from the seated operator. (Courtesy Japan Electron Optics Labs. Ltd.)

Figure 16.4 The column of the experimental Toulouse 6-lens 3 MV electron microscope. The pillars support the generating, stabilizing and electron-accelerating equipment mounted above the column. The increased size of the lenses can be gauged by comparison with the Toulouse 1 MV instrument shown in Fig. 16.1. (Courtesy Prof. G. Dupouy and CNRS, France)

100 kV. The theoretical resolving power is not significantly different from that at 1 MV due to the increased focal length of the objective lens. The overall sharpness of images of metals is significantly better, however, due to the reduced chromatic effect. Contrast in biological specimens is even less than at 1 MV, and the dark-field contrast-stop method is almost essential.

16.8 Conclusions

Because of its potential practical value in examining thick specimens of constructional materials such as alloy steels, nuclear reactor fuel cans, semiconductor materials, cements, carbon fibres, ceramics, polymers, sintered materials, natural minerals and so forth, intensive HVEM research and development are taking place at several centres throughout the world. The most important of these are the CNRS laboratories at Toulouse, France, under Prof. G. Dupouy; the Cavendish Laboratories in Cambridge, England, under Dr. V. E. Cosslett in collaboration with GEC–AEI; and in the commercial laboratories of Hitachi and JEOL in Japan.

So far, only three manufacturers have entered this field, all of whom offer instruments costing around one dollar U.S. per volt. These are GEC–AEI with the EM–7 (U.K.), the JEOL JEM–1000 (Japan) and the Hitachi HU–1000 (Japan). Hitachi are also currently designing a commercial 3 MV instrument.

The cost of a megavolt HVEM plus its building and ancillary equipment is such as to put the purchase of an instrument out of the reach of all but the very wealthiest institutions. Most are owned by Governmental departments or industrial consortia. They are generally operated on a time-sharing basis, frequently in collaboration with local Universities, and some instruments work 16 hours a day, seven days a week.

In spite of the cost disadvantage, there are about 40 known HVEMs in the world operating at above 500 kV. There are 12 in Japan, 11 in the U.S.A., 8 in the U.K. and 4 in France. The total investment involved must be around £30 million, a sum which would have purchased 1,000 conventional top-class 100 kV TEMs. Some of the information in the fields of metal and materials technology which has so far been published could not have been obtained by any other means. The amount of significant biological information which has been published is disappointingly little, and could in the main have been obtained at 100 kV, though with the expenditure of greater effort. One wonders what lies in the future.

CHAPTER 17

Special Methods in Biological Electron Microscopy

17.1 Introduction

As has already been pointed out in Chapter 3, the conventional transmission electron microscope (TEM) which we have so far been discussing in this book is only one of a family of electron microprobe instruments which can be classified under various combinations of fixed beam or scanning beam, transmission imaging or reflection imaging, point probe or flood probe. The TEM is a fixed-beam, transmission-image, flood probe electron microanalyser. We will first discuss at greater length the other instruments in the family and their relative advantages and disadvantages to the biologist, and then discuss particular special methods which can be applied to the TEM instrument itself or to its image in order to achieve particular results.

17.2 The Scanning Transmission Electron Microscope (STEM)

In the STEM (Fig. 17.1) a focused probe spot of the smallest obtainable diameter and the highest obtainable brightness is scanned in a raster across a thin specimen identical with a TEM specimen. The transmitted electrons are simply captured in detectors placed beneath the specimen, there being no imaging lenses behind the specimen. The electrical output from each detector modulates the brightness of a display spot on a separate cathode ray tube. The display spot is scanned in exact synchrony with the probe spot across the face of the display tube. Resolving power is determined ultimately by the diameter of the probe spot, and image magnification is determined simply by the ratio of the length of the display line on the face of the image tube to the length of the probe line on the surface of the specimen. Overall magnifications of the order of millions can be obtained without difficulty and without the use of imaging lenses.

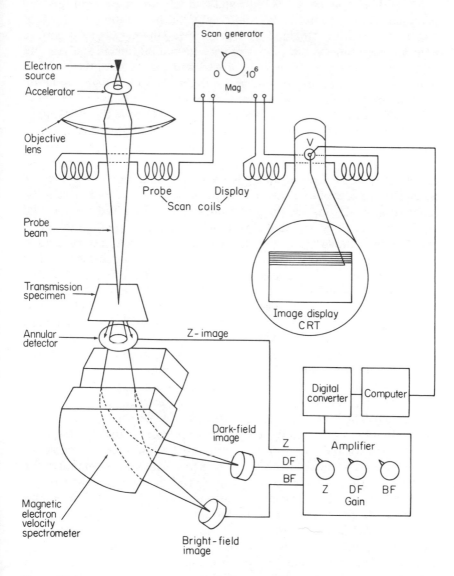

Figure 17.1 A block diagram showing the principle of the scanning transmission electron microscope (STEM)

It might be thought at first sight that building up an image in this way is a crude technique not to be compared with the 'continuous' form of picture information obtained with the TEM. This may have been true a decade ago, but the techniques of spot formation, image analysis and image processing have been so improved and refined over the past few years that almost all the advantages in the field of ultra-high-resolution electron microscopy now lie with the scanning mode of image formation.

Fig. 17.1 shows a diagram of the optics of the STEM in the very sophisticated high-resolution form which is currently under intensive development. The electron probe spot is formed from a cold cathode, point-source triode gun and is focused on the specimen by one or possibly two magnetic lenses. The actual electron source is a very fine tungsten point as used in an emission electron microscope (see p. 70), etched to a radius of a few hundred Å. Electrons are dragged from the tip by an electrostatic field formed by an electrode placed close to it. Because the tip radius is so very small, a quite moderate voltage (3 kV) forms a colossal electric field at the point. The relative brightness of this type of source measured in amperes per unit solid angle can be 10^5 times as great as that of a conventional hot tungsten filament (thermionic source) and 1,000 times as small (see p. 398). The electrons are then accelerated to 50–100 kv. A single high-quality TEM objective lens suffices to focus an intensely bright electron spot of 2–5 Å dia. on the specimen. The disadvantage of this type of source is that it requires a vacuum a million times better than that of the conventional TEM, due to instability caused by ionization of residual gas. The necessary vacuum hardware in the form of ion, sorption and turbomolecular pumps has only recently become available (see p. 399). The lens used is a conventional TEM objective lens used as a diminishing lens; it is equipped with a stigmator to make the spot as circular as possible and also has a set of deflection coils to enable the probe beam to be moved from side to side to generate the scanning raster.

The ultimate resolving power of the STEM depends primarily on the size of the probe spot. This cannot be made infinitely small, first because of diffraction effects leading to the formation of an Airy disc (see p. 38) and secondly because of spherical aberration in the probe-forming lens. This lens appears from the standpoint of conventional optics to be a condenser lens, because it focuses an image of the electron source onto the specimen. In point of fact, it is the objective lens of the STEM, because all the morphological information is obtained from the illumination system. Helmholtz's principle of reciprocity states that all beam paths in an optical system must be reversible (see Figs. 1.11, b and c). If source and image positions are reversed in the TEM, the condenser becomes the objective; this is the situation in the STEM. Therefore, the resolving power of the STEM is limited by the aberrations in the probe-forming system. Just as in the TEM, this is around 2Å, imposed by spherical aberration.

Let us now consider what happens to the electrons transmitted by the specimen. These may be classified into three broad categories, as shown in Fig. 5.1. Firstly, a proportion of incident electrons passes straight through the specimen without interacting with it. These are called 'undeviated primaries'. Secondly, some incident electrons interact with the electron cloud or shell around one or more atoms in the specimen. These incident electrons share their energies with specimen atom electrons, which may be raised to orbits of higher energy or else be ejected as secondary electrons. The resulting energy loss causes the incident electron to be deviated from its path but it is also reduced in velocity due to the loss of energy. These transmitted electrons are called 'inelastically scattered primaries'. Thirdly, a small proportion of the incident electrons passes close enough to the nuclei of specimen atoms to be deflected through very large angles, even through 180°. These electrons share no energy with the massive nucleus but are deviated through much wider angles than inelastically scattered ones. These electrons are called 'elastically scattered primaries'. Electrons scattered through more than 90° are said to be 'backscattered'.

The transmitted electrons, bearing information about the specimen in the form of angular deviation and velocity loss, are now passed through two filters in sequence. The first separates wide-angle electrons from narrow-angle simply by their angular deviation from the axis. The electrons which pass through the angular filter pass into a velocity filter or 'energy spectrometer'. This is a sector-shaped magnetic field which spreads the narrow-angle electrons into a velocity spectrum. The undeviated primaries are therefore separated from the inelastically scattered electrons. Three signals emerge from three suitably placed detectors, corresponding to the undeviated, inelastic and elastic transmitted electrons. Each signal is amplified separately, and modulates the brightness of the recording spots on the faces of three separate synchronous display tubes. The significance of the three display images is as follows. The undeviated electrons are those which pass into the image-forming lenses of the conventional TEM; this image is therefore the conventional or 'bright-field' image which the eye can interpret straightaway. The inelastic signal is composed of the electrons which have been subtracted from the bright-field signal; they therefore form an image of reversed contrast called the 'dark-field' image (see p. 258). It is at very much higher resolution than the dark-field image which can be obtained from the TEM because chromatic aberration has been eliminated. The elastic signal corresponds to the distribution of electrons which have interacted with atomic nuclei. The higher the Z-number of a specimen atom, the greater will be the probability of this occurring. This signal therefore corresponds to the distribution of heavy metal atoms in the specimen, and is called the 'Z-signal'.

These three simultaneously displayed images are capable of yielding far more information about the specimen than the single TEM image. The signal from the detectors is in the form of a voltage varying with time—a so-called 'analog' signal. This analog signal can be 'digitized' by dividing the range from full black (zero signal) to peak white (maximum signal) into a number of 'grey levels' (the eye can distinguish 10 to 12 steps of contrast), each of which is assigned a numerical value. The resulting digital signal can then be processed by a programmed computer. It is then possible to combine the dark-field and bright-field signals so as greatly to reduce the signal due to the substrate on which the specimen is mounted, and thus to increase the contrast of the bright-field image. It is possible to 'clean up' the signal and to remove random pulses from meaningful ones. Contrast can be increased, decreased or reversed; and the signals can be recorded on magnetic tape and played back over and over again, thus enabling transient phenomena to be recorded and analysed at a later date. The signal from one scan line or part of a scan line can be extracted and displayed as a curve on another display tube, thus enabling direct measurements to be made.

The field emission electron source has several advantages over the conventional hot filament source. As well as being much smaller and brighter, it is much more highly coherent (see p. 27). Chromatic aberration or 'energy spread' arising from random fluctuations in the energies of the emitted electrons is much less, which contributes to the attainment of high resolution in the electron spectrometer. From a practical point of view, the life of a field emission tip is far longer than that of a conventional hot filament, provided that it does not become contaminated. Minor contamination can easily be removed by heating or 'flashing'. Source life is estimated to be months rather than hours. The STEM has a number of minor advantages, mainly in ease of operation. As magnification is changed, there is no variation in brightness of the image on the display screen, there is no change in image focus and there is no image rotation or inversion. Due to the elimination of the imaging lenses, the effect on resolution of the chromatic aberration introduced by the specimen is greatly reduced and much thicker specimens may be examined.

On the debit side, however, the high-resolution STEM suffers from much greater complexity. The cold cathode field emission source requires a vacuum of at least 5 orders of magnitude better than the hot filament gun: 10^{-12} of an atmosphere as against 10^{-7}. This necessitates the use of a special pumping system for the gun part of the column, involving heaters for outgassing the metal parts, plus ultra-high vacuum pumps such as ion pumps and sorption pumps. The specimen chamber does not require this ultra-high vacuum; it can be isolated from the gun chamber by a very small aperture just big enough to allow the passage of the electron beam. Gas molecules which diffuse back from the specimen chamber are then dealt with by the ultra-high vacuum pumping system. The electronic system is far more complex than the TEM; as well as stabilizers for the accelerating voltage and objective lens current it requires all the scanning, processing and display electronics, which can be as complex as the user requires and could involve 6 or more display screens and several computers.

The principle of the STEM was first demonstrated in a crude form by von Ardenne in 1938, but the modern version with velocity analyser was developed in the late '60s by Professor A.V. Crewe and his associates at the University of Chicago. Results presented so far have shown that it is possible to image single atoms of high Z-number mounted on a very thin (20 Å) carbon substrate. The specimen used was a long-chain polymer of benzene tetracarboxylic acid linked by thorium or uranium atoms spaced 10 Å apart. Single silver atoms on a carbon substrate have been successfully imaged and have been shown to be labile under electron bombardment. Crewe and his colleagues have also shown the immense potential of the STEM in imaging specimens of extremely low inherent contrast. This is made possible by the digital nature of the signals from the detectors. It is possible to combine the elastic, energy-loss and unscattered signals from the energy spectrometer in such a way as to provide optimum specimen contrast with minimum background noise from the carbon substrate, and thus unstained biological macromolecules can be imaged. Preliminary results on unstained native DNA clearly show information on strandedness, helicity, base separation and flexibility. By reacting the DNA with other molecules, e.g. polypeptides and enzymes, the sites of activity can be demonstrated. The potential of this instrument in biology is immense; it may well prove to be the greatest single step forward since the original TEM work of Ruska and his colleagues in the '30s.

The great advantage of the STEM for high-resolution biological work is its ability to analyse the image into separate signal channels, all of which are free from chromatic aberration and which can be processed and displayed separately and simultaneously to yield information which it is impossible to obtain from the single signal channel of the TEM. It appears probable that the resolving power of the STEM will be limited ultimately by degradation of the chemical bonds in the specimen due to interaction with probe electrons—an interesting application of Heisenberg's principle of uncertainty. High-energy probe electrons break chemical bonds irreversibly by the energy-sharing process involved in inelastic scatter; the more complex the molecule, the more sensitive it is to irreversible damage. The STEM probe has the advantage here over the flood-beam TEM, since it is possible to obtain an image after one single passage of the probe spot, which limits specimen damage to the irreversible minimum.

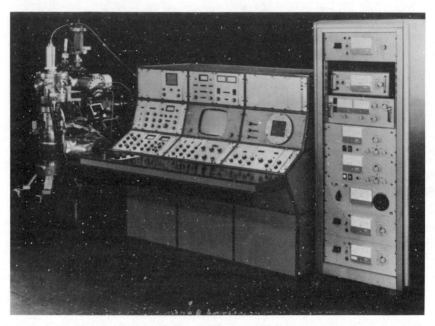

Figure 17.2   The prototype AEI 'STEM-1'. The column is inverted as in this manufacturer's 'Corinth' TEM

A high-resolution STEM with field emission source, electron spectro-meter and image analysis facilities is not commercially available at the time of writing (1975). Only two such instruments have so far been constructed and described in the literature, both being in the Chicago laboratories of A. V. Crewe and J. Wall. However, all the major manu-facturers in Britain, Europe and Japan are now heavily engaged in STEM development work, and commercial models should shortly be on the market. The price will undoubtedly be very mich higher than that of a high-resolution TEM. Fig. 17.2 shows the prototype British AEI–STEM–1 with its control console and power supply cabinet, and Fig. 17.3 shows a block diagram of the two-lens column. STEM modifications of conventional high-resolution TEMs are now available as described below, but in the absence of an electron spectrometer and image analysis equipment, it is debatable whether the rather minimal advantages justify the cost.

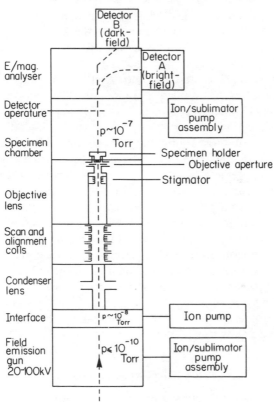

Figure 17.3 A block diagram of the AEI 2-lens STEM column, UHV pumping system, electromagnetic analyser and detectors

It is possible to modify a conventional high-performance TEM with two condenser lenses to give a STEM image. To reduce the spot size from a diameter of around 2 μm to around 100 Å, a third condenser lens (sometimes called a 'mini-lens') is placed close to the objective lens, between the second condenser lens and the specimen. The objective is switched off when the STEM mode is in use. The spot is displaced to form a raster on the specimen by adding a sweep circuit to the normal beam deflection coils, which are used in the TEM for centring the beam about the microscope axis. The sweep circuit also drives the deflector coils of the display tube in synchronysm. An electron detector, which is generally a simple Faraday cage, is placed behind the objective lens to intercept the transmitted electrons. This gives a digital signal which can be amplified, processed and used to modulate the brightness of the display tube spot. The advantages of a comparatively low-resolution bright-field STEM image (without a velocity analyser) are then available. The most important advantage is in the electronically amplified and reversible image contrast, which enables a high-contrast image to be obtained from a specimen which would otherwise give very little contrast in the TEM. Final image contrast and brightness are readily adjustable, using controls identical to those found on a conventional television set. Another advantage is in the reduction of damage to the specimen by the electron beam, since it interacts with a given area of specimen only once per scan. Such an instrument (see Fig. 9.4) is a half-way house to the high-resolution STEM, and applications are being found in the field of biology. Improvements are under continuous development, notably a new form of high-brightness point-source gun using a fine point of lanthanum boride as the electron emitter. This material operates at red heat (1600 °C) and does not require the ultra-high vacuum of the cold cathode emitter. It should be possible to obtain spot sizes down to 10 Å in a normal TEM in this way. If an X-ray detector is placed in the specimen chamber of a TEM converted to STEM, a ready-made electron probe X-ray microanalyser results (see p. 389).

17.3 The Scanning Electron Microscope (SEM)

The SEM is used to examine the surfaces of specimens at high resolution. It utilizes a probe-forming electron optical system and probe-beam scanning mechanism identical to the STEM (see Fig. 17.4.) The probe spot, which is made as small and bright as possible since ultimate resolution depends on this, is scanned in a raster over the surface of the specimen. The incident primary electrons may themselves be reflected from the surface under examination, (backscatter), or they may excite the emission of secondary electrons from just below the specimen surface. The image-forming electrons are attracted into a detector placed close to the surface. Each captured electron gives rise to a flash of light in a solid scintillator. The light output, corresponding to the secondary electron signal, is amplified in a photomultiplier and the resulting electrical signal modulates the brightness of a display spot scanning synchronously with the probe spot. A direct image of the surface in the 'light' of emitted

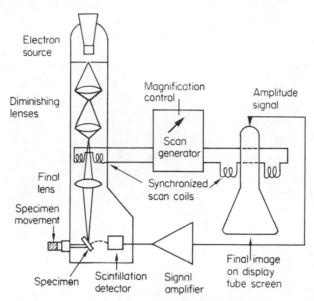

Figure 17.4   A diagram of the scanning electron micro-scope. A description of the method of operation is given in the text. This instrument can only be used for examining surfaces. Its resolving power depends on the smallness of the spot of electrons focused on the specimen surface and the speed at which it travels across the surface. A high-resolution (200 Å) micrograph requires an eamina-tion time of a minute or more

secondary electrons is thus produced on the face of the display tube, and may be photographed in the usual way to give a 'scanning electron micrograph'. Secondary electrons are generally used rather than back-scattered primaries, since these are emitted from beneath the surface and one is able to 'see into' deep holes and re-entrant depressions in the surface. In either case, the number of electrons emitted is a function of the nature of the surface and its angle to the incident beam. By placing an electrically charged wire mesh grid in front of the detector and biasing it suitably, either reflected primaries or emitted secondaries can be collected. A picture of the surface is then built up on the face of the display tube by scanning the probe spot in a raster of 600–1,000 individual lines. Magnification is simply the ratio of the length of the scan line on the specimen to the length of the display line on the final image tube. The resolving power of a SEM depends primarily on the diameter of the probe spot, but the signal-to-noise ratio in the final image also affects it. This depends on the speed with which the probe spot is scanned across the surface. The slower the scan, the greater the signal corresponding to

Figure 17.5 A typical modern high-performance SEM, the Philips PSEM–500. This compact instrument combines SEM, STEM, EXMA and cathodolumines-cence facilities with a minimum spot size of 80 Å. On the left below the column are the controls for the eucentric goniometer stage, and a range of detectors for electrons, X-rays and light may be fitted on the right. The control panels to the left of the column house the display and photographic facilities. (Courtesy Philips, Eindhoven)

each image point and the less the spurious signal due to random electrons or 'noise'. A micrograph at maximum resolution requires a total scan time of a minute or more, therefore the image must be recorded photographically by means of a camera (generally a Polaroid) focused on the display tube screen. A visual image of the whole field at reasonable resolution can be obtained on a separate long-persistence display tube with a scan time of a few seconds. This enables the operator to select a field at a suitable magnification, focus the probe spot and set the correct brightness for the final photograph. Since X-rays characteristic of the specimen atoms are also given off at the specimen surface (see the following section), an X-ray detector placed in the specimen chamber close to the electron detector enables a simultaneous X-ray emission image to be built up on a second display screen. Fig. 17.4 is a schematic diagram showing the principle of the SEM, and Fig. 17.5 shows a commercial SEM fitted with an X-ray spectrometer (see below).

One of the great advantages of the SEM is the relatively short specimen preparation time. Metals or conducting specimens can be examined straightaway after simply mounting them on a metal stub with an electrically conducting adhesive. Most biological specimens, being nonconducting, must first be given a conducting surface coating. This is done by placing the specimen on its stub in a vacuum evaporator (see p. 476) and condensing a very thin film (less than 100 Å thick) of an easily evaporated heavy metal, such as gold, evenly on the surface. It is necessary to rotate the specimen to avoid 'shadows' (see p. 478). After the apparatus has been set up, it takes only a matter of minutes to coat a specimen. However, it is generally necessary first to dehydrate the specimen or serious artifacts may be introduced when the specimen is introduced into the vacuum. Dehydration can be simple air drying, but it is generally best to use liquid CO_2 'critical point drying' in order to avoid distortions due to surface tension effects, which cause the collapse of delicate surface structures such as microvilli and pseudopodia. Even when relatively complicated dehydration techniques are necessary, the time and labour involved in the preparation of a specimen are considerably less than the embedding and ultramicrotomy (see Chapter 18) involved before a biological specimen can be examined in the TEM.

Resolving power as with the STEM is limited ultimately by the diameter of the probe spot. Conventional instruments using thermionic tungsten cathodes have resolving powers of around 200 Å, which limits top magnification to × 10,000. However, the new lanthanum boride point-source guns, which give a smaller, higher brightness virtual source under normal vacuum, are now being applied to commercial SEMs, with resulting

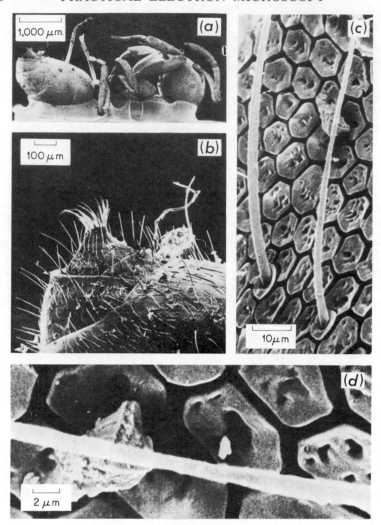

Figure 17.6   A series of scanning electron micrographs taken at 20 kV on a Cambridge Steoreoscan instrument of a garden ant mounted on the object stub with 'Durofix' and coated with an 800 Å thick layer of gold to render the surface conducting. (a) the whole animal, approx. × 12·5 (b) the tail area and anal sternite, approx. × 70. (c) the bristles and surface structure of the anal sternite, approx. × 1,150. (d) the fine structure of the surface, × 3,450. The series shows the remarkable magnification range obtainable by varying the ratio of length of scan line to length of recording trace line. Note the tremendous depth of focus and absence of image scan lines. The instrument can only be used to examine surfaces. (Courtesy Cambridge Scientific Instruments Ltd.)

improvement in claimed resolving power. 50 Å is becoming commonplace, and there seems to be no reason why the cold cathode field emission gun used on the STEM should not be applied to the SEM, in which case the resolving power could be less than 10 Å.

There is no reason why the SEM should not be used as a simple STEM. It simply requires a deeper specimen chamber and a special specimen holder to enable the electron detector to be placed below the transmission specimen. Commercial adaptations for this purpose will shortly be available from SEM manufacturers.

The SEM has already shown itself to be of great value to the biologist. Specimens ranging from whole insects (see Fig. 17.6) down to the surfaces of red blood cells have been examined. The depth of field is extremely great due to the absence of imaging lenses; the size of the specimen is limited only by the space available in the specimen chamber. It is only possible to examine surfaces with the SEM, but resolution is at least 10 times as great as that of the light microscope and is being improved continually. This, combined with the immense depth of field, is rapidly making access to an SEM essential for the biological electron microscopist.

17.4 The Electron Microprobe X-ray Microanalyser (EXMA)

The principle of this instrument has been discussed on p. 67. Since X-rays are emitted whenever a high-energy electron beam interacts with matter, any electron microprobe instrument, be it TEM, STEM or SEM, may have an X-ray detector, analyser and display unit attached to it.

The major problems in X-ray microanalysis are the detection, identification and quantitation of the individual constituents in the heterogeneous beam of X-rays emitted by an excited specimen when an electron probe spot is focused on it by the electron gun of a conventional SEM or TEM. There are two basic ways of attacking these problems; both have their advantages and disadvantages. The first, and older, method is to disperse the X-ray beam into a wavelength spectrum, using a crystal as a diffraction grating, and then to scan the spectrum with a gas-flow proportional counter. Wavelength is given by the angle of the counter to the spectrometer crystal, and the signal can be quantified by measuring the output current from the counter. This is called a 'wavelength dispersive' (WD) spectrometer (see Fig. 17.7). The second, more modern, method is to use a lithium-drifted silicon semiconductor detector which gives out electrical signals proportional to the energies of the incident X-ray quanta. In this way, the detector acts as its own energy spectrometer, forming an 'energy dispersive' (ED) spectrum. Quantitation is achieved by counting

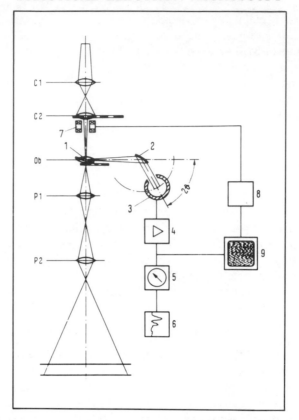

Figure 17.7   A diagram of the beam paths of the Siemens wavelength dispersive TEM X-ray microanalysis equipment. The double condenser focuses a fine intense electron spot (1) on the specimen, which is held in a special tilting holder. The electron beam is transmitted and the specimen is imaged on the screen in the normal way, but the emitted X-rays are directed on to an analyser crystal (2) which reflects an X-ray spectrum into the counter (3). The counter can be rotated in an arc 2θ about the crystal; this angle is plotted on the x-axis of the following figure. The output from the counter is amplified (4) and can be read on a ratemeter (5) or else is the ordinate of the pen recorder (6), giving the y-axis of the trace in the following figure. To obtain a 'map' of the X-ray output corresponding to a given element over the surface of the specimen, the angle 2θ is set to correspond to the characteristic emission line, and the electron beam is scanned across the specimen with the coils (7) by signals from the scan generator (8). The amplified output from the counter modulates the brightness of the beam of the cathode ray tube monitor (9), which is scanned in unison with the electron spot, giving an 'X-ray map' of the scanned area. A transmission image on the final screen of the microscope is continuously available during both these procedures. (Courtesy Siemens Ltd., Berlin)

the number of pulses corresponding to a given energy coming out of the detector in unit time.

Each method has particular advantages and disadvantages. The WD method is more sensitive to light elements (Z-number less than Na $= 11$) of primary interest to biologists, and has a higher spectral resolving power for heavy elements. The main WD disadvantage is the inherently time-consuming nature of the analysis. The X-ray spectrum has to be traversed by the detector, and the output drawn out as a series of peaks on a pen chart recorder (see Fig. 17.8). To cover the whole range of elements in an unknown specimen, 3 or 4 different crystals must be used to cover different parts of the spectrum, and the total analysis time is of the order of hours, during which the specimen suffers marked deterioration under the intense electron beam. The ED spectrometer, on the other hand, has the immense advantage of possessing a digital output signal; this advantage was discussed in the previous section on the STEM (see p. 380). This enables a spectrum to be displayed on a cathode ray tube, peak heights being built up in a matter of seconds. It is very much more sensitive in the medium-Z-number range, because the ED detector can be placed much closer to the X-ray source than can the WD crystal with its associated mechanical components. It can therefore intercept X-rays from a far greater solid angle—up to 10^3 times as large. The lithium-drifted silicon crystal comprising the ED detector is also far more efficient in converting X-rays to signal pulses than the WD diffraction crystal plus proportional counter. On the debit side, the ED detector has to be kept at liquid N_2 temperature in order to reduce thermal noise, and has to be shielded from the microscope vacuum, otherwise pump oil and other contaminants would condense on it and render it inoperative. This means that a window, generally of beryllium, must be placed between the X-ray source and the ED detector, which renders it almost completely insensitive to the very low-energy X-rays from light elements. This disadvantage will doubtless be overcome in the near future.

The X-ray microanalyser can be operated either in the probe mode or the scan mode. In the probe mode, the X-ray spectrum from a comparatively small volume of the specimen (depending on its thickness) may be analysed quantitatively for all the detectable elements present. The WD spectrometer draws the peaks on a chart; the ED presents the information almost instantaneously on a display tube. In the scan mode, the detector is set to an X-ray spectral line (WD) or an energy (ED) corresponding to a selected element, and the probe beam is scanned in a raster across the desired area of the specimen. A 'map' of the specimen in the 'light' of the selected element is generated on a synchronous display tube and is recorded photographically. The electron gun is then set to give a wide-angle flood beam, and the conventional TEM image is focused on the final screen and photographed in the usual way. The two photographs can subsequently be superimposed to show the distribution of the selected element. Accurate quantitative elemental analysis can then be made of any desired area of the specimen in the probe mode. Fig. 17.7 shows a diagram of the beam paths of the Siemens version, and Fig. 17.8 shows an elemental analysis recording.

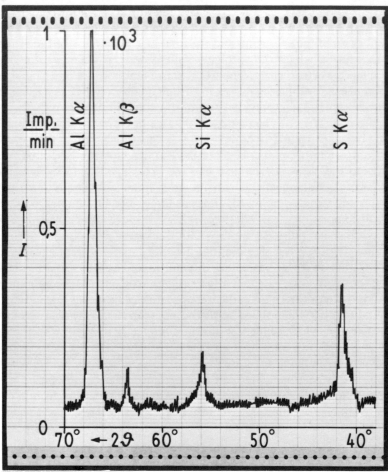

Figure 17.8 A scan trace from the pen recorder of the Siemens X-ray spectrometer attachment fitted to the Elmiskop 101. The specimen was a sample of the etching residue on a transistor surface, and shows a high aluminium content in the silicon. (Courtesy Siemens Ltd., Berlin)

In the present state of the art, WD spectrometers can detect and measure all elements above $Z = 8$ (oxygen). ED spectrometers operate above $Z = 11$ (sodium). The claimed limits of detection range from 10^{-14} g for simple ED–TEM accessory spectrometers to 10^{-19} g for the special-purpose AEI–EMMA instrument for the mid-range elements such as iron. Spectrometer resolving power hinders detection for high-Z elements; the very low X-ray yield unfortunately makes the low-Z elements of biological significance very difficult to detect and quantitate. Due principally to the lateral spread of the electron probe as it passes through the specimen, X-rays are generated from a greater volume of the specimen than would be expected simply from consideration of probe spot size. Lateral resolution within the specimen in practice is in the range 100 µm to 1 µm, depending on thickness. In biological terms, only fairly large cell organelles are capable of elemental analysis. In any case, any specimen preparation techniques involving dehydration prior to insertion into the microanalyser vacuum are bound to alter very significantly the distribution within biological specimens of soluble elements such as Na^+, K^+, Mg^{2+}, Ca^{2+}, P, S, Cl^- etc. which are of particular biological interest. It is necessary to prepare ultrathin frozen sections by the technique of 'ultracryotomy' in order to study their distribution. However, the development of this potentially extremely sensitive and powerful technique is only in its infancy, and great strides forward are bound to be made in the near future. It is already becoming almost indispensable to metallurgical electron microscopists, and most TEM and SEM manufacturers are offering ED or WD spectrometers which are fitted through a port in the specimen chamber together with third condenser 'mini-lenses' to reduce the size of the probe spot. The AEI 'EMMA–4' TEM fitted with WD X-ray microanalysis equipment is shown in Fig. 17.9. For maximum sensitivity, it is necessary to design a special EM for microanalysis such as the GEC–AEI 'EMMA–4' which combines WD and ED facilities. Such an instrument is inevitably more expensive than a simple TEM plus microanalyser. Some results of biological significance have already been forthcoming, such as the distribution of calcium in bone and tooth; the localization of iron in haemopoietic tissues and the sites of incoporation of metal-containing drugs in tissues, but the largest field of application has so far been in materials technology and metallurgy.

17.5 Ultra-high Resolution

Transmission electron microscopists will not be happy until they have achieved their ultimate aim: to image individual atoms. This has already been achieved with the field-ion microscope and the STEM with metals. The atoms of the lighter elements in biochemical compounds have inter-atomic distances of about 1·5 Å. The difficulties in imaging them in trans-mission are firstly that they do not perturb an electron beam sufficiently to

Figure 17.9   The AEI 'EMMA-4' TEM X-ray microanalyser. An EM-801 is fitted with two WD X-ray spectrometers on either side of the modified specimen chamber. Four separate motor-driven dispersion crystals enable the spectrum of mass numbers from sodium to uranium to be covered. An ED spectrometer is fitted at the rear of the column. A normal transmission image of the specimen may be obtained on the final screen at any time during the course of the analysis.
(Courtesy GEC–AEI Ltd., Manchester)

give rise to an image, i.e. they have insufficient contrast; and secondly that dimensions as small as interatomic distances can hardly yet be resolved. The second problem, that of ultra-high resolution, was the first to be tackled. Specimens of suitable contrast will no doubt present themselves as resolving power increases.

The resolving power of electron microscopes is limited, as we have seen in Chapter 4, both by spherical aberration which is inherent in the objective lens; and by chromatic aberration, which arises both in the specimen and from instabilities in the accelerating voltage of the illuminating electrons. Knoll and Ruska as far back as 1932 estimated the theoretical ultimate resolving power of an electron lens operating at 75 kV as 2·2 Å. They based their calculation on inherent uncorrectable spherical aberration. It is only recently that this resolving power has been achieved, and that by lattice rather than by point resolution. The attainment of such a very high resolving power does not depend only on the instrument; it calls also for a very high degree of skill and knowledge on the part of the microscopist. Since electron optical magnification and image brightness are not in general high enough for really accurate focusing at these levels, the attainment of such resolution is also to a certain extent a matter of luck. It requires specimen stability within ±1 Å while the photographic exposure is being made. The stabilities of lens current and accelerating voltage must also be of the highest attainable order.

Spherical aberration can only be reduced by improvements in lens design. It can be reduced by increasing the magnetic field and thus reducing the focal length of the objective lens, but as we have already seen in Chapter 4, this follows the law of diminishing returns. Practical difficulties with the magnetic saturation of iron limits the flux density of the lens field using the conventional form of construction to about 25,000 gauss, corresponding to a focal length of about 1 mm. Experiments are being made using iron-free lenses having superconducting coils cooled by liquid helium to generate the necessary intense fields of 50,000 gauss or more. The rate of gain in resolving power falls off as focal length decreases (Fig. 4.7), and the effort to get below 1 mm seems to be hardly worth while.

Spherical aberration can in theory be reduced more easily by using a somewhat different form of lens, the 'single-field condenser-objective' first proposed by Glaser in 1941. This is a very powerful symmetrical lens with the specimen placed at the centre of the bell-shaped field instead of a little above it as is the case with the conventional type of lens. The upper polepiece acts as a condenser, focusing the incident beam on the specimen. The lower polepiece acts as an objective of short focal length. This combination has been shown to have a spherical aberration constant of about one-tenth that of a normal lens, together with a chromatic aberration constant of about one-half. These lenses are at present under development by Ruska and his coworkers among others. It should be possible in practice to resolve about 1·2 Å at 100 kV using the single-field condenser–objective lens. It must be emphasized however that the routine attainment of such high resolution demands a better stabilization of high tension and lens currents and better specimen stability than are generally available at present.

The path of development of light microscope lenses is also being followed, using combinations of lenses to reduce aberrations. The superposition of electrostatic negative and electromagnetic positive lenses is one possibility. Electrostatic 'electron mirrors', which have reverse aberration characteristics to those of lenses, might be used in combination with lenses, but the practical difficulties appear to be very great. A further possibility is the use of combinations of cylindrical electrostatic lenses with round electromagnetic lenses. By varying the relative strengths of the individual components of such 'octupole' lenses, spherical aberration can in theory be completely corrected. The necessary corrections to be applied to the various components are worked out by a computer during the operation of the lens, and correcting signals are fed back. The apparatus appears to be formidable, but experiments are in progress in the U.S. on the construction of such a lens system.

Chromatic aberration can be reduced by making the heterochromatic electron beam leaving the objective lens approximately monochromatic once again, as it was when it left the electron gun. This is done by filtering out electrons deviating in velocity on either side of a chosen mean by more than a certain amount. The electron velocity filter works on the same principle as a mass spectrometer, which uses a magnetic field to separate charged particles of different mass travelling at the same velocity. A velocity filter separates electrons of the same mass travelling at different velocities. The principle of the apparatus used is shown in Fig. 17.10. The electron gun produces an almost monochromatic beam, since the accelerating voltage is stabilized to within one part in 10^6. After passing through the specimen, the beam becomes heterochromatic and passes through the objective lens in the usual way. A wedge-shaped magnetic field is now placed between the objective lens and the rest of the imaging system. This 'magnetic prism' causes faster and slower electrons to follow different paths, so that they can be sorted out by causing them to fall on either side of a slit aperture. The electrons passing through the slit will emerge as an approximately monochromatic beam once more, and can be imaged by a conventional intermediate and projector system, the final image being greatly reduced in chromatic aberration. The method improves both resolution and contrast.

To sum up, the history of the conventional transmission electron microscope shows that it was a relatively simple matter to improve resolving power from 100 Å to 10 Å. It has since proved very difficult indeed to improve this further to 3 Å. It will be a task of phenomenal magnitude to increase resolving power to 1 Å, but while the problem remains, it will be tackled. It will no doubt be overcome eventually, and the goal of imaging individual atoms of low atomic number with a transmission electron microscope will be achieved. This goal now appears to be in sight, using the STEM (see Section 17.2).

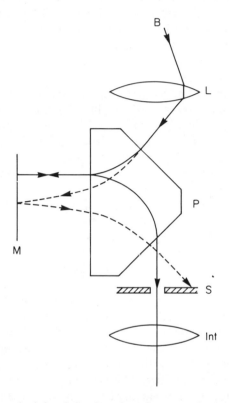

Figure 17.10   The principle of the Castaing electron velocity filter for reducing chromatic aberration. The beam B emerging from the objective lens is focused into the entry face of the magnetic prism P by the auxiliary lens L. Fast electrons follow paths having greater radii of curvature than slow electrons, and emerge at a different point on the back face of the prism. They then strike the electrostatic electron mirror M and are reflected back into the prism. Once again, fast and slow electrons follow different paths and emerge at different points. A selection slit S located at the conjugate plane of the lens L allows only electrons covering a small range of velocities to enter the intermediate lens and form the final image. A range of ± 1 V at 75 kV is claimed, giving enhanced resolution and contrast. (Courtesy Dr. Castaing and Société d'Optique, Précision, Électronique et Mécanique, Paris, France)

17.6 High-brightness Electron Guns

In terms of energy emission per unit solid angle, the conventional heated tungsten hairpin filament in a triode accelerator (thermionic gun) can be made about 10^5 times as bright as a light source. This type of electron source is very satisfactory in the conventional TEM provided that ultra-high resolution is not sought. However, it has a number of disadvantages which become especially important in the scanning mode, where resolution is limited by the diameter and brightness of the probe spot. It is very difficult to design a thermionic gun having both the desirable scanning characterisitcs of high brightness and small crossover diameter (see Fig. 6.9). The thickness of the tungsten wire and its hairpin shape give rise to an elliptical crossover of roughly 20 μm optimum diameter. The intrinsic brightness of the crossover is limited by space charge effects around the emitting tip. To reduce the crossover diameter significantly, a pointed filament (see Fig. 15.4) must be used, and to increase brightness the operating temperature must be raised, leading to a greatly reduced filament life of only an hour or so.

There are at present three methods of overcoming the two major disadvantages of the tungsten source. The first is to use as emitter a substance of lower work function (see p. 112) in a conventional triode gun. The most promising material so far investigated is lanthanum hexaboride (LaB_6). A thin rod of this material is ground to as fine a point as possible and is affixed to a tungsten hairpin heater. The operating temperature is about 1,500°C (1,000°C lower than pure tungsten) and the source can be made 2–5 times as small and bright as a tungsten filament gun. The emitter has a lifetime of weeks rather than hours provided it is not poisoned by contaminants, a malady to which all rare earth emitters are particularly prone. The radius of the emitting point increases with time, and it requires a warm-up period before use. The vacuum requirements are about the same as those for a tungsten filament gun. At present, it offers rather minimal advantages over a well-designed tungsten source and is unlikely to come into general use in its existing form.

The second and most obvious solution is to use the cold cathode field emission source as used in the high-resolution STEM (see Section 17.2 above) with all its attendant ultra-high vacuum problems (see the following section). This type of source allows of a probe spot diameter of 2 Å with sufficient brightness to image individual silver atoms. As ultra-high vacuum technology develops and becomes commonplace, this will be the obvious solution to the problem.

A third possible method, at present under investigation by le Poole and Philips, is to use the molten end of a very fine tungsten wire as the emitter. The tip is kept molten by a laser beam focused on to it through a glass window in the gun casing. The tungsten point evaporates in use, but the wire is automatically replenished by feeding it in continuously from a reel contained in the gun vacuum. High brightness, small spot size and long life under normal vacuum conditions are claimed.

17.7 Ultra-high Vacuum

The attainment of the full resolving power of an electron microscope is plagued by the deterioration of the specimen whilst under examination. As we have seen in Chapter 7, specimen detail is obscured by the deposition of material derived from the interaction of the electron beam with the residual gases in the column. This is 'contamination'. Specimen detail can also be destroyed by electron-induced combination with hydroxyl ions which are derived from residual water vapour in the column vacuum. This is 'stripping' or 'beam damage'. Both are clearly due basically to the gas pressure in the specimen chamber being too high, and therefore the effects of both can be reduced or eliminated by improving the vacuum in the space immediately surrounding the specimen.

Specimen chamber pressure may easily be ten times higher than that registered by the high vacuum gauge, which is generally situated much closer to the pumps. Insufficient attention has been paid in the past to proper specimen chamber pumping by some instrument designers, who have been content to allow the specimen chamber to be pumped most inadequately through the bores of the objective polepieces. Both contamination and stripping can be reduced to negligible proportions by proper pumping of the specimen chamber. The normal operating vacuum in the column (10^{-4}–10^{-5} torr) is not good enough to reduce contamination and stripping to negligible proportions, even if it can truly be achieved at the specimen. It is necessary to reduce the pressure at the specimen to 10^{-6}–10^{-8} torr for such a reduction to be possible. This very high vacuum can be achieved in two ways: by the use of special sorption ('getter-ion') pumps attached by very wide-bore leads to the specimen chamber; or by the use of very large cooled (down to 20 °K) surfaces ('cryopumps') close to the specimen. In either case, the specimen chamber must be sealed off as effectively as possible from the main vacuum, which inevitably contains hydrocarbon molecules from the main oil diffusion pump and water vapour molecules from the photographic material, by the use of small apertures located in or near the bores of the objective polepieces.

As we have already seen in the previous section, much of the present progress in electron microscopy is centring about the development of high-brightness, cold cathode field emission electron guns. The very high brightness of this type of source depends on the ability of the very concentrated electric field at the tiny emitting point to overcome the space charge which forms about any electron source, repelling fresh electrons and thus limiting the emission. Such very high-intensity electric fields are only stable at gas pressures some 10^4 times less than that normally attainable in a TEM emission chamber using conventional vapour diffusion pumps backed by rotary pumps. Even with the use of cooled baffles between the diffusion pump and the vacuum chamber, which reduce pumping speed drastically, the ultimate vacuum is still limited by backstreaming of pump vapour. A pump using a quite different principle is therefore necessary to achieve a pressure of less than about 10^{-7} torr.

There is no rigorous definition of an 'ultra-high' vacuum; it is generally thought of as being a pressure less than that attainable by a conventional diffusion pump. Pressures between 10^{-8} and 10^{-12} torr are regarded as being ultra-high vacuua. It is of interest to note in passing some kinetic data for gases in this pressure range. At a pressure of 10^{-10} torr, there are still no less than 3×10^6 gas molecules present in each centimetre cube. In spite of the presence of this apparently enormous number of molecules, the mean free path—the mean distance any molecule travels before colliding with another—is no less than 500 km! This gives some idea of the smallness of molecules. Some idea of the velocity of the molecules is given by the fact that at this pressure and at room temperature, each cm^2 of surface is bombarded by no less than 4×10^{10} molecules/sec. In spite of this, if the gas is nitrogen, it takes some 10 hours to build up a monolayer of gas on the surface.

The source of residual gas in a leak-free system is simply the 'outgassing' or desorption of nitrogen, hydrogen, oxygen, water vapour etc. from the metal and ceramic components of the gun chamber itself. Most of the absorbed gas can be liberated by heating or 'baking-out' the components to a temperature of about 350 °C while pumping continuously with a cryogenically trapped vapour diffusion pump. This necessitates the use of special metallic vacuum seals and heat-resistant electrical components. After cooling, the baked-out system is connected to the ultra-high vacuum (UHV) pumps.

There are two different types of UHV pump at present commercially available. The first is the turbomolecular pump, sometimes called a 'molecular drag' pump. The second is the sorption pump, of which there are several varieties. The turbomolecular pump removes gas from the system, transfers it to a backing pump and thence to the atmosphere. A sorption pump traps gas molecules and holds them firmly in a completely closed system.

The turbomolecular pump is a kind of vapour-free diffusion pump. It consists simply of a number of circular metal discs about 20 cm in diameter fixed to a revolving shaft. In between the rotating discs are annuli which are fixed to the casing. This labyrinth construction closely resembles that of a small steam or gas turbine engine working in the opposite sense, i.e. the rotor is rotated at 20–50,000 r.p.m. by an electric motor and gearing and the device acts as a pump. It will only work in the molecular flow region, i.e. at pressures such that the mean free path is much greater than the clearance between each rotor and the walls of the stator compartment. Gas molecules diffuse from the vessel to be evacuated and strike the rapidly moving rotor blades. This gives them a component of velocity towards the next stage of the labyrinth. The gas molecules are thus bounced from the rotating surface to the stationary surface, gaining momentum as they approach the high pressure end of the rotor. The number of molecules and hence the pressure increases from state to stage of the labyrinth. An 8-stage double-ended turbomolecular pump allows a pressure build-up from 10^{-10} to 10^{-3} torr along the length of the rotor. Pumping speed depends on the design, but commercial pumps are available which operate between 200 and 500 l/sec, a factor of 10 slower than a sorption pump of comparable size. A great advantage of the turbomolecular pump is the fact that the pumping speed for noble gases, notably helium and argon, is not significantly different from that of chemically reactive gases. Also, there is no magnetic field associated with it. However, due to the very high-class engineering necessary in the construction of such a pump, with its high rotor speeds and fine clearances, the cost is very much greater than that of a sorption pump.

Sorption pumps are very simple and have no moving parts. The molecules of residual gas in the vacuum chamber are caused to strike a reactive surface to which they are adsorbed or with which they combine chemically. The residual gas is generally ionized before being attracted onto the adsorbing surface, which is usually renewed continually by the evaporation or sputtering of the adsorbing or 'getter' metal onto it. The whole pump can be placed within the vacuum chamber, in which case it is called a 'nude' pump, or it can be connected to it by a short, wide tube. Sorption pumps are classified into simple sortion pumps, diode and triode ion pumps, sublimation pumps, getter-ion pumps and sputter-ion pumps.

Simple sorption pumps have a very long history. They were the only means by which the early workers on gas discharges could reach pressures below those attainable with hand-operated piston pumps. They consist simply of a closed vessel made of glass or metal and connected to the vacuum vessel by a short, straight pipe. The pump vessel is filled with fresh charcoal, and is surrounded by a Dewar flask filled with liquid N_2. For continuous pumping, two pumps must be used side by side, one connected to the vacuum vessel and adsorbing while the other is closed off, warming-up and desorbing. The two pumps are used alternately to absorb and desorb. Modern sorption pumps use 'molecular sieve' fillings made from artificial zeolites with pore sizes in the 10 Å region. These will work at room temperature and have a heater incorporated to regenerate the sieve material. They will pump down to 10^{-2} torr, and are used either as backing pumps or as foreline traps between rotary vane pumps and turbomolecular pumps. This ensures that no contamination from the rotary oil pump can reach the UHV system.

Ion pumps are almost as simple as sorption pumps. A Penning gauge (see p. 152) acts as an ion pump. Positive ions formed by multiple collision strike the cathode and are adsorbed onto it. A simple diode ion pump such as this is not very efficient, because the cathode soon becomes saturated. The answer is to use a more efficient adsorbent, and to renew it continuously. The pumps most frequently used on STEM emission chambers are some variant of the sublimation or 'getter' principle. A highly active layer of a reactive metal such as titanium or zirconium is evaporated continuously from a heated metal filament or a refractory surface onto an inert cathode presented to the gun emission chamber. The principle is again a very old one; the mirror-like patches seen on the inside of the glass bulb of the old-fashioned thermionic valve or vacuum tube are the evaporated getter. For continuous pumping, the getter surface must be renewed continually. This combination forms a diode getter-ion pump. This can be improved by the addition of a third electrode held at about 1 kV, which causes the initial ionization; we then have a triode getter-ion pump. A large commercial triode getter-ion pump evaporates titanium onto the cathode surface from a spool of titanium wire carried inside the pump at a rate of 5 mg/min, and pumps 3,000 l/sec of hydrogen, 2,000 l/sec of nitrogen, 1,000 l/sec of oxygen but only 5 l/sec of argon.

If the gas ions are accelerated to a sufficiently high energy, they can be made to strike a cold titanium surface and cause titanium to be sputtered onto the anode, which then adsorbs unionized gas molecules from the system. This type of pump is called a 'sputter-ion' pump.

Sublimation pumps simply evaporate titanium from a heated surface onto a large area of cold surface. Gas molecules which diffuse into the space between the filament and the cooled surface combine with the titanium and are thus removed from the system. Nude sublimation pumps can reach extremely high pumping speeds for reactive gases.

Sorption pumps are in general extremely robust, silent, fool-proof, disaster-proof and can be left running unattended over long periods of time. Their disadvantagges for electron microscopy are twofold: the high magnetic fields associated with all forms of ionization pump, and the very low pumping rate for noble gases. Nevertheless, they are at present the only practical method for obtaining ultra-high vacuua at a reasonable cost.

17.8 Environmental Specimen Chambers (Wet Cells)

The use of very high accelerating voltages (one million volts or more) offers the possibility of examining the inner structure of living bacteria or tissue culture cells, and observing the dynamic interaction of the cell organelles. The very high penetrating power and reduced specimen damage due to ionization effects should allow organelle movements in living cells to be studied, at least for short periods of time, provided means can be found to introduce living cells in the wet state into the electron microscope and to obtain adequate contrast. This involves the construction of pressure chambers which will not burst open when introduced into a vacuum, and which have sides or ends which are transparent to high-voltage electrons. Fortunately, such chambers are relatively easy to construct, since evaporated carbon films have been found to have a remarkably high tensile strength. They have been developed by Stoganova in Russia, Heide in Germany and Dupouy in France, and consist of an air slab some 5 to 20 μm thick bounded at opposite faces by carbon–Formvar membranes about 600 Å thick. The volume of the chambers is about 0·1 ml, and the diameter of the membrane window about 1 mm. The electron beam must therefore traverse the two windows plus the thickness of the living cells, some water and some humid air. Experiments by Dupouy on bacteria at a million volts have shown a certain amount of initial success. The published micrographs show much less internal detail than can be seen in micrographs of sections, due in part to the superposition of detail consequent upon the great depth of focus of the electron microscope, but the authors claim that the bacteria which were exposed to the beam were subsequently cultured and were found to divide normally.

A major problem has been found to be the very poor contrast shown by living, unstained material. Dupouy has shown with bacteria that this can be improved by the use of a darkfield method which he describes as a 'contrast screen' (see p. 257). A very thin wire is placed across the objective aperture, allowing only scattered electrons to form the final image, the main beam being intercepted. The contrast and detail of bacterial flagellae are greatly enhanced (see Fig. 11.7).

Although the results so far obtained by the combined use of ultra-high voltages and pressure chambers to study living cells are somewhat disappointing, the technique obviously has tremendously exciting possibilities for the biologist, and will develop rapidly as more very high voltage electron microscopes become available.

A later development is the differentially pumped gas reaction chamber for studying chemical reactions such as oxidation, reduction, nitriding etc. *in situ*. Solid windows are not used; very small apertures through which the electron beam passes the high-pressure chamber from the specimen chamber itself. The small gas leak passing through the apertures is collected in a subsidiary chamber which is independently pumped with its own pumping system. In this way, the gas leak into the main specimen chamber is made small enough to be dealt with by the main column pumping system. This type of chamber enables a gas–solid reaction to be observed in the HVEM. It is a valuable technique in metals technology, but does not appear to be particularly applicable to biology.

17.9 Image Intensifiers

If a specimen is particularly sensitive to damage by the electron beam, very low beam intensities must be used to image it. The image may not be visible on the final screen. It would be possible to photograph such an image using very long exposure times, but it is first necessary to find and focus the specimen. It is necessary therefore to increase the brightness of the final image in some way. This is the function of an image intensifier.

The basic principle of the image intensifier is to form the final electron microscope image on a transparent transmission phosphor screen mounted beneath the normal fluorescent screen and camera, the latter generally being removed from the column. Since the first, dim image is formed on a transparent screen, all the subsequent operations can be performed out of the electron microscope vacuum. The primary image is now coupled optically to a photocathode (Fig. 17.11), which emits electrons when excited by light. The optical coupling must be as efficient as possible, so that as little of the light from the primary image as possible is lost in the coupling process. A lens of very large aperture or glass fibre optics must be used. The electrons emitted by the first photocathode are accelerated by a high potential of several thousand volts and focused on to a second cathode, which emits several secondary electrons for each primary electron striking it. This process can be repeated a number of times, each multiplier stage increasing the total number of electrons available to excite a final transmission screen similar to the first transmission screen. The light intensity pattern on the final screen corresponds to the pattern on the first transmission screen in the electron microscope, but the light

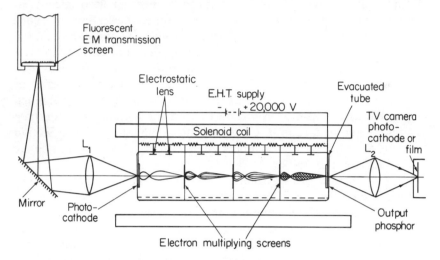

Figure 17.11 A schematic diagram of the electron microscope image intensifier and image train developed by GEC–AEI Ltd. The primary electron image is formed on a transparent phosphor screen below the camera in the microscope column, and is focused on to the photocathode of an electron multiplier. Four stages of electron multiplication produce an identical image of greatly increased brightness on the transparent output screen, which may be examined by eye, photographed or focused on to the photocathode of a TV camera for further amplification and display on a monitor. (Courtesy Dr. Charman, GEC–AEI Ltd. and *J. Roy. Micr. Soc.*)

output is greatly increased. The final screen may then be examined and photographed directly, or, as is more usual, it can be optically coupled to the photocathode of a television camera, which is scanned in the usual way. The signal from the camera is further amplified and is finally displayed on the screen of a closed circuit TV monitor tube.

17.10 Television Display Systems

The television display system, which may also be used to view the primary image directly without the addition of an image intensifier, has a number of advantages. Firstly, the signal from the dim image can be amplified electronically, resulting in further image intensification. Secondly, a long-persistence screen can be used in the final display tube, so that the random 'noise' generated by the individual electrons forming the weak primary image may be ironed out by the addition of a number of display tube scans. Thirdly, contrast can be reversed electronically, so that light parts in the image can be made to appear dark. Fourthly, contrast can be increased or decreased electronically as required. Fifthly, the picture signal can be stored on video tape, so that transient phenomena can be recorded, and can be examined subsequently many times over to assist in their interpretation. Sixthly, a further stage of magnification of about five times is introduced, since the display screen is about that much bigger than the transmission screen. Finally, the picture stored on tape as an analogue signal can be converted into a digital signal (see Section 17.2) and the data can then be processed in a variety of ways by computers to yield further information.

The simplest form of TV system is a camera mounted outside the column and focused on the fluorescent screen, operating a closed-circuit monitor mounted away from the column. This simple system has many disadvantages, the main ones being lack of sensitivity and loss of resolution due to the granularity of the primary fluorescent screen. If, however, the camera is mounted so as to view the transmission screen of an image intensifier, immense gains in sensitivity are possible, the image remaining easily visible on the TV monitor screen as it disappears into the 'noise'.

The system can be made so sensitive that very high direct electron optical magnifications can be used, since the resulting dim image can be made readily visible. Detail of about 100 μm can be displayed on a conventional 625 line television picture, so that a primary instrumental magnification of 100,000 followed by the TV magnification of 5 is adequate to display detail of 5 Å.

The minimum current density at the electron microscope final image transmission screen which can at present be intensified to give a comfortable, useful, albeit rather noisy, TV image is about 10^{-14} amp/cm^2. For comparison, the comfortable viewing screen luminosity for the fully dark adapted unaided eye is given by a current of about 10^{-11} amp/cm^2. The equipment therefore gives about a thousandfold increase in image brightness.

Image intensifiers combined with television display systems are now available from all the major manufacturers. Some use lens coupling, others use fibre optics. Various types of television camera photocathode are used: videcon, plumbicon and image orthicon. Each has certain advantages over the others. It is possible to increase overall sensitivity and resolution by placing the image intensifier and the TV camera photocathode directly inside the vacuum of the electron microscope, but the photocathode surfaces are extremely sensitive to contamination and require a vacuum of at least 10^{-8} torr. This improvement is therefore very much in the developmental stage at the present time, but may become available in the future.

17.11 Optical Transforms

This is a' light optical rather than an electron optical technique, but it is becoming of considerable interest in extracting morphological information from electron micrographs of specimens which have any form of regular, repeating structure which can be made visible in the electron microscope. It has been applied to the solution of the structure of some viruses, but can be applied to any other regular semicrystalline structure seen either in sections or in particulate preparations. Resolving power and astigmatism on a micrograph may be accurately measured.

The principle of the method is first to make an optical diffraction pattern, using a transparent electron micrograph as the diffraction grating or 'subject'. This pattern of diffraction spots is then carefully examined, and all the possible forms of symmetry are noted. Any random spots are then treated as 'noise', and are blanked off by cutting a mask which will only let light through certain areas of the diffraction pattern. This 'cleaned-up' pattern is then reconstituted into an image by the reverse process of recombination, and a simplified, noise-free image of the original object results. The amount of detail included in the recombined image depends entirely on the subjective assessment of the original diffraction pattern, and spurious images can arise.

The optical apparatus used is basically simple, but requires very accurate alignment. The component parts are shown closely spaced in Fig. 17.12. The light source must be as coherent as possible; a low-power gas laser is ideal, but pinhole sources were originally used. The light beam from the laser (a) is spread by the negative lens (b) and passes to the long focal length collimating lens (c). This forms a parallel light beam which passes through the structure in the subject in the holder (d), which is a suitably masked electron microscope plate. A second very long focal length lens (e) forms a diffraction pattern at the plane (f), where a shutter for photographic exposures is placed. The greater the distance between source and diffraction pattern (the camera length), the larger is the resulting pattern; hence the very long focal length lenses. An enlarging lens (g)

forms an enlarged image of the primary diffraction pattern at the screen (*h*), and can be recorded on a high-speed plate or Polaroid film. This diffraction pattern is then examined and masked, and is then replaced in the plane (*h*). A fifth lens behind (*h*) (not shown) is focused on the diffraction pattern and forms a reconstituted image, which is photographed. This apparatus was developed by Prof. Roy Markham and his colleagues, and can be obtained commercially from Polaron Ltd., London, N.3.

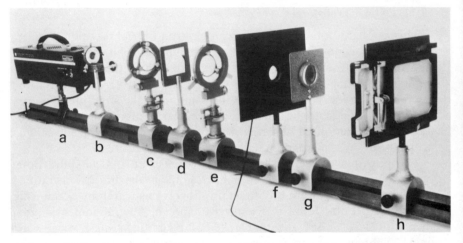

Figure 17.12 An optical diffractometer. The gas laser (*a*) produces a highly coherent beam of light which passes through an electron micrograph of a regularly repeating object held in the holder (*d*) and forms a diffraction pattern on the plane of the photographic plate (*h*). (Courtesy Prof. Roy Markham and Polaron Ltd.)

An example of the use of the optical diffractometer is shown in Fig. 17.13. An electron micrograph of the surface of a negatively stained particle of cowpea chlorotic mottle virus (*a*) displays very ill-defined repeating patterns at several levels of organization. An optical diffraction pattern of this area was then made. This is a cruciform shape (*c*) surrounded by six inclined spots and a great deal of random 'noise'. The important areas of the diffraction pattern are decided upon with the help of drawings of hypothetical models of the virus, and with this information the irrelevant parts of the diffraction pattern (*c*) are masked off, giving the 'cleaned-up' diffraction pattern (*d*). This is then reconstituted to give the structure shown in (*b*), in which the two levels of organization are now made much clearer. The use of this method obviously requires the exercise of considerable intellectual honesty by the user, since 'unwanted' features can so easily be got rid of. However, it is obviously of value in the interpretation of repeating structures such as virus surface structures. The masking technique can also be used to separate the 'back' structure from the 'front' structure on 3-dimensional particles.

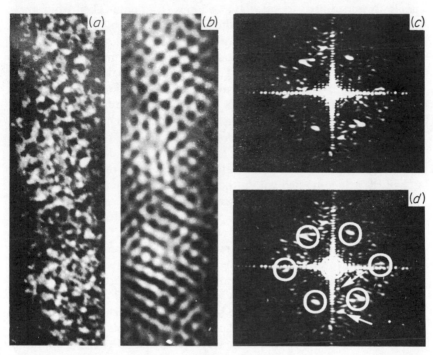

Figure 17.13   An example of an optical transform. The repeat patterns of the electron micrograph of a virus surface (a) are obscured by 'noise'. The diffraction pattern (c) reveals which are the repeating elements and the part of the pattern due to noise. The unwanted noise is masked out in the 'cleaned-up' pattern (d), which is then resynthesized into the transform (b), which shows clearly the various orders of repeating structures. (Courtesy Prof. Roy Markham and Academic Press Inc., New York)

17.12 Stereology

Stereology, like optical diffractometry (see above), is a technique which is normally applied only to the final finished electron micrograph. It is a method for extrapolating from areas to volumes; from two-dimensional space (the micrograph of the section) to three-dimensional space (the biological tissue or other solid material from which the section was made).

The most important parameters which stereology enables us to calculate are the volume fraction and the surface area/unit volume of structures (e.g. mitochondria) distributed at random throughout a solid (e.g. a cell). The mathematical derivations used are based on concepts taken from statistics, dimensionality and topology. The rigorous proofs of the final simple formulae used to calculate volumes from data taken from areas involve the highest realms of mathematics, but, like all the best mathematical results, they are of the most satisfying elegance and simplicity.

The technique used to obtain the areal data from the electron micrograph is called 'point counting'. The method is to make a grid of intersecting lines by photographing a ruled grid (e.g. a 'Letratone' sheet) and printing the resulting negative on $8'' \times 10''$ sheets of ester-base fine grain ordinary film. The resulting transparency is laid at random over the micrograph, and all the line intersections which lie over the required structure type (e.g. mitochondria) are counted by hand, preferably using a push-button laboratory enumerator. The total area of the cell (or that area of the cell contained within the micrograph) is then measured in the same way. The volume fraction can be proved to be simply the ratio of the first point count to the second point count, provided the distribution of the required object is random within the matrix. To ensure random sampling, a number of micrographs taken at random from randomly chosen sections must be made. The required parameter, e.g. the volume fraction, is calculated after each point count, and a progressive mean is taken. When the results fall steadily within about $\pm 10 \%$ of the progressive mean, sufficient micrographs have been counted. In general, it suffices to count only 5–10 micrographs. Results obtained by this simple, powerful and accurate analytical technique have given rise to radical rethinking about the structure of the liver and the lung, for example.

Readers who require more information about the techniques and theoretical background of stereology are referred to the publications of Elias and Underwood given in the bibliography.

PART 3

SPECIMEN PREPARATION

CHAPTER 18

Electron Microscope Histology

18.1 Introduction

The final two chapters of this book will deal with the most commonly used of the various methods for the preparation of biological materials prior to examination in the electron microscope. It is not the intention to give lists of recipes; these can all be readily found in the various books dealing exclusively or almost exclusively with specimen preparation, and which are listed in the bibliography. It is essential for the biological electron microscopist to possess at least one of these books; there is therefore little point in repeating their detailed contents in these chapters. The aim here is to discuss the principles and rationale behind the methods used where these are known. Specimen preparation is still very much of an empirical art based on experience and to a certain extent on intuition, and the modes of action of some of the reagents used and the physical processes involved are still very obscure. The basic aim of the electron microscopist is to obtain morphological information, and to obtain it by reproducible methods which can be repeated anywhere in the world, so that the results obtained by one laboratory are strictly comparable with those obtained by another. Certain methods have been found to give what are generally agreed to be 'good results'; the criteria for these 'good results' will be discussed later in this chapter. The budding electron microscopist must study carefully the published results of the acknowledged masters of the art of the partcular method he intends to use, and strive to repeat them exactly before attempting to improve on them. The almost indefinable characteristic called 'quality' in electron micrographs is slowly and almost imperceptibly improving all the time as new methods spread and old ones are cast aside, and the microscopist must follow the published literature with diligence if he is not to fall behind.

The methods used at present to prepare biological specimens for electron microscopical examination fall basically under two heads: histological and

413

biochemical. The basic aim of the histological or 'cook and look' approach is to preserve tissues faithfully and exactly as they were in life and to render clearly visible all the inter-relationships between tissues, cells, organelles and macromolecules; in fact, to preserve structure at all costs, generally at the expense of function. The basic aim of the biochemical or 'grind and find' approach is to separate out comparatively large quantities of pure components of cells or tissues, and to discover and measure their functions; in fact, to preserve function at all costs, frequently at the expense of structure. The electron microscope is used by the biochemist mainly as a tool for recognizing separated particles by their shape and size, but it comes into its own in discovering certain ultrastructural features which may be masked or not preserved in ultrathin sections. The stalked particles which project from the surface of mitochondrial membranes and which have been christened 'lollipops' are a case in point. The structure of membranes is particularly difficult to decipher in sections, and workers are increasingly studying isolated fragments of membranes by the techniques which will be described in the last chapter. The two techniques, histological and particulate, require different resolving powers. Histological ultrathin sections rarely allow of resolutions better than about 20 Å, but the surfaces of isolated particles can be studied with the maximum available resolving power of the microscope.

18.2 Electron Histology

The remainder of this chapter will discuss the methods used in the preparation of histological ultrathin sections for the electron microscope. It will be recalled from Chapter 4 that the main characteristic which limits the choice of specimen is the 'mass-thickness', the product of physical thickness and density. About 50 % of a 50 kV beam of electrons will traverse a section of material of density 1 and thickness 0·1 μm (1,000 Å). This is the practicable upper limit of specimen thickness. It is not therefore possible to examine whole cells, either living or dead, in the normal 50–100 kV instrument, although as we have seen in Chapter 16 this may be possible in the future by using accelerating voltages of a million or more. It is necessary therefore to slice cells into sections which are only 500 Å thick or even thinner. Before this can be done, the cell must be stiffened so that it is capable of bearing against a sharp cutting edge sufficiently well for such very thin slices to be cut, and before this it must be killed in such a way as to preserve faithfully every detail down to the molecular architecture just as it was in life. The latter process is called 'fixation' and the former is called 'embedding'; the sequence is generally referred to as 'fixation and embedding'.

The necessary steps in electron histology are as follows:

1. Fixation: killing the tissue but at the same time preserving faithfully all tissue fine structure from the relationships between the cells down to the molecular architecture of the cell organelles.

2. Block staining: rendering the preserved fine structure visible in the electron microscope by causing certain components of it to attract heavy metal ions and thus to scatter electrons differentially. Stages 1 and 2 can be combined.

3. Dehydration: removing all the water from the tissue and replacing it with an inert fluid miscible both with water and with the infiltrating fluid.

4. Infiltration: replacing the dehydrating fluid with another fluid or mixture of fluids which can easily be hardened into a solid which is elastic enough to be sliced into 500 Å thick sections very thinly and evenly. Stages 3 and 4 can be combined.

5. Polymerization: hardening the infiltrating fluid permeating the tissue so that it forms a solid matrix supporting it rigidly without disturbing any of the spatial relationships.

6. Sectioning: slicing the solid matrix plus supported tissue (called the 'block' or sometimes the 'embedment') into 'ultrathin' sections less than 600 Å thick.

7. Mounting: transferring individual sections or ribbons of sections to copper specimen support grids for insertion into the electron microscope.

8. Section staining: increasing the existing differential electron scattering power (contrast) of the tissue constituents by reacting the mounted sections with solutions of heavy metals (section stains).

The sections of tissues are now ready for examination in the electron microscope without any further preparation. Prints of electron photographs of such sections are the familiar 'electron micrographs' published in the literature.

Each of the above steps will now be considered in greater detail.

18.3 Fixation

The aim of fixation is to preserve every detail of cellular ultrastructure right down to the molecular level exactly as it was in life the instant before the cell was killed. This ideal situation demands that all the processes of animation (enzymatically catalysed metabolism) be suspended in an instant of time, that the semiliquid contents of the cell and its surrounding tissue be instantaneously solidified without disruption, that the relationships of every organelle and every molecule be faithfully preserved exactly as they were at the instant of fixation, and that any process likely to destroy the structure after fixation, such as autolysis (self-digestion) or attack by microorganisms be prevented. The structures must also be preserved in such a way that the various processes involved in dehydrating and embedding do not remove any components, add unwanted ones, or distort relationships.

The first problem in fixation is to obtain cells in their normal living condition. Most tissues have to be obtained from a living animal. It must always be remembered that the very act of taking an animal out of a cage and placing it on a laboratory bench causes complex nervous and hormonal stimulation of many tissues which may well be reflected in changes in the ultrastructure of particular cells. Anaesthetizing the animal is no solution; anaesthetics are known to change the structure of certain cells, especially those of the central nervous system. As soon as the animal is dead and the heart ceases to beat, anoxia immediately begins to affect practically all the tissues, and the dying cells begin to release their autolytic enzymes. This problem is one which has been recognized since the inception of histology, and it has been greatly intensified by the electron microscope, which can detect changes far too small to be noticed by the light histologist.

The techniques employed for tissue removal and fixation must be carefully designed to cover each individual case. Obviously, if collagen is to be studied in tendons by fragmentation techniques, it matters little how or even when the animal was killed. On the other hand, preservation of brain tissue for example demands the technique of perfusion—substituting a fixative solution for the circulating blood in a living animal. This brings its own problems, for although the method will obviously give excellent preservation of the endothelial cells lining the larger blood vessels, it may also cause vasoconstriction and thus prevent the fixative from even reaching other tissues.

After having brought the fixative solution to the tissue, the second problem is to ensure that it penetrates through the extracellular connective tissue and the plasma membranes of the cells and spreads through the cytoplasm and nucleoplasm as soon as possible. Some fixatives such as osmium tetroxide solution are known to penetrate very slowly. The tissue must therefore be cut into very small pieces indeed—preferably cubes of not more than 0·5 mm side. Other fixatives such as aldehydes penetrate more quickly, but are still relatively slow. The tissue pieces can be somewhat larger in this case, but should not exceed 1 mm or at the most 2 mm in thickness.

As the fixative passes through the cell, it transforms the viscous colloidal protein sol which is the cytoplasm into a solid elastic gel. Effects due to differences in osmotic pressure between the fixative solution and the cytoplasmic sol may cause disruptive effects to the organelles, as may changes in pH due to the chemical reactions taking place at the advancing fixative–cytoplasm front which may release hydrogen ions. In theory, therefore, the fixative solution should contain in addition to the substance which is actually performing the fixation by 'denaturing' or gelling the protein sol two other constituents: neutral salts or other inert substances to make the solution isotonic with tissue fluid; and a buffer to 'mop up' released hydrogen or hydroxyl ions and take care of any dilution of the fixative caused by its mixing with the tissue fluid. The salts in the buffer may in themselves increase the tonicity of the fixative sufficiently, so the buffer itself may perform both functions.

The most suitable and generally used method for obtaining fresh tissue from a small live animal is to place it in a reasonably large covered glass vessel having the base covered with cotton wool, and then to allow it to quieten down. Anaesthetize it lightly with CO_2 gas, ether or chloroform, introduced through a tube, remove it and pin it down on a cork covered dissecting board as quickly as possible. Cut the skin over the required tissue with very sharp scissors, expose the area containing the tissue and flood it immediately with fixative solution from a Pasteur pipette. Remove a block of tissue with stainless steel instruments, and transfer it to a large drop of cold fixative placed on a piece of clean white card or dental wax. Dissect out the required tissue beneath the surface of the fixative under a × 5 stereoscopic dissecting microscope, and cut it carefully using a new scalpel or razor blade into fragments which do not exceed 1 mm at the most at the thickest part. Quite large flat sheets or strips of tissue can be fixed, provided the thickness does not exceed 0·5–1 mm. Transfer the required fragments to fresh cold fixative contained in a vial closed with a polythene stopper, using the tip of a wooden toothpick or a Pasteur pipette. 2″ × $\frac{5}{8}$″ sample tubes with polythene closures are very suitable; only about 1 ml of fixative need be used, thus economizing on expensive chemicals such as osmium tetroxide. Subsequent processing is carried out in the same vial by removing one solution and substituting another. Mark the tissue number on the vial with a diamond pencil. After the required tissue has been removed, do not forget to despatch the experimental animal by severing the spinal cord, or the aorta.

Fixation is the most important single procedure in the whole gamut of processes involved in biological electron microscopy. The highest resolving powers are of no avail on poorly fixed material. The essentials of fixation are speed and certainty. The reasons for speed have been discussed above. The reasons for certainty should be obvious, but are only really brought home when for instance a piece of connective tissue or fat is discovered to have been embedded in mistake for a piece of testis or thymus. The mistake is only too easy to make; even the most highly skilled amongst us have been known to make it from time to time. The resulting disappointment is enormous, but it is nothing to the waste of time and effort that went into the preparation of the wrong material. Such mistakes can only be prevented by the closest attention to the job in hand.

18.4 Fixatives

The primary purpose of the fixative is to solidify the protein sol which flows between the meshes of the network of phospholipoprotein membranes which form the framework of the cell, and thus to preserve the spatial relationships of the various organelles as they were at the instant of fixation. It should also render insoluble all the other chemical constituents of the cell, such as nucleic acids, nucleoproteins, carbohydrates and lipids. It is a great advantage if the fixative also provides electron contrast. A fixative may be very good for gelling, but very poor at preservation. Aldehydes, for instance, preserve structure excellently, but allow lipids to be completely extracted by the dehydrating alcohol. Osmium tetroxide is very good for rendering lipid insoluble, but has no effect on carbohydrates such as glycogen. Aldehydes preserve glycogen well, but are useless for providing electron contrast. Good preservation is of little value if the structure preserved cannot be seen. Osmium tetroxide on the other hand is the best substance known for providing electron contrast.

It is clear therefore that no one substance is likely to be able to perform all the functions required by a fixative. Osmium tetroxide (which should not be called 'osmic acid' because its aqueous solution is neutral) comes very close to being this ideal substance. It stabilizes protein sols by combining chemically with them and forming cross-linkages, not by precipitating them. It is therefore an almost perfect preservative of fine structure. It preserves lipids firstly by forming addition compounds with unsaturated fatty acid chains, and secondly by its solubility in triglycerides. It is very poor at preserving carbohydrates, but extremely good at preserving the phospholipoprotein membrane skeleton of the cell. It preserves nucleic acids poorly, but nucleoproteins well. But, above all, it is the best known producer of electron contrast. Since it combines chemically with practically all the constituents of the cell, osmium metal remains behind in the fixed cell, attached firmly to the structures it stabilizes, and delineating them almost perfectly. Electron contrast is produced by physical density, and osmium is the densest substance known. Osmium tetroxide has additional advantages: it does not harden and embrittle fixed tissue, so ultrathin sectioning is possible; and it neither shrinks nor swells the fixed tissue. Osmium tetroxide is therefore unlikely to be superseded as a primary fixative for electron microscopy.

In spite of its general excellence, osmium tetroxide has a few disadvantages. It allows glycogen to escape during the subsequent procedure, it does not react significantly with nucleic acids, and it penetrates tissue very slowly due to the low diffusion rate of its large molecules. These disadvantages can be overcome by using an aldehyde as a primary fixative, washing the aldehyde out, and then using osmium tetroxide as a secondary fixative. Aldehydes penetrate rapidly, have the property of stabilizing glycogen so that its subsequent loss is prevented (although aldehydes do not in fact combine chemically with glycogen), preserve certain protein structures such as microtubules which osmium tetroxide preserves poorly or not at all, and stabilize nucleoproteins better than osmium tetroxide. Most modern work is therefore on material which has been fixed first in an aldehyde (glutaraldehyde is generally used) followed by a second fixation in osmium tetroxide solution.

One other substance is occasionally used as a primary fixative in electron microscopy, although it is falling out of favour. This is potassium permanganate solution. It is a particularly good fixative for phospholipids, and preserves spatial relationships well, but fixes practically no other constituent of the cell. It is therefore very useful for cell membrane structures, and in particular for myelin. It also acts as an electron stain by precipitating coarsely (50 Å) granular electron dense material on the membranes. It has been used to a great extent on botanical material for its excellent preservation of chloroplast membranes, and on nerve material for the preservation of the myelin sheath. It is worth remembering if membrane structures are being primarily investigated.

18.5 Buffers and Additives

The purpose of buffers and other additives is to maintain the pH of the fixative solution at the physiological value in spite of the chemical activity of the fixative, and to prevent shrinkage and swelling of the tissues by osmotic pressure effects. Additives must obviously be free from disruptive effects on the cell, and must be non-toxic. Also they must not react chemically with the fixative, and must not precipitate out with change in pH.

The most commonly used buffer is a modification of Michaelis's barbiturate buffer, introduced into electron microscopy by Palade, who used it in conjunction with 1 % osmium tetroxide solution as a primary fixative. The original buffer contained sodium chloride in order to increase the tonicity; this is omitted in Palade's mixture. Michaelis's buffer is specially designed to cover a very wide range of pH, from 2 to 9. It contains two components: sodium acetate to provide buffering capacity in the acid range, and sodium barbiturate (Veronal or barbitone sodium) for the alkaline range. Baker has pointed out that since the buffer is used at pH 7·3 (slightly alkaline), the sodium acetate component provides no buffering capacity at this pH. However, it may well be that this component increases the tonicity of the fixative or has other unrealized beneficial effects. The use of Palade's fixative has the merit that it is used almost universally when osmium tetroxide is used as the sole primary fixative, and therefore results from laboratories all over the world are strictly comparable.

Other buffers may be used, provided they do not react chemically with the fixative and are not toxic to tissues. Phosphate, citrate, cacodylate, acetate, maleate and borate have all been used with apparently satisfactory results.

The necessity for using additives to increase osmotic pressure is a controversial matter. Some workers maintain that they are essential, others maintain that they are useless. Sucrose was added to Palade's fixative by Caulfield, and this mixture enjoyed a vogue for a while. Sodium, potassium, calcium and magnesium chlorides have also been used. The fact remains that no marked beneficial or deleterious effect has been reported from their use. Indeed, some workers maintain that osmium tetroxide alone in distilled water gives as good results as any of the fixative mixtures containing osmium tetroxide. The best fixative for a given material can only be determined by trial and error, comparing the results obtained and judging them by the accepted criteria of good preservation (see p. 470).

18.6 Fixation Technique

The two-stage fixation procedure introduced by Sabatini, Bensch and Barrnett using glutaraldehyde buffered with phosphate (Sorensen's dibasic–monobasic sodium phosphate mixture) or cacodylate followed by osmium tetroxide is now almost universally used. The primary glutaraldehyde fixative is used at a concentration of between 1–6 % in a 0·05–0·1 M buffer at around pH 6·8 to 7·6. It must be used cold (the vials are immersed in melting ice in the refrigerator during fixation) and the time for primary fixation varies from 1–3 hours depending on the size and penetrability of the blocks. After primary fixation, the glutaraldehyde must be washed out of the blocks very thoroughly, since it combines with and reduces osmium tetroxide, and would cause unwanted precipitation in the tissue. It is washed out with several changes of the buffer solution alone, to which sucrose may be added to increase tonicity and prevent extraction. It has been found that fixed tissue can be kept in the cold buffer wash for months without deterioration, although this may not be universally applicable.

Since aldehyde fixation does not blacken tissue and make it impossible to distinguish the various components, as does osmium tetroxide primary fixation, it is very convenient to microdissect the original blocks of tissue at this stage under the surface of the buffer wash. Osmium tetroxide penetrates very slowly indeed into fixed, gelled tissue, and the blocks must be made as small as possible before secondary fixation. They should never exceed 0·5 mm in one dimension, and should preferably be thinner. After microdissection, the tiny tissue blocks are returned to the vial with a Pasteur pipette and are now treated with osmium tetroxide solution. This may be buffered or unbuffered according to taste; a 2 % solution in distilled water gives excellent results. The secondary fixation may be carried out at room temperature, as are all the steps of the following dehydration and embedding procedures.

A word of warning will not be out of place concerning osmium tetroxide solutions. Osmium tetroxide is very volatile, and the vapour is intensely and unpleasantly toxic. Since it is an excellent fixative, the vapour kills any epithelial cells it comes into contact with. It is especially damaging to the corneal epithelium of the eyes, and the mucous epithelia of the nose and mouth. If the vapour is allowed to come into contact with the eyes, dead corneal epithelial cells will be sloughed off for days, interfering with vision. Osmium tetroxide, both solid and in solution, must always be handled with rubber gloves in a fume cupboard with the glass front of the hood drawn down as far as possible and the extractor fan turned on full. If it is used on the bench, a fan must be arranged to blow air across the operator's hands to prevent the vapour from coming in contact with the face.

Glutaraldehyde is not as volatile and consequently is not so dangerous. Care must be taken however not to get too close to the fixative solution and inhale the vapour. Glutaraldehyde may be used safely on the open bench; another good reason for using it as a primary fixative.

The single stage fixation procedure using Palade's fixative (1 % osmium tetroxide solution made up in 0·1M sodium barbiturate–sodium acetate buffer) is identical with the procedure for primary fixation in glutaraldehyde. The fixative is used in ice at around 4 °C. The tissue pieces must be cut up very small, since they are difficult to cut up after fixation owing to tissue blackening.

Many other primary fixatives have been described; the saline veronal–acetate buffered osmium tetroxide solution of Rhodin and Sjostrand; a similar mixture described by Zetterquist; Dalton's potassium dichromate–osmium tetroxide mixture, to mention only a few. Most have been compounded for some special purpose, and may not give good results if used indiscriminately. The beginner especially should stick to the two almost universally accepted methods (those of Palade and of Sabatini, Bensch and Barrnett) and strive to get acceptable results by using them before trying any variations.

The fixation methods so far described are chemical methods carried out in solution. Other methods are possible, notably the physical methods of heat and cold. Heat denatures protein, but in the process the spatial relationships of the cell organelles are so disrupted that the method is useless for electron microscopy. Proteins are also coagulated by cold, but the chief disadvantage is that ice crystals may form in the cytoplasm as the water separates out, and the fine structure is also disrupted. This can be prevented to a certain extent by soaking the tissue before freezing in protective agents such as glycerol, ethylene glycol or dimethyl sulphoxide. Freezing is then accomplished by plunging the soaked tissue into isopentane cooled to liquid nitrogen temperature. The tissue must be in very small pieces and the freezing must be very rapid if ice crystal formation is to be avoided. One of the advantages of this method is that the water can then be removed by sublimation, the specimen being kept cooled (at about − 20 °C) and under vacuum. The water is removed by condensation on to a cold finger cooled to liquid nitrogen temperature, or by a dish of phosphorus pentoxide. Special apparatus for freeze drying can be purchased commercially. The frozen-dried tissue is then embedded by infiltrating it with a plastic monomer (under vacuum or penetration may be poor) and polymerizing. Sections are then cut and stained, and the structures thus preserved can be compared with the structures preserved by chemical fixation. It has the further advantage that enzymatic activity is preserved to a far greater extent than with chemical fixation, and if the tissue is embedded in a water-soluble plastic (see p. 427), histochemical reactions can be performed on the sections to localize the sites of activity of certain enzymes.

18.7 Dehydration

The aim is to replace all the free water in the specimen with a fluid which is miscible both with water and with the embedding monomer. The dehydrating fluid may be ethyl alcohol, methyl alcohol, isopropyl alcohol, acetone, methyl cellosolve or the monomer of a water-soluble plastic embedding medium. Of these, ethyl alcohol is by far the most widely used, because it does not harden the tissue and make it brittle for subsequent ultrathin sectioning. Acetone especially hardens the tissue, and may be used in special cases for this purpose. The procedure is simple. A series of mixtures of water and dehydrating fluid of decreasing water concentration is made up. The fixative is poured out of the vial, care being taken not to pour the tissue blocks out with it, and the remainder is carefully removed with a finely drawn Pasteur pipette. The most dilute dehydrating fluid is then poured on. This is usually 50 % or 75 % ethyl alcohol. Since the tissue has generally come straight from a 1 % osmium tetroxide solution, a certain amount of osmium tetroxide will remain in the tissue. This is reduced by the alcohol to black, insoluble osmium dioxide, which forms a fine granular or colloidal suspension in the first dehydrating alcohol. This occurs over the first ten minutes or so. The first rinse of 75 % alcohol is then poured off, pipetted out, and another few ml poured on. There should be no further precipitation in this second fluid, which is left in contact with the tissue for 30 minutes. It is in turn poured off, pipetted out and replaced by 95 % alcohol, which is left in contact with the tissue for a further 30 minutes and is then substituted by 100 % alcohol. Two further changes of 'absolute' ethanol are given, followed by one change of ethanol which has been specially dried by standing over anhydrous copper sulphate. This final 'absolute absolute' alcohol must be kept carefully stoppered, and the copper sulphate must be changed when it begins to turn blue. It is essential that all the free water be removed, or a poor embedding will result, since the resin monomer is in general immiscible with water. A further refinement of dehydration is to pass the tissue from the last alcohol into an 'antemedium', which is a fluid completely miscible with both alcohol and the resin monomer. The fluid generally used is 1,2-epoxypropane (EPP), otherwise known as propylene oxide, which is completely miscible with the epoxy resins almost invariably used nowadays for general embedding purposes. EPP is extremely volatile and has a very low viscosity, so it penetrates tissue very rapidly, carrying with it the resin monomer, and then evaporates out, leaving the monomer behind. It is highly inflammable and is a fire risk. After the alcohol has been substituted by EPP by a further 30 minute soaking, infiltration begins immediately.

It is essential that the processes of dehydration and infiltration be carried out as rapidly as possible, because all the reagents used are powerful lipid solvents, and remove a significant amount of lipid even after it has been fixed with osmium tetroxide.

18.8 Block Staining

Further differential electron contrast can be added after osmium fixation and before the tissue blocks are embedded. The washing and dehydrating fluids can be used to add further stains. Uranium, iron, bismuth, and potassium permanganate solutions have been used following osmium fixation and before dehydration. 1 % phosphotungstic acid (PTA) dissolved in the final dehydrating alcohol gives very intense staining of collagen. If this procedure is used, care must be taken not to mix PTA-containing alcohol with epoxy-propane. PTA catalyses the reaction between alcohol and epoxy-propane, and an explosion followed by a fire can result. Indium trichloride in acetone is claimed to enhance the contrast of nucleic acids.

18.9 Infiltration

The procedure for infiltration depends on the choice of resin for the final embedding. In the early days of electron microscopy, the resin used was butyl methacrylate. This has exceptionally good cutting properties when polymerized, but produces gross artifacts in the tissue due to the shrinkage which it undergoes during polymerization. These 'explosion artifacts' drag the cell components apart, giving a characteristic vacuolated appearance to the embedded tissue. The polymer is also unstable in the electron beam, and evaporates when irradiated. This is advantageous for the production of contrast in the section, because the electron-scattering embedding medium is removed, leaving only the osmium-stained cell components behind on the supporting film. It is therefore unnecessary to section stain methacrylate sections before examining them in the electron microscope. But evaporation leads to further artifacts, due to the collapsing of structural features in the 600 Å thick section one on top of the other, causing gross distortion of ultrastructure below about 100 Å. The evaporated plastic has to go somewhere, and deposits on the polepieces of the objective lens. This leads to very rapid increase in astigmatism in the microscope, which soon becomes uncorrectable, and the column must be dismantled for cleaning. Because of its disadvantages, butyl methacrylate is hardly ever used nowadays as an embedding resin, and will not be considered further. It has been replaced by cross-linked epoxy resins, which hardly shrink on polymerization and are so stable in the electron beam that sections may be examined without a supporting film if mounted on 200-mesh grids. These resins are more difficult to cut into ultrathin sections than butyl methacrylate, due to their less elastic mechanical properties, and also need to be section stained to provide increased contrast, but their advantages in ultrastructural preservation are so great that they are now almost universally used.

18.10 Embedding Media

The disadvantages of the methacrylates arise mainly from the fact that they form linear polymers which do not crosslink to form a stable three-dimensional structure. The long chains of the linear polymer intertwine to form a stable solid, but it is very readily attacked by solvents and heat. A cross-linked polymer having similar elastic properties and similar compatibility with dehydrating alcohols was needed. Epoxy resins were first suggested by Maaløe and Birch-Andersen in 1956, but the resin they chose had limited compatibility with alcohol. A more compatible epoxy resin, 'Araldite', was tried by Glauert, Rogers and Glauert, also in 1956, and was found to have excellent properties as an embedding medium for electron microscopy. It is manufactured in large quantities as an adhesive and for various other purposes such as protecting the windings of transformers from attack by moisture and microorganisms, and consequently is cheap and readily available.

18.11 Epoxy Resins

This name is given to a family of thermosetting synthetic resins which have a characteristic light yellow or honey colour and syrupy viscosity, and which, when mixed with suitable 'curing agents' and heated, polymerize irreversibly into cross-linked, yellow-brown solids. The mechanical properties of the cured resins are functions of the resin monomer and the curing agents. The substance called the 'resin' has two types of chemically reactive group: epoxide end groups, and hydroxyl groups spaced along the length of the chain, with the following general formula:

The epoxy end-groups are highly strained three-membered rings, which open very readily and attach themselves to other groups containing reactive hydrogen atoms, notably amines. The addition of an amine to an epoxy resin will therefore cause the molecules to join up end-to-end, forming long-chain polymers. The intermediate hydroxyl groups are also capable of reacting with other reactive substances, notably acid anhydrides, to form cross-bridges between the resin molecules. If therefore a mixture of resin, amine and anhydride is heated together, polymerization will take place in 3 dimensions—lengthwise, and across the chains—forming a very stable, inert substance consisting of polyesters and polyethers very resistant to heat and solvents. The mechanical properties of the resulting polymer, which affect its cutting properties when it is being cut into ultrathin sections on a microtome, are governed by the length of the hydroxyl-containing central part of the resin chain, the chain length of the acid dianhydride, and the proportion of amine.

Commerical epoxy resins consist of mixtures of diepoxy resin molecules having variable numbers of intermediate hydroxyl groups, varying between 1 and 5. The greater the number, the more viscous the resin.

The cross-linking agent is called the 'hardener' or sometimes the 'curing agent'. It is usually the anhydride of a substituted dibasic organic acid, succinic or phthalic acid generally being chosen. The most commonly used hardeners are 'DDSA' (dodecenyl succinic anhydride) and 'MNA' ('methyl nadic anhydride' or more correctly 'methyl endomethylene tetra-hydro phthalic anhydride'). These are also yellow, syrupy liquids whose viscosity depends on the molecular weight.

The end-to-end linking agent is called the 'catalyst' or 'accelerator'. It is generally 'BDMA' (N-benzyl, N-N-dimethylamine), although other diamines are used. These are limpid liquids of low viscosity and characteristic, repellent odour of rotting fish.

The usual proportions used to form the cured solid resin are 50 parts of resin, 50 parts of hardener and 1–2 parts of catalyst. These components are mixed very thoroughly, and are then polymerized or 'cured' by heating to 60 °C for 48 hours.

The cutting properties of the final polymer depend on the components of the resin monomer mixture. The Araldite resin CY 212 (manufactured by Ciba, Duxford, Cambridge) yields an extremely hard, brittle block which is quite incapable of being sectioned if it is polymerized with MNA. If polymerized with DDSA, the block is still very hard; too hard for most tissues. It is therefore usual to make the cured block more elastic by adding a 'plasticizer' to the mixture. Dibutyl phthalate (DBP) is generally used in the proportion of about 0·5–2 % of the monomer mixture. The greater the proportion of DBP, the softer the block. On the other hand, if the resin Epon 812 (Shell Chemicals, known as 'Epikote Resin 812' in Europe) is used with DDSA, too soft a block is produced. If used with MNA, too hard a block results. This resin is therefore very convenient to use, because a cured block of any required hardness to match the embedded tissue can be made, simply by varying the proportions of MNA ('hardener') and DDSA ('softener').

The two types of embedding medium described above are generally known as 'Araldite' and 'Epon' respectively. Another epoxy resin, 'Maraglas', has been described by Freeman and Spurlock. The resin, Maraglas 655, is marketed by the Marblette Corp. of New York. It is used in conjunction with a hardener, Cardolite NC 513, and dibutyl phthalate. The resin has the disadvantage of being immiscible with ethyl alcohol, and has not attained the popularity of the other two, in spite of other claimed advantages.

18.12 Other Embedding Media

Another embedding medium which has been used with success on very hard materials such as bacteria and plant tissues is a polyester resin, 'Vestopal-W'. It has the disadvantages of being very viscous and immiscible with ethyl alcohol, but the sections are very stable under the electron beam. The method has been described by Ryter and Kellenberger.

Water-soluble embedding media have certain potential advantages, notably that reactions using aqueous media can be carried out on the sections. These embedding media have therefore considerable potential applications in the field of electron histochemistry which are only just being realized. They have the additional advantage that the use of drastic solvents such as alcohol and epoxy-propane as dehydrating agents is unnecessary; the tissue is simply infiltrated stepwise with mixtures of embedding medium and water containing successively less water.

Some low molecular weight epoxy resins, due to their polar nature, are miscible with water. About 30 % of Epon 812 is miscible with water, and can be removed by shaking the resin with water, separating the water-miscible part in a separating funnel or centrifuge, salting-out the water-soluble resin and finally drying it in a vacuum desiccator. The water-miscible resin can be obtained commercially under the name 'Aquon'. The resin itself is used as the dehydrating agent (it is a powerful fat solvent), and the infiltrated tissue is then further infiltrated with the usual mixture of resin, hardener (DDSA) and catalyst and polymerized. The resin polymer is not soluble in water. A water-soluble resin based on the Araldite family is called 'Durcupan' and is obtainable from Ciba. It is used in a similar way to Aquon.

A water-soluble embedding medium which appears to have had more success in histochemistry is based on methacrylate. Instead of the non-polar methyl and butyl esters, the polar 2-hydroxyethyl methacrylate commonly known as 'glycol methacrylate' (GMA) is used. Unlike the water-soluble epoxy resins, the polymer of GMA is affected by water, and sections swell in contact with it. Water-soluble histochemical reagents will therefore react with embedded tissue components, and GMA is finding increasing applications in the field of electron histochemistry. Like butyl-methyl methacrylate mixtures, it forms only non-crosslinked straight chains on polymerization, and hence requires no hardener. The catalyst used to promote end-to-end polymerization is benzoyl peroxide, used at a concentration of about 1 % in the final water-free infiltrating medium. It shows similar 'explosion artifacts' to butyl methacrylate, but these can be reduced with cross-linking agents such as divinyl benzene.

18.13 Block Hardness

Much of the success or otherwise in cutting ultrathin sections lies in matching the hardness of the embedded tissue to the hardness of the block of embedding resin. Some fixed tissues, notably those containing a high proportion of collagen such as dermis and tendon, are very hard, and will tear out of a soft block when an attempt is made to section them. Other tissues, notably embryonic tissues, tend to be very soft, and section unevenly if embedded in a hard block. It is necessary therefore to try to match the hardness of the polymerized embedding medium to that of the tissue. Epon 812 is very convenient here, because the hardness of the final block is controlled by the proportions of the two hardeners used in the monomer mixtures. The hardness of polymerized Araldite CY 212 is controlled by dibutyl phthalate, and glycol methacrylate is hardened by the addition of methyl methacrylate.

18.14 Embedding Procedure

The tissue blocks remain in their vial and the embedding fluids are changed, as in the dehydrating procedure. It is usual to replace the final dehydrating alcohol with a 50–50 mixture of alcohol and resin mixture before infiltrating with the full-strength resin mixture. A stock mixture of resin, hardener and plasticizer (Araldite) or resin and the two hardeners (Epon) is first made up in the correct proportions, leaving out the catalyst.

It is best to warm the resin and hardener before mixing to reduce the viscosity. Mixing must be very thorough. The resin-hardener mixture is stable at room temperature, and does not increase appreciably in viscosity for some months. Before embedding a vial of tissue blocks, a small quantity (about 25 ml) of resin mixture is catalysed by adding the required amount of the amine, and mixed very thoroughly indeed. The 50–50 mixture is made *in situ* simply by leaving 2 ml or so of absolute alcohol in the vial and adding an equal amount of catalysed mixture. 30 minutes soaking should suffice. Pour off the 50–50, and replace it with full-strength resin. Soak for some hours, then change the resin. Repeat this several times, depending on the size of the tissue blocks (the smaller the block, the shorter the infiltration time; another very good reason for using the smallest practicable tissue pieces). The resin becomes appreciably more viscous after a few hours at room temperature; therefore keep the small amount of catalysed monomer in a tightly stoppered bottle in the refrigerator. Care must be taken that water does not condense from the atmosphere into the cold resin when the stopper is removed. The final infiltration should be overnight, after which the infiltrating medium will have become almost too viscous to pour.

The choice of mould used to make the final block is determined by the chuck on the microtome it is proposed to use. Most microtomes have a chuck taking a cylindrical block of nominal 7·5 mm diameter by about 15 mm long, which is cast in a mould made of gelatin. The Size No. 00 gelatin drug capsules used in veterinary work (made by Parke, Davis) are generally used. These are placed on end in a simple holder made of a piece of hardwood or polythene drilled with a number of blind $\frac{5}{16}$″ diameter holes. The capsule lids are removed, and the tissue pieces are lifted carefully one at a time with the tip of a wooden toothpick from the viscous overnight infiltrating fluid, and the surplus embedding fluid is removed by rolling each piece carefully on filter paper. Each tissue piece is then placed precisely centrally at the rounded base of each capsule. A small label typed on thin tissue, about 3 mm by 10 mm, is then slid into the capsule to act as a permanently embedded designation of the block number. Each capsule is then filled to within 2 mm of the top with fresh low-viscosity catalysed resin mixture, and then the lids are replaced. The racks bearing the filled, closed capsules are then placed in an oven at 60 °C and are left for 48 hours to polymerize or 'cure' completely.

The catalysed embedding medium is best made up in throw-away plastic containers. Any glassware of any value must be cleaned immediately with alcohol, or the resin will polymerize in it. The small glass vials are also thrown away after being used once.

Special polythene capsules with the ends already pointed into truncated pyramids can be used ('BEEM' capsules, Polaron Ltd., London). These are more expensive than gelatin capsules, but the work of block trimming (see below) is simplified.

After the blocks have been polymerized, the racks are removed from the oven, the gelatin capsule lids removed, and the capsules are placed in hot water on a magnetic stirrer for an hour or so to dissolve the gelatin capsule away from the block. After drying and sorting, the blocks are immediately filed away in numbered pill boxes to await examination in the electron microscope.

It is not always convenient or possible to cut tissues into 0·5 mm cubes, and thin slices of much greater area may have to be embedded, for instance in studies of the retina, arterial endothelium and similar sheets of tissue. In this case, the flat tissue is best pinned down on to small pieces of 2 mm thick polythene using the smallest entomological pins obtainable. Processing then takes place on the polythene mounts until the final embedding, when the sheet of tissue is very carefully cut away from the pins and embedded flat in shallow polythene pill-box lids which can be obtained in various diameters. The most convenient size yields a polymerized block 4 cm in diameter and about 2–3 mm deep. The polythene lid mould can be pulled away from the block and used over and over again. These transparent blocks are then examined under a low-power dissecting microscope at about × 20, and the areas of interest are marked out by scratching round them with a mounted needle. The required areas are then sliced out with a very fine-toothed modelmakers' saw or else by pressing the block against the edge of a razor blade in a vice. The slices can be held in a vice chuck in the microtome, or the embedded tissue can be orientated exactly by trimming with a razor blade under a stereo binocular microscope. The trimmed orientated blocks are then fixed to blank resin blocks cast in gelatin capsules specially for the purpose. The catalysed resin mixture left over from an embedding is used to maintain a supply of ready-cast blank blocks. The end of a blank block is sliced off to present a flat surface, and the trimmed tissue block is affixed to it with a drop of catalysed resin mixture whicn is then polymerized in the oven. The composite block thus fabricated is then trimmed as described below.

18.15 Epoxy Resin Dermatitis

Some people have a strong allergy to epoxy resins and their components. The polyamine hardeners and catalysts seem to be the causative agents rather than the epoxide resin itself. Actual contact between the skin and these substances may bring on highly irritating skin rashes, and breathing in the vapours may cause wheezing, running at the eyes and nose, and sneezing. The utmost cleanliness must be maintained in the working areas where epoxy resins are used. If possible, they are best confined to the fume cupboard, especially the unpleasant-smelling BDMA. People who find that they are allergic to resin monomer mixtures should wear disposable rubber or plastic gloves and always work with the resins beneath a fume hood with the draught at maximum. Always wash the hands very thoroughly after using epoxy resins, even after touching the stock bottle. Use plain soap and water and not organic solvents such as chloroform to remove the resin from the skin. Some people may find it beneficial to use barrier creams. The resin manufacturers supply detailed instructions describing the precautions to be taken, and these should be strictly adhered to.

18.16 Ultramicrotomy

The cutting of even, undamaged sections free from scratches, holes, tears, chatter marks and other defects, and which adhere together in straight ribbons is undoubtedly the most difficult operation in the whole field of electron microscopy. Success in sectioning depends on the success of the embedding (correct choice of block hardness, thorough mixing of resin components, proper infiltration, correct polymerization), a reliable microtome, a really sharp, correctly adjusted glass knife, and last but not least, the skill of the operator. It cannot be too strongly emphasized that skill in cutting really first-class ultrathin sections comes only with long practice and considerable thought. Some people are far more adept than others; it is a manual skill of the highest order requiring dexterity in the highest degree. The essentials of the procedure will first be described, followed by a discussion of some currently available commercial microtomes.

18.17 Block Trimming

The block is first mounted in a chuck under a stereo binocular × 5 microscope. The rounded end of the embedded block is then trimmed freehand to a four-sided truncated pyramid of about 45° angle and about 0·5 mm square face (Fig. 18.1(a)). The square face is the one from which the sections are cut by the glass knife on the microtome. The trimming is done by first cutting horizontally across the top of the block with a new razor blade from which the oil has been wiped away, about 200 μm deep into the tissue (Fig. 18.1(*b*)). The outer part of the embedded tissue is almost invariably discarded, due to damage inflicted by the scalpel when it was originally removed from the animal and in subsequent cutting processes, and to the fact that the outermost part of the tissue block is almost invariably over-fixed or otherwise poorly preserved. It is often

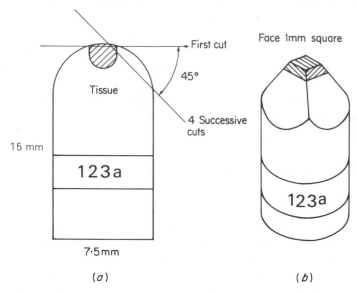

(a)   (b)

Figure 18.1(*a*)   Side view of a piece of tissue embedded in a No. 00 gelatin capsule after removal of the gelatin, showing the stages of trimming

(*b*) Perspective view of the trimmed block

instructive to examine the first freehand slice (if sufficiently thin) with a low-power microscope to confirm that the correct tissue is being examined —this can sometimes save a lot of time. When the required tissue has been exposed horizontally, make four further cuts at 45° to the block axis to give a square top face of approximately 1 mm side. The block is now 'rough trimmed' for cutting 'thick' 0·5 µm sections for examination in the light microscope.

The cutting quality of the block and the hardness match with the embedded tissue can be accurately judged while trimming. Too hard a block will shatter with conchoidal fractures and the tissue will tear out. Too soft a block will 'give' before the razor blade cutting edge with a characteristic 'rubbery' feel. Individual blocks in a batch may vary somewhat, and if all are trimmed at the same time, the best can be chosen for ultrathin sectioning.

Blocks should be allowed to mature if possible before sectioning. The resin improves in cutting properties over a period of weeks, if not months, after initial curing. Several days at least should elapse before the blocks are trimmed and sectioned.

18.18 Block Trimming Machines

For those constitutionally incapable of the accurate freehand trimming of blocks with a razor blade, machines are now commercially available. Such machines cost almost as much as a hand-operated ultramicrotome, and are thus an expensive luxury, but they are capable of greater accuracy in trimming than a freehand operator can attain. While this may not matter overmuch if only one single perfect section is required (this probably covers 90 % of biological section work), it becomes very important when long ribbons of sections are required for stereological investigations where three-dimensional information has to be built up from the virtually two-dimensional image obtainable from a single section. If the top and bottom edges of the block face are not exactly parallel, the ribbon of sections will curve over in one direction or another, and cannot be mounted on a straight slot grid or holder. Trimming machines ensure that the four faces of the truncated pyramid are exactly perpendicular to each other (or are at any other chosen angle). They may also act as convenient 'thick' (thicker than 0·5 µm) section microtomes for examination of the embedded material in the light microscope before cutting ultrathin sections for the electron microscope. This technique is described later in this chapter.

Three block-trimming machines are at present on the market: the 'Pyramitome', manufactured by the Swedish firm of L.K.B.; the 'Block Shaper', manufactured by the British firm of Cambridge Instruments; and the 'TM 60' manufactured by the Austrian firm of Reichert. The first two are simple hand-operated thick-section microtomes using glass knives and equipped with accurately and reproducibly orientable collet block holders which can easily be set to trim the block pyramid prior to sectioning. Both are excellent 'thick-section' microtomes.

The Reichert 'TM 60' is in effect a micro-milling machine, using a high-speed motor-driven end mill. The block is mounted in a chuck which can be swivelled to any angle. It is advanced horizontally towards the revolving cutter which is passed backwards and forwards across the block until the required amount of resin has been removed to expose the tissue.

This forms the block face. The chuck is then swivelled to the required pyramid angle, and each pyramid face in turn is shaped with the milling cutter. This instrument has the disadvantage that the tissue removed while trimming proceeds cannot be examined in a light microscope; it is not a 'thick section' microtome.

18.19 Ultramicrotomes

It is now about 20 years since the first ultramicrotomes which were genuinely capable of cutting sections thinner than 500 Å first appeared on the commercial market. Many designs have appeared since then, some of which have disappeared. By this process of the survival of the fittest, the prospective purchaser can be assured that the ultimate performance of all the instruments at present holding a share of the market is as good as it is possible to achieve in the present state of the art. All will routinely cut sections 500 Å thick, and will cut occasional sections as thin perhaps as 300 Å. Greatly exaggerated claims were made for microtomes in the early days—all were alleged to be able to cut sections less than 100 Å thick—but accurate measurements have since been made (see below) which show that these claims were for various reasons inaccurate by a factor of 4 or 5. Manufacturers are now more cautious in their claims. Most do not guarantee that their instruments will cut down to a stated thinness, since so much depends on the quality both of the block and the glass knife, to say nothing of the skill and experience of the operator. Microtomes differ mainly in their convenience, the number of adjustments provided, the degree of automation, flexibility for doing things other than simply cut ultrathin sections (e.g. block trimming and thick sectioning) and in the amount and quality of the accessory equipment provided (e.g. zoom magnification stereo binocular magnifiers). Just as with electron microscopes, you pays your money and you takes your choice.

Two of the most important steps forward in the cutting of ultrathin sections were the discovery by Claude in 1948 that the sections would adhere in ribbons and float away from the knife edge on a water surface, and the discovery by Latta and Hartmann in 1950 that the fracture edge of a piece of broken glass provided an almost perfect cutting tool. In consequence, the knives used are generally triangular pieces of $\frac{1}{4}''$ thick plate glass of about $1''$ base prepared in a fashion to be described later in this chapter, to the back of which a trough or 'boat' made of adhesive tape or metal is affixed with melted wax (see Fig. 18.8). The trough is filled with water or water plus a surface tension reducing agent, and the sections as they are cut float on the water surface, forming a continuous ribbon moving back from the cutting edge. The whole ribbon of selected sections detached from it are picked up by submerging a grid held in forceps behind the ribbon and bringing it up underneath.

The basic principle of an ultramicrotome is shown in Fig. 18.2. The essential ideals of a good design are extreme rigidity, perfect reproducibility and complete absence of sliding bearings. All the necessary movements are allowed by means of flexible strips as shown, point pivots or ball pivots. The embedded specimen E in its pyramidal trimmed block is gripped by the collet chuck C with as little overhang as possible. The chuck is attached to the free end of the very rigid arm A, and is constrained to move with three degrees of freedom: (a) up and down in an arc (1), flexing about the spring steel strip S_1; (b) from side to side in an arc (2), flexing about the spring steel strip S_2; (c) linearly (3), along the axis of the arm. Movement (1) is called the 'cutting stroke', and brings the block face down in contact with the cutting edge of the glass knife K, thus cutting the section. Movement (2) is called the 'by-pass stroke', and moves the

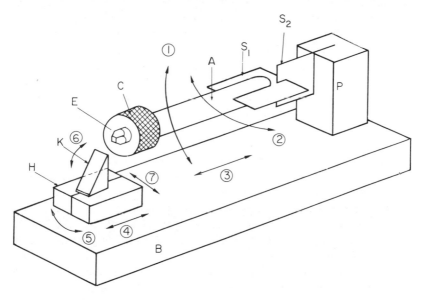

Figure 18.2   The basic parts of an ultramicrotome. The specimen embedded in the block E is held in the collet chuck C at the end of the massive arm A, which is free to move in the arc (1) about the spring leaf S_1, and in the arc (2) about the spring leaf S_2, which is held on the post P which in turn is fastened to the massive baseplate B. The triangular glass knife K is clamped in the holder H, and can be advanced, and swivelled. The specimen is moved towards the cutting edge of the knife either by thermal expansion of the arm A, or by a mechanical advance (not shown)

block face laterally so that when the arm lifts to return to its starting point (the 'return stroke'), the block face is moved well away from the back of the knife. This is to prevent electrostatic attraction from pulling the freshly-cut section back on to the block face. Movement (3) is called the 'advance', and moves the block face forward by an amount equal to the thickness of the desired section at each stroke.

The knife K is gripped firmly in a holder H, which can be moved axially (4) (knife advance) independently of the arm, which is fixed firmly by means of the strip S_2 and the post P to the very rigid base B. The knife angle can be adjusted both horizontally (5) and vertically (6) so that the cutting edge of the knife can be made exactly parallel with the block face, and so that the block face clears the back of the knife after the cutting action. The knife holder can also be moved bodily sideways (7) so that the best part of the cutting edge can be selected.

All microtomes are alike in having an arm moving in arc past a glass knife clamped in an adjustable holder, both firmly clamped to a very rigid base. They differ in the methods used to move the arm, the type of flexible mounting used, the method of advance, the method used to ensure adequate clearance between the block face and the knife edge on the return stroke, and whether the knife or the specimen chuck or both can be swivelled. Designs vary in complexity from the simplest hand-driven instrument derived from the rotary paraffin microtome to fully automatic motor-driven, push-button controlled instruments which will cut a complete ribbon of sections in the absence of the operator.

Probably the most important point of design difference between microtomes is the method used to advance the block towards the knife edge. This can be classified into mechanical advance and thermal expansion advance. Mechanical advance gives incremental steps of a certain given minimum thickness, which should be fully reproducible and independent of the time taken for a complete cutting cycle. Thermal advance on the other hand is continuous, and since expansion takes place continuously, the longer the time taken between two consecutive cycles, the thicker will be the section. Section cycle time must be absolutely regular on a thermal advance microtome, and therefore all such instruments are driven by a motor or other reproducible electrical device. Hand-operated instruments, since the cycle time is bound to be somewhat irreproducible, are fitted with mechanical advance using a micrometer lead screw and some kind of pawl and ratchet device to turn a nut by multiples of a fixed minimum increment on each cycle. Either the nut or the lead screw is coupled to the arm through its rear flexible mounting, and so advances the specimen by a chosen fixed amount on each upward by-pass stroke.

18.20 Representative Commercial Ultramicrotomes

The ultramicrotomes to be described are those of which the author has had first-hand experience. The list is not intended to be exhaustive; other instruments are on the market. No disrespect is implied if a particular microtome is not mentioned; the reason is simply that the author has not been able to operate it personally for long enough to assess it adequately. Simple microtomes are described first, followed by more complex ones.

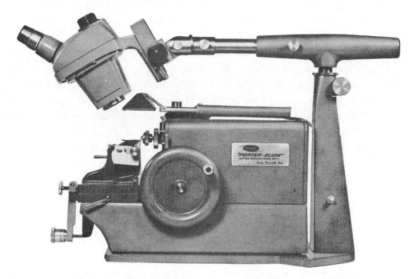

Figure 18.3   An example of a hand-operated ultramicrotome, the Sorvall 'Porter–Blum' Mark 1

18.20.1 Porter—Blum MT–1 (Hand-operated, see Fig. 18.3)

Made by Ivan Sorvall Inc., Norwalk, Conn., U.S.A. Designed by Prof. K. R. Porter and J. Blum at the Rockefeller Institute, New York. This was the first simple, practicable hand-driven microtome to be marketed. The design dates from about 1957, but has persisted with only minor modifications ever since. The arm is moved up and down on the cutting stroke by means of a crank turned by hand with a flywheel mounted at the side, and sideways on the by-pass stroke by means of a face cam. The fully orientable collet specimen chuck (a vice chuck is available) is mounted on a ball at the end of the arm, and can be swivelled in any direction. The knife clamp has vertical angle, fore-and-aft and sideways movement. The arm is supported at the rear on gymbals with spring-loaded ball contact pivot bearings, and is advanced forward by means of a compound lever of very high mechanical advantage. The lower end of the lever is advanced by a nylon half-nut on a lead-screw, which is turned by a pawl and ratchet. One tooth of the lead-screw ratchet gives an advance of 250 Å, so the microtome can only be used seriously at two settings: two teeth ($=500$ Å) and 3 teeth ($=750$ Å). It is at present impossible to section regularly at 250 Å, and sections thicker than 750 Å can only give rather low resolution, although the contrast is high. Because the instrument is hand-driven, it has possibilities of immense flexibility, but only if the operator is very highly skilled in its use. Cutting speed (the rate at which the block face moves across the cutting edge, measured in mm/s) can be varied simply by varying the speed of rotation of the handwheel; a really skilled operator can even cause the cutting speed to vary at different parts of the block face. The cycling time or 'cutting rate' can be varied immediately at will from about 4 c/s (as fast as the wheel can be turned evenly) down to as long as the operator chooses. About 50 sections/minute are normally cut. It also means that a block face can be pretrimmed by cutting approximately 0·5 μm sections far more rapidly than on any other instrument. Its great virtues of speed, simplicity and flexibility endear it to microtomists of the older school, who were mostly brought up on it. It can also be used as a thermal expansion microtome, simply by setting the mechanical advance to minimum, taking off the cover and placing an Anglepoise lamp close to the arm. The prototype version operated in this way. The maximum setting of the advance is 19 teeth, so it will cut approximately 0·5 μm thick sections for examination under the light microscope without any trouble other than resetting a single knob. There is also a multiple by-pass attachment so that the arm can be by-passed an integral number of times, allowing sections as thick as 2 μm to be cut. The instrument is one of the cheapest on the market, but it must be emphasized that very considerable manual skill is needed to operate it. A very satisfactory optical system for illuminating and examining the sections as they are cut on to the water surface, which includes a Bausch and Lomb zoom stereo binocular microscope, is supplied as an accessory.

18.20.2 Porter–Blum MT–2 (Automatic)

This is a refined and motor-driven version of the original MT–1, retaining most of its features. The bearings are spring-loaded knife edges. The mechanical advance system is modified so that a very much larger continuously variable range of section thickness from zero up to 4 μm can be obtained. Specimen retraction is by retraction of the specimen arm, not by sideways movement. A stepless range of cutting speeds from 0·09 to

3·2 mm/s is provided, although the cycling time at the lowest cutting speed is $2\frac{1}{2}$ minutes. The motor drive can be overriden by a handwheel at the side, which can be used instead of the motor drive to allow for manual operation. The knife holder can be swivelled through a large angle for the cutting of block pyramids, and a push-button allows alternate thick and thin sections for light and electron microscopy to be cut. The mechanical complexity of the instrument is considerably greater than that of the MT–1. It is provided with the same excellent Bausch and Lomb zoom binocular and fluorescent lamp as is fitted to the MT–1. The instrument can be set up more quickly than other automatic microtomes, and it is capable of cutting excellent ribbons of even serial sections.

18.20.3 *Cambridge–Huxley Mark 1* (Hand-operated)

Designed by Prof. A. F. Huxley and made by the Cambridge Instrument Co. Ltd., Cambridge, England. This is a hand-operated, mechanical advance instrument; the cutting stroke is gravity powered and the cutting speed is regulated by an oil-filled dashpot. The rate of fall of the arm can be controlled over wide limits by adjustments of the oil dashpot by-pass screw, and cutting speed can be reduced to 0·1 mm/s or less, the normal cutting speed being about 1 mm/s. This can be an advantage when cutting very hard blocks. The arm is recycled by lifting an operating lever at the side with the left hand, which automatically moves the arm aside for the by-pass stroke. The recycle time is about 1 second, and the normal time for complete descent is about 8 seconds. Thus, about 6 sections per minute can be cut. In consequence, pretrimming before the block begins to section over the whole of its area can be very time-consuming compared with the Porter–Blum. The mechanical advance is by means of a micrometer, which moves the double leaf spring arm suspension forward through a long lever. The micrometer spindle is moved around by a ratchet wheel operated by a pawl attached to the operating arm. The smallest automatic advance increment is 50 Å, one tooth of the ratchet, and the maximum section thickness is 1,500 Å, 30 teeth. Thick sections are conveniently cut by opening a small door at the side and advancing the ratchet wheel manually; one quarter turn gives 0·5 μm. The knife holder can be swivelled and angled but the specimen chuck (a vice chuck is available) is fixed. Knife feed is by a large micrometer mounted on the knife holder. This instrument is also one of the cheapest on the market.

18.20.4 *Cambridge–Huxley Mark 2* (Automatic)

This is a refined and motor-driven version of the original Mark 1. Basically, it is mechanically identical to the Mark 1 and can be hand operated in exactly the same way. In addition, a small electric motor is fitted in the base plate, which drives a cam lifting the chuck arm on the by-pass stroke. The motor then stops, and the arm falls by gravity on the cutting stroke, the cutting speed being controlled by the oil dashpot. At the end of the free fall, a microswitch is actuated, starting the motor again and returning the arm to the top of the stroke. Press-button single-stroke operation is provided by electrically by-passing the lower micro-switch. Advance is mechanical by pawl-driven micrometer and lever. The

Figure 18.4　An example of an automatic motor-driven ultramicrotome, the LKB 'Ultratome' Model III

oil dashpot is now externally adjustable by means of a 'cutting speed' control, giving speeds between 0·18 and 1·8 mm/sec. Cycle time cannot be adjusted independently of cutting speed, but varies between 3 and 10 cycles/min at minimum and maximum cutting speeds respectively. The automatic advance has now been provided with two immediately selectable and independant ranges. 'Macro' is for thick sectioning, and can be set to give sections of 0·5, 1·0 or 2·0 μm thick. 'Micro' is·for normal ultra-thin sectioning, and can be set from zero to 1,500 Å in 50 Å steps. The amount of micrometer advance is indicated on a scale on the front, and is reset by a handle on the macro knob. Specimen height is set by a control similar to that on the Mark 1 version. The knife holder is basically similar, and can be swivelled $\pm 20°$, traversed back-and-forth and sideways by micrometers, and the knife rake can be set between $0°$ and $15°$. All knife movements are rigidly clamped while sectioning. The illuminating and viewing system is mounted on a completely separate stand with a built-in power supply for the swivellable 4 W fluorescent strip lamp. A choice of Bausch and Lomb, Beck or Olympus stereo zoom binoculars is offered, the range being $\times$ 7 to $\times$ 40. The microtome can be remotely controlled both in serial and single-section modes by a hand-held control box. Optional accessories include a vice chuck and a fully orientable specimen holder, either of which simply replaces the standard fixed-orientation 7 mm collet chuck. A tissue-freezing device for cryo-ultramicrotomy is also available.

The instrument is very robust and simple to use. Untrained personnel can cut excellent sections after only an hour or so of instruction. It is the cheapest, fully automatic ultramicrotome available.

18.20.5 LKB Ultratome (Fig. 18.4)

Designed by Algy Persson and manufactured by LKB, Stockholm, Sweden. This was one of the first really successful fully automatic ultramicrotomes. It is basically an arm carrying a fully orientable chuck as shown in Fig. 18.2 but which can only move up and down. There is no by-pass on the return stroke; instead, the knife holder is retracted about 30 μm away from the block face by means of an electromagnet, which is excited on the up-stroke and flexes the base of the knife holder. The arm is moved up and down by means of a thin Terylene cord attached to its mid-point, which in turn is attached to a lever which is attached to the shaft of an electrical 'torque motor'. This is basically a giant permanent magnet moving-coil galvanometer movement. When a current is passed through the coil, the operating lever (which corresponds to the indicating needle of the moving coil galvanometer) moves up and down, carrying the arm and chuck with it. Knife advance is manual; automatic advance is thermal. Three heating coils are mounted along the length of the specimen arm, and current is fed into them in such a way that the heating power increases with time, and so the effect of heat loss with temperature is compensated for. Expansion is claimed to be linear with time over a period of about 15 minutes. When the arm is hot, it can be cooled down rapidly by sucking a current of air down its hollow centre; the instrument is ready for a subsequent run in about 10 minutes. Section thickness is determined by the current through the heaters, which is measured by an ammeter calibrated in section thickness. Maximum thickness is 1,000 Å. The time for the cutting cycle is fixed at 3 seconds for the fixed cutting speeds of 2, 5, 10 and 20 mm/s. A slow cutting speed of 1 mm/s can be arranged, when the cutting cycle time is doubled, and hence the heater current must be halved for the same section thickness. The cutting power, like the Huxley, is gravity, damped in this case electromagnetically by a current passing through the coil of the torque motor. The return stroke is electrically powered. The lifting and damping currents for the torque motor are generated by transistorized pulse-generating circuits housed in a remote control box. The operation of the microtome can thus be remotely controlled, and once it begins to cut sections, it is completely automatic, and will cut a ribbon of up to 200 sections over a period of 10 minutes in the complete absence of the operator, whose breathing and movements do not perturb the microtome. Like all automatic machines it is greatly superior to any hand-operated instrument for cutting ribbons of serial sections of almost completely even thickness. The instrument is superbly made and finished, and most of the features are very well thought out. All the transistorized circuits are on plug-in printed circuits, so servicing is simple. The controls provided allow the cutting range level to be adjusted, to compensate for different knife heights; single-stroke, automatic or manual control; cutting speed control; two-range section thickness control; and controls for cooling air blower and lighting. A NIFE stereo binocular microscope is provided, together with a swivelling combined tungsten and fluorescent lamp which can be set in the correct position to view the knife edge in darkfield (see below) and thus to judge the quality of the cutting edge before sectioning begins. It can also be used for block trimming in conjunction with a special holder. The instrument is available in two versions; the Model I as described above, and the Model III, which has additional refinements such as an extended cutting speed range from 0·1 mm/s up to 60 mm/s. The low speeds enable very large block faces (in excess of 1 mm square) to be sectioned, while the higher speeds enable very soft embeddings (e.g. the gelatin–water embeddings now being tried for electron histochemistry) to be sectioned. One drawback of all fully automatic instruments is the length of time they take to set up, compared with the hand-operated Porter–Blum and Huxley. Both patience and skill are necessary to get the microtome cutting regularly, but once this has been achieved, the results are as good as can be obtained in the present state of the art.

18.20.6 Reichert Om U3

Designed by H. Sitte and manufactured by C. Reichert A.G., Vienna, Austria. This is the most fully automated ultramicrotome available so far, and is probably the simplest to use. The principle again follows Fig. 16.2. The arm, carrying a fully orientable chuck, is mounted on two steel bearing balls running on hardened steel tracks, allowing the arm to move up and down and sideways. The arm is lifted by a motor-driven cam and connecting rod with a handwheel over-ride through two further bearing balls, and at the same time is displaced sideways to by-pass the knife. The arm drops on the cutting stroke, driven by gravity but supported by the connecting rod. The two-speed motor drive gives a slow cut, adjustable in height, and a rapid by-pass return. The advance feed is by thermal expansion, using a separate expansion unit of very low thermal inertia mounted between the arm post and the baseplate, which is claimed to be completely linear with time and to operate without delay. An automatic, disengageable mechanical advance feed to the knife is also provided. The three important variables in microtomy—cutting speed, advance rate and cutting rate—are set independently simply by altering the settings of 3 dials. The dials each carry a red dot marking corresponding to an 'average' setting which will give approximately 600 Å thick sections from a resin block of average cutting properties. If suitable sections are not forthcoming at this setting each parameter can then be adjusted independently as cutting proceeds. Cutting speed is continuously variable from 0·5–10 mm/s; cutting rate from 6–12 sections/minute, and section thickness from zero to 1 μm. The actual operation of the instrument is by using a group of 4 push-buttons for main switch, light, thermal feed and motor drive. The knife holder design is based on that of the tool post and saddle of a lathe, having two screw-operated dovetail slides for fore-and-aft and sideways movement. The automatic knife advance is used to cut thick sections. The knife clamp can be swivelled vertically and horizontally to adjust the knife angle, and the chuck is fully swivellable to adjust the specimen angle. The chuck can be removed from the arm and mounted in place of the knife clamp, so that freehand trimming can be done on the microtome, using the stereo binocular microscope (Reichert with 3-power lens turret or A–O Zoom). The complete instrument is mounted on a convenient console table. Because of the number and precision of the adjustments provided, the instrument is reasonably quick to set up, and, because of the automation and logic of the controls, excellent sections can be cut by a novice after a short acquaintance with the instrument.

Like all automatic microtomes, the Reichert excels at cutting very long ribbons of even serial sections, which can be cut in the absence of the machine-operator. The instrument has few drawbacks other than price and the results are as good as can be obtained in the present state of the art.

18.21 Cryo-ultramicrotomy

The advantages of rapid freezing of fresh tissue as a method of fixing and preserving cell and tissue components are well known in light microscopy. Recently, many attempts have been made to apply the method to electron microscopy. Early attempts showed drastic structural disruption at the ultrastructural level due to ice crystal formation during freezing. This can be partly overcome by extremely rapid freezing and by sectioning at very much lower temperatures than those used in light microscopy. Very small pieces of tissue are plunged into isopentane cooled with liquid N_2, the freezing rates being of the order of hundreds of degrees per second. The frozen tissue is sectioned immediately on a conventional ultramicrotome which has the knife and block surrounded by a special insulated chamber cooled with circulating liquid N_2. Sections are collected on grids either from a dry knife or from a flotation medium such as cyclohexene, and are then freeze-dried before being examined either stained or unstained in the electron microscope. Unstained contrast is very low, but special histochemical staining methods are being worked out. The method seems promising for localizing soluble cell components which are removed during conventional fixation and embedding procedures. In consequence, most ultramicrotome manufacturers are now offering accessory apparatus for 'cryo-ultramicrotomy', as this technique is now becoming known.

18.22 Section Thickness

All modern ultramicrotomes have automatic advance mechanisms which should in theory be able to cut sections down to a few Ångstroms in thickness. Yet it seems to be impossible to cut sections evenly of less than 300–400 Å in thickness. Why should this be? Unfortunately, at the present time nobody seems to know the answer to this. The physics of the cutting process appears to be very imperfectly understood. Whatever the embedding material, whatever the knife material, whatever the microtome, however apparently perfect the knife edge, whatever its angle, sectioning seems to stop at around 300 Å, except for odd wedge-shaped areas found in sections of uneven thickness in which it may be possible to find areas which may be thinner than this, using internal standards of reference of known size. The ability to cut sections 50 Å thick routinely may be the next great stride forward in the examination of biological ultrastructure at high resolution. At the time of writing, nobody seems to have any very clear idea as to how this will be accomplished.

18.23 Section Thickness Measurement

Ultrathin sections are cut on to the surface of water or other liquids on which they float until picked up or 'harvested' on an electron microscope specimen grid. When viewed by reflected light, the difference in optical path length between the light reflected from the surface of the water beneath the section and the light reflected from the upper surface of the section gives rise to interference (see Chapter 1) between the two light rays, which, if white light is used, causes reinforcement at certain visible wavelengths and hence makes the sections appear to be coloured. Because the path-length difference is very small (about one-fifth of a wavelength for a 500 Å thick section) the colours are very pale and barely discernible, let alone judged with any degree of accuracy by the eye. The colours to be expected can be calculated from a knowledge of the optical properties of the resin; they correspond fairly well with the section thickness actually measured by using other methods. The colours seen by the eye can be described as follows. The very thinnest sections can barely be made out; they appear just a little lighter than the surrounding water, and are called 'dark grey'. Slightly thicker sections are easier to see (see Fig. 18.14, p. 461), and are called 'grey'. The faintest tinge of yellowish colour can next be discerned, which looks very much like the light reflected from the surface of polished silver, so these sections are called 'silver'. As the section thickness increases, the yellow colour becomes more predominant, and sections are called 'gold' and then 'bronze'. A bluish tinge now begins to creep in, and the sections look 'purple'. The colours now become more saturated and obvious, passing through blue, green and yellow. Sections which are thicker in the colour scale than purple are in general useless for electron microscopy, being far too thick.

The correlation of these interference colours with actual section thickness exercised microtomists from the very earliest days of ultrathin sections, and the first estimates were made by shadowing the edges of methacrylate sections with metals (see Chapter 19). Due possibly to the evaporation of the methacrylate in the vacuum and under the heat of the evaporating metal, some of the first estimates were considerably in error, silver sections being estimated as about 100 Å thick. As time went on, it became increasingly obvious that sections were in fact considerably thicker than these early estimates. Although few physical methods are applicable to the measurement of such extremely thin objects, microtomists in various parts of the world began to devote much energy to the accurate determination of this very important dimension. Bachmann and Sitte in 1960 used multiple beam interferometry; Peachey in 1958 used ellipsometry; Williams and Meek in 1965 used radioactive sections of known specific activity; and Rösch in 1959 published a theoretical optical treatment. All these workers obtained results which were in very good agreement, in spite of the very diverse methods and treatments used. One very important observation has emerged: in the 500 Å thickness region, the thickness of a section has to change by at least 50 % before the change can be appreciated by the eye observing section colour. This is because the eye is here trying to differentiate light intensity (the difference between 'dark grey' and 'light grey'), a function for which it is a notoriously poor performer, rather than differences in actual colour. Also, autoradiographs made from the radioactive sections have conclusively shown that there is a very great *intra*-section thickness variability; a single section of one apparently uniform interference colour can vary in thickness over its area by as much as 50 %. The most characteristic conformation of such sections is wedge-shaped, or thinner at one end than at the other.

The colour sequence for sections of epoxy or methacrylate resins floating on water is now generally accepted as being approximately as follows:

Dark grey	less than 400 Å
Grey	400–500 Å
Silver	500–700 Å
Gold	700–900 Å
Purple	greater than 900 Å

Purple sections are virtually useless for electron microscopy, and gold sections give rather poor resolution. Is there an optimum section thickness? Obviously, as section thickness decreases, so contrast and hence visibility both decrease. There is very little point in having the possibility of very high resolution in a section if the embedded material is incapable of forming an image. So far, even the thinnest sections of reliably reported thickness have yielded excellent contrast after suitable section staining. The section stain is probably a surface phenomenon, and contrast in a stained section is not as dependent on section thickness as might be expected. Whether 50 Å thick sections will obey this rule when and if they become possible is a matter for conjecture. As things stand, the general rule is: the thinner the section, the better. However, epoxy resin sections much less than 500 Å thick are very delicate indeed, and will not in general stand up to the electron beam when mounted unsupported on bare grids, as is the modern practice. Such sections need to be supported on carbon films and nets (Chapter 17), and this involves a great deal of preparative work. The tendency therefore is to work mainly with 'barely silver' sections, about 600 Å thick. These can give resolution of around 25 Å, and are entirely adequate for 95 % of thin section biological work, where instrumental magnification very rarely exceeds × 50,000.

18.24 The Microtome Knife

The simple edge of a piece of broken glass, which goes back to the Stone Age, appears to be the sharpest cutting edge the wit of Man has so far been able to devise. The polished fractured edge of a diamond is harder and more wear-resistant, but is not a better cutting edge. All the attempts which have so far been made to sharpen steel knife edges have yielded results very inferior to the glass edge. For these reasons, and because of the extreme cheapness of the raw material from which glass knives are made, glass knives are used almost universally in ultra-microtomes.

The exact way in which the glass is fractured is of the greatest importance in making a satisfactory glass knife. The starting material is a $12'' \times 8''$ sheet of $\frac{1}{4}''$ (or 7 mm) thick common shop window plate glass, obtainable from any glass merchant, who will supply sheets ready cut. The kind of glass is important, since this determines its freedom from internal stresses. All plate glass was until recently made by a polishing process, which ground away the irregular top and bottom surfaces of a thick sheet extruded from a vat of semi-molten glass. In this way, the cooling stresses in the outside layers were removed, and the glass (known as 'polished plate') was to a great extent relieved of internal strains, and would crack very predictably and reproducibly. Unfortunately for the electron microscopist, this process has now been almost universally replaced by a flotation method of manufacture, where the sheet of molten glass is extruded on to the surface of a bath of molten tin. Here it flows to give almost perfect, optically flat surfaces which do not need polishing, but the cooling stresses remain in the glass. This 'float glass' is not as satisfactory for making glass knives as the old 'polished plate'. The difference is not great, but the knives made from float glass appear to be softer, more friable and more difficult to make reproducibly. The microtomist is therefore urged to find a supply of polished plate glass if possible. One sheet the size of a good shop window will last for many years, and is still to be found in the stocks of glass merchants.

The sheets of glass are first cut into strips $12'' \times 1''$. This is done by scoring down the centre of the sheet as lightly as possible with a new wheel cutter; diamonds such as are used by glaziers give very unsatisfactory results. The sheet is then cracked in two by placing it on a folded cloth with the score line uppermost, placing a throat swab stick beneath the score, and then leaning with the palms of both hands evenly on both sides of the score. If the scoring has been done correctly and to the very edge of the sheet, a crack will run slowly along, following the score, cutting the sheet neatly in two. The fracture should not show any significant 'frilling', and the surface of the glass exposed by the fracture should have an almost mirror finish. The 12″ × 4″ sheets are next divided down the centre to give 12″ × 2″ strips, which are in turn cracked down the centre to give the 12″ × 1″ strips from which the microtome knives are made.

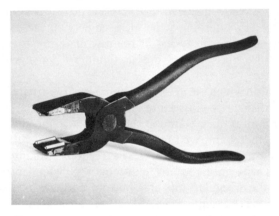

Figure 18.5 A pair of glass pliers modified for cracking glass knives by taping 3 hardwood sticks to the jaws

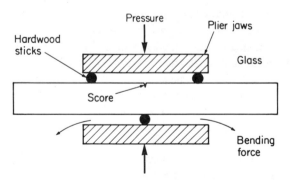

Figure 18.6 The application of the modified glass pliers for cracking plate glass strips and squares to make glass microtome knives

Suitable 1″ wide strips can be purchased ready-made from some microtome manufacturers, but this not unnaturally increases the cost of a glass knife by a large factor. After the 1″ wide strips have been made, they are then further divided into 1″ square pieces, which are the raw material from which the triangular knives are made.

Cracking glass by hand as described above is greatly facilitated by the use of a pair of modified glass pliers (Fig. 18.5). In order to crack glass cleanly, a line of pressure (A, Fig. 18.6) must be applied exactly under the score, and two lines of pressure (B) must be applied from above exactly equidistant on either side of the score, giving rise to a bending force tending to open out the crack formed by scoring. The pliers are modified to apply this bending force simply by taping short lengths of circular cross-section hardwood throat swab sticks to the inner surface of the jaws, two sticks on the upper jaw and one on the lower (Fig. 18.6). The piece of glass is then gripped in the plier jaws, two sticks on the upper scored surface and one on the lower unscored one, and the plier handles are grasped with both hands together. Force is then applied to close the pliers (considerable force is necessary, hence the need for both hands). It must be applied at a slowly increasing rate and in a carefully controlled fashion. The crack will run out from the pliers quite slowly; it should take about a second to run from one end of a 12″ strip to the other. The more slowly and evenly the crack runs, the more perfect will be the flat surface thus produced. Large pieces of glass must be supported on the cloth on the bench top, the plier jaws being pressed against the side of the bench. The ability to crack large pieces of glass is acquired only by practice, and is best demonstrated.

The actual triangular glass knives are made from the 1″ square pieces of glass using the modified pliers in the following way. First, a light wheel score is made on one surface of the glass square which is slightly displaced from the diagonal in the direction shown in Fig. 18.7(a). The score must stop short of the edge to give a free break of several millimetres. This free break will curve over into the surface to the left of the diagonal, and should meet it in the previous free break area about 1 mm below the intersection of the two perpendicular edge faces (Fig. 18.7(b)). The edge formed at the termination of the free break path and the smoothest flat edge is the actual cutting edge of the knife. The scored squares are cracked using the modified pliers as shown in Fig. 18.7(c). The plier jaws must be placed so that the hardwood pressure points are exactly parallel with the score and not with the diagonal.

The finished glass knife has the appearance shown in Fig. 18.8(a). The cutting quality of the edge is determined mainly by the position of the intersection of the free break with the free break of the clearance face, '*d*' in Fig. 18.8(*b*). *d* should be 0·5–1 mm. The length of the cutting edge can be divided roughly into 3 regions: (1) the best cutting edge; (2) the 'frilled' region, useful for rough trimming or for cutting thick sections for light microscopy; and (3) the 'spur' or 'peak' region on the lower side of the knife as it is broken, which is useless. The aim is to get the greatest length of usable cutting edge. If *d* is too great (Fig. 18.8(*b*)), the spur will be large, and the 'frills' or 'whiskers' which form on it will run for half the length of the edge or even further, yielding a very short usable cutting edge (1). If *d* is too small, the score side of the edge will curve away as shown in Fig. 18.8(*d*), yielding a curved edge which is useless for sectioning. A reasonably long good straight cutting edge may result (about one-third the total length), but the trimming region (2) will be short. The ideal

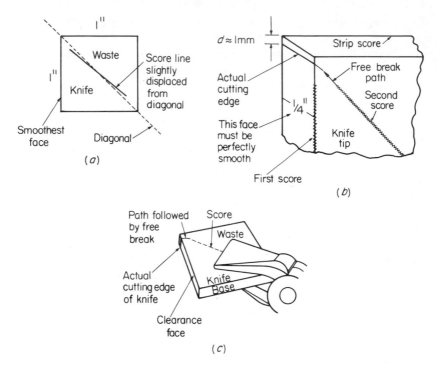

Figure 18.7(*a*)   How to score a 1″ square piece of glass to crack a microtome knife

(*b*) The path which should be followed by the free break on cracking the 1″ glass square with the modified pliers

(*c*) Holding the glass square in the pliers to crack the knife

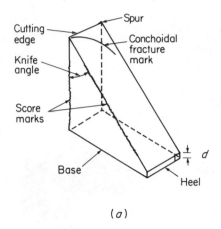

(a)

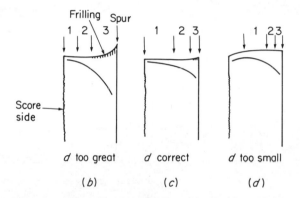

Score side

d too great   d correct   d too small

(b)   (c)   (d)

Figure 18.8(a) The appearance of a correctly made glass knife

(b) The appearance of the cutting edge if the second score mark was angled too far from the diagonal. A large 'spur' and a long, 'frilled', useless length of knife edge (3) results

(c) The appearance of a correctly cracked knife. About half the edge is usable (1), one-third can be used for trimming (2), and one-sixth is frilled (3) and useless

(d) The appearance of the cutting edge if the score was angled too close to the diagonal. The edge is curved, and a short length of usable edge (1) results

(a)

(b)

Figure 18.9(*a*)   The apparatus used to examine a knife edge for flaws, using darkfield illumination. (Photo M. Turton)

(*b*) The appearance in darkfield of a good knife edge. The really good part of the cutting edge (1) can scarcely be seen. The brighter the line made by the edge, the poorer are its cutting powers. Visible frilling begins at (2), and is bad close to the spur (3). (Photo R. D. Hardy, approx. × 25)

knife (Fig. 18.8(c)) has about one-half its length usable as the cutting edge, one-third usable for trimming, and only one-sixth 'frilled' below the hardly visible spur, which is unusable in any case.

18.25 Evaluating Glass Knives

After having cracked the knife and before going on to the next stage of waxing on the trough (see below), the edge should be inspected by dark-field illumination at a magnification of about × 20. It is best if a special piece of apparatus such as is shown in Fig. 18:9(a) be made up. This consists of a lamp and transformer set with the lamp filament focused on the back edge of the knife from a low angle. The knife is placed on a matt black surface, and the microscope is focused from above directly on the edge. The appearance through the microscope should be as shown in Fig. 18.9(b). The more perfect the cutting edge, the more difficult it is to see, as at A. If the edge can be clearly seen, as between B and C on the right, it is frilled, and is useless for ultrathin sectioning. The illuminating and binocular systems of the LKB Ultratome are very conveniently arranged for knife examination, as the lamp can be switched to give a focused spot and can be swivelled about the knife edge to give the correct darkfield illumination, while the binocular can be swivelled upright about the same axis to the correct viewing position.

18.26 Knife Making Machines

The art of making really first-class glass knives reproducibly every time by the pliers and hand break method described above is only learned by long practice and experience. Making good glass knives is the most important single step in obtaining good sections. However skilled the operator and however perfect the microtome, a bad knife cannot possibly yield good sections. A working laboratory with two electron microscopes and three microtomes may use 50 or more knives per day, and a regular supply of first-class knives is essential. To crack first-class knives reproducibly, every parameter must be accurately reproducible: length and position of score, depth of score, position of bending force, amount of bending force and the rate at which it is applied. These requirements are so critical that even the best hand knife-cracker using pliers can only guarantee about 80 % of usable knives, and a learner may only obtain 1 in 10. To overcome these difficulties, a machine called the 'KnifeMaker' (Cat. No. 7800A) has been designed by the Swedish firm of LKB. This is essentially an accurate scoring jig combined with a precision bending machine which is fed with 12″ × 1″ glass strip. It first scores and bends the strip to produce the inch-square blanks, and then scores and bends these squares to produce the knives. When the jigs have been correctly set up, a complete novice can produce perfect knives within an hour, so little skill being needed to operate it. It is ideal, if not essential, for the larger laboratory, where a junior technician can make the day's requirement of knives in an hour first thing in the morning. It may be a luxury for the lone, highly-skilled worker, but can save even him time and frustration.

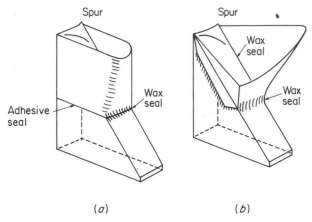

(*a*)   (*b*)

Figure 18.10(*a*)   An adhesive tape trough waxed to the knife
(*b*) A metal trough waxed to the knife. The metal trough
has a greater water surface area for manipulating ribbons of
serial sections

18.27 The Knife Trough

Ribbons of ultrathin sections can only be collected by floating them on the surface of water or other suitable liquid. It is necessary therefore to arrange a small water container behind the cutting edge of the knife. This is called the 'trough' or 'boat', and can be made from a variety of materials. The most commonly used trough is made simply by cutting a length of about $1\frac{1}{2}″$ of $\frac{1}{2}″$ wide black plastic adhesive insulating tape and passing it around the back of the knife as shown in Fig. 18.10(*a*). The tape is pressed firmly to the glass, which must be dry and grease-free, and the free ends are carefully trimmed with a sharp razor blade flush with the clearance face of the knife. Great care must be taken not to touch the cutting edge with the fingers or the razor blade. The lower rear part of the tape must then be sealed to the knife back with a scrap of dental wax, which is applied to the gap and then carefully and neatly melted with a hot glass rod. Care must be taken in choosing the type of adhesive tape, as some have a water-soluble adhesive which contaminates the water in the trough and dirties the sections.Metal troughs (Fig. 18.10(b)) can also be used; they can be made with a much greater water surface area, which is an advantage when long ribbons of sections are required. They are best made from bent 0·005″ thick stainless steel shimstock. They must be removed and dewaxed in xylene after use; the tape troughs are of course thrown away with the knife after use. Metal boats also require more care when waxing them on to the knife back, which is best done by placing a scrap of wax at the junction of trough and knife back. The back of the trough is then heated with a very small blue flame, when the wax melts and flows by capillarity up the sides. Care must be taken not to get wax on the cutting edge. The wax seal is easily broken if a metal trough is accidentally knocked.

18.28 Diamond Knives

The cutting edge of a glass knife is very fragile and subject to rapid wear. A really good edge under ideal conditions will cut a ribbon of sections possibly 2 cm long. This represents anything from 20 to 200 sections, depending on the dimensions of the block face. Attempts have therefore been made to replace glass with a harder material. The obvious choice is polished diamond. A diamond knife is made by splitting a diamond down a crystal plane so as to obtain a wedge-shaped piece of diamond with an included angle of about 45°. The two intersecting faces of the wedge are then polished with diamond dust in a direction parallel to the cutting edge so formed until microscopical examination shows the edge to be apparently flawless. The width of the cutting edge is very short compared with that of a glass knife; 0·5–1 mm is normally available. Wider edges require a larger diamond, and become almost prohibitively expensive. The finished knife, containing only a few cubic millimetres of diamond, is then mounted in a strip of brass about 1 cm long by 5 mm wide by 2 mm thick. This con-stitutes the actual knife, and is in turn mounted in a metal holder, usually made of die-cast aluminium, the same size and shape as a triangular one-inch glass knife and having the trough integral with it. The assembly is interchangeable with a glass knife in all microtome knife holders.

Diamond knives are very expensive, due to the rarity of suitable raw diamonds and the labour involved in polishing the edge. In the author's admittedly limited experience, a really good glass knife will give better sections initially than the best diamond knife, but the diamond will go on cutting slightly second-class sections indefinitely. Diamond knives seem to have an enormous individual variation in their cutting properties. Obtaining a really good one is a gamble, and a very expensive one at that, since knives cost at least £150 each. The edge, although extremely hard, is also very brittle. One injudicious move of the microtome advance resulting in a blow on the knife edge from a metal part of the microtome chuck chips the edge and the expensive knife is then useless. Another disadvantage is that the surface of the diamond is very difficult to wet with water, and tends to repel the meniscus as the trough is filled. Rubbing the surface of the knife with a detergent (or even saliva), using the broad end of a soft wooden toothpick and working from side to side, allowing the cutting edge to split the wood, will in time make the surface wettable. Great care must be exercised during this procedure.

In spite of the great expense and other disadvantages of diamond knives, several highly experienced workers claim that certain individual samples have truly remarkable cutting properties and durability, but they admit that they have only acquired such gems by persistently ordering and replacing knives—an extremely expensive procedure beyond the resources of most laboratories. A good diamond knife, if the worker should have the great good fortune to acquire one, is a pearl of great price, and should be jealously guarded by, and used by, one worker only.

18.29 Microtomy

When the block has been trimmed and a supply of glass knives complete with troughs has been made, section cutting can begin. First, the microtome must be set up. The setting-up process demands about 90 % of the skill required in microtomy; once a microtome is set up and has begun to section, only the hand-operated types require any skill for the actual process of sectioning.

18.29.1 Setting-up the Microtome

First mount the trimmed block in the microtome chuck. Most have collet chucks, which grip the block evenly around its circumference. Do not tighten the chuck too much, or the plastic will be squeezed out and will flow towards the knife, causing ultrathick sectioning. On the other hand, if the block is too loose it may vibrate and cause 'chatter marks' on the sections. Set the block as far into the chuck as possible, leaving the minimum of overhang. This also reduces chatter. Set the top and bottom edges of the face exactly horizontal. Try not to handle the microtome parts more than necessary, as this causes thermal disturbances. Do not mount blocks when they are hot or cold, but only at room temperature.

Next mount the knife in the clamp. Make sure the clamp is well withdrawn back from the chuck, or the cutting edge may foul the metal parts and become chipped. Make sure the knife fits snugly against the end stop, and the base is resting on the floor of the clamp. Take care when tightening the knife clamp not to damage the wax seal between knife and trough. Do not overtighten. For the first rough trimming, use a knife which has already been used for ultrathin sectioning.

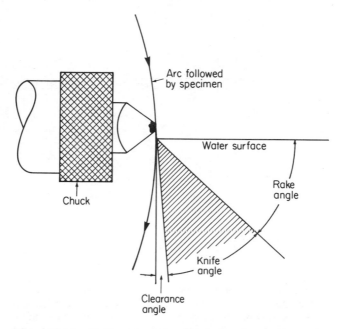

Figure 18.11   Knife and cutting angles. The clearance angle is tangential to the arc followed by the specimen. The knife angle does not necessarily bear any relationship to the actual cutting angle at the specimen (see next diagram)

The action of the cutting edge is the same as that of a lathe tool, and the same nomenclature is used for the angles involved. In setting up the knife, the most important angle is the clearance angle (Fig. 18.11). The angle of the knife itself is already fixed by the cracking process. It is usually a nominal 45° since the knife is made from a square. Other angles may be used occasionally, in which case the knife must be cracked from a rhombus. In general, a more acute angle is used with soft blocks and a more obtuse one with hard blocks. In practice, the knife angle makes little difference to the cutting, as will be seen from Fig 18.12. The actual cutting edge of the knife cannot at molecular level be a straight line formed by the intersection of two planes; it must be rounded. In this case, what in fact is the true knife angle? Obviously, no answer can be given, which explains why in practice the nominal knife angle is of little significance.

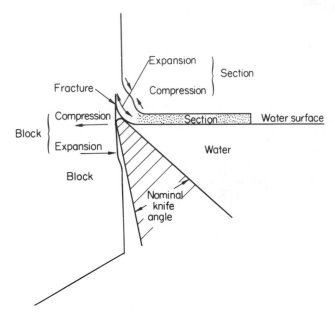

Figure 18.12   The stresses suffered by the section and the block during cutting. The actual cutting edge is curved, and fractures the block before paring the section off. The section may be considerably compressed, especially if the knife edge is blunt

Since the nominal knife angle is fixed, the only angle which can be adjusted is the clearance angle. This must be made as small as possible. Great care must be taken to see that the block face does not contact the front face of the knife after the section has been cut, or the block will become compressed, and regular sectioning will be impossible. In practice, an angle of between 1–5° is used; the softer the block, the greater the clearance angle needed. The knife holder on most microtomes can be rotated to set the clearance angle, which is shown on a scale of degrees. Do not forget that this scale applies only to knives in the form of a right-angled isosceles triangle, and does not apply when the knife angle has been changed from 45° by cutting it from a rhomb. The clearance angle can be set if necessary by looking at the side of the knife and gauging the clearance face against a plumb line or a vertical line scribed on the wall beside the microtome. If standard 45° knives cut from inch-square glass are always used, the clearance angle, once set, need never be altered provided the knife is always carefully set on its base in the holder.

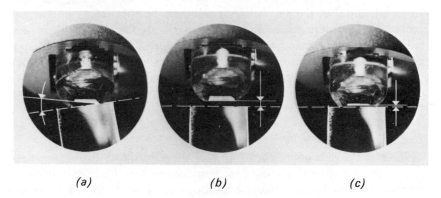

(a)   (b)   (c)

Figure 18.13   Advancing the knife edge towards the block. First, the edge is made parallel with the block face (*a*). Next, the block is swivelled until the face remains parallel with the knife edge as the microtome arm moves up and down (*b*). Finally, the knife edge is advanced towards the stationary block, observing the reflection of the edge in the block face, until all clearance has been taken up (*c*). (Photographs courtesy Reichert)

Having set or checked the clearance angle, next set the knife edge parallel with the block face. For this procedure, it is essential that the block face has a smooth, reflective surface; hence the necessity for using a brand-new, unscratched razor blade for block trimming. Bring the centre of the block face level with the knife edge by adjusting the microtome arm height. Next, switch on the light and advance the knife with the coarse knife advance control, watching down the binocular magnifiers. The block face should be brightly reflecting the light reflected from the knife holder; if not, place a piece of white card or a small mirror in front of the clearance face so that it does. There must be no water in the knife trough during this procedure. As the knife edge approaches the block, the appearance shown in Fig. 18.13(*a*) will be seen. Next alter the swivel angle of the knife to make it parallel with the block face. The horizontal angle of the knife is now correct (Fig. 18.13(*b*)). It is now necessary to make the block face exactly vertical so that it remains parallel to the knife edge as the microtome arm moves up and down. Therefore, move the arm up and down by an amount equal to the height of the block face, and watch the clearance between knife and block carefully to see if it increases or decreases. If it does, make the necessary adjustment with the chuck vertical swivel adjustment until the clearance between the block face and knife is the same at all arm levels. Next, advance the knife towards the block, using the knife advance fine control. Watch the block face and knife edge very carefully; it should be possible to see the reflection of the edge in the face (Fig. 18.13(c)), and thus the gap between them can be adjusted with twice the accuracy. This adjustment must be made with the greatest care and patience. One false move will force the block face into the cutting edge, gouging the face and chipping the edge, and the whole procedure of block trimming, knife making and setting up will have to be repeated. With practice, all these adjustments can be made to an accuracy of better than 10 μm.

18.29.2 Thick Sectioning

The block face is now cleaned-up and a few thick (0·5 μm) sections are cut for examination in the light microscope to check that the material is correct and the fixation and embedding are good. First, fill the knife trough with water or other fluid, using a dropping pipette. The water surface must be exactly level with the knife edge with no meniscus. The lamp and binoculars must be set so that a pool of light is reflected straight into the binocular from the water face when the water level is exactly horizontal, as shown in Fig. 18.14. On a new microtome, this adjustment must be made, which, once set, will remain correct. The most convenient way to make the final adjustments to the water level is to use a clean plastic hypodermic syringe with an angled needle. Be careful not to spill water over the clearance face of the knife or the block face; a wetted block face effectively prevents sectioning. If water is spilled, it should be soaked up with rolled-up filter paper.

Next, set the mechanical advance to about 0·5–1 μm per stroke, and set the horizontal knife traverse so that the block face is opposite the 'trimming' part of the cutting edge (Fig. 18.8(c)), 2). Now set the microtome going (preferably using the hand operation if fitted), and watch very carefully through the binocular. If no coarse mechanical advance is fitted, the knife holder must be advanced by hand in 0·5 μm increments, each increment being given on the by-pass or upstroke. Thermal expansion advance if fitted may under no circumstances be used here.

This process of initial trimming or 'facing' of the block is where the greatest skill in microtomy comes in. The greatest care and patience must be exercised as the knife advances towards the block. The block face will first begin to section at one corner, giving rise to triangular sections.

These become larger until they cover the whole area of the block. Now stop sectioning, and gather up all the fragments of section which are floating on the water surface with a strip of lint-free paper or a very fine (No. 00) sable hair brush, such as is used for retouching photographic negatives. Be very careful not to bump the trough or break the wax seal. Relevel the water surface with the syringe if necessar₍. Next cut about 3 complete sections, each 0·5 μm thick, and stop sectioning again.

These sections are for staining and examination in the light microscope. If they are adherent in a ribbon, they should be parted. This is done with 'eyelash manipulators', which are made very simply by mounting a human eyelash (long, curved and delicately tapering; some people's lashes are far better than others) into the end of a wooden toothpick. Split the narrow end of the toothpick with a razor blade and insert the root end of the hair into the slit. Fix it with a minute scrap of dental wax melted with a thin, hot glass rod. These manipulators can be used to move thick and ultrathin

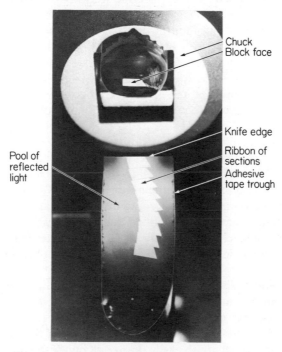

Figure 18.14 The appearance of the sections floating on the water surface in the trough when illuminating conditions are correct. (Photograph courtesy Reichert)

sections about on the surface of the water in the trough. Using two eyelash manipulators, one in each hand, part the sections. Harvest them singly on to strips of microscope slide cover slips which have previously been made by scoring inch square cover slips with a diamond into $1'' \times \frac{1}{8}''$ strips. Sieze a strip of cover slip by the end in a pair of No. 4 watchmakers' balance forceps, and lock the forceps by sliding a small rubber O-ring towards the points. Holding the forceps in one hand and a manipulator in the other, slide the free end of the cover slip beneath the surface of the water in the trough, and bring it up behind a freely-floating section. Brush the section on to the upper surface of the glass strip, observing carefully through the binocular magnifier. Remove the strip plus mounted section and place the strip, section upwards, on a piece of filter paper and let it dry naturally. The section will adhere firmly to the glass without adhesive, and when dry it is stained by the toluidine blue method given on p. 465, mounted and examined with a light microscope.

Examine the stained light microscope section first with the low power and then with the oil immersion objective. The cytological preservation should be so perfect and faithful that the high-power light image looks almost like a low-power electron micrograph. A great deal of preliminary information on tissue preservation can be extracted in this way, and a decision is now made whether to continue with the block, or whether to discard it and try another. If it is desired to continue, note the position of the particular features required, and bring the whole of the section into the field with a × 4 objective lens (Fig. 18.15, 3). Next, remove first the knife and then the block complete on the chuck from the microtome. Be careful not to alter the chuck swivel angle. Mount the chuck beneath a stereo binocular microscope, and using an absolutely new razor blade, cut away the unwanted tissue from the block face (Fig. 18.15, 4), maintaining the sides of the pyramid, while comparing the image with that of the stained section. Localize the wanted tissue as closely as possible. Be ruthless; cut the block face down to an area no greater than 0·2 mm side. Really good ultrathin sections are very difficult to cut from larger block faces.

Fine trimming can be done on some microtomes using the built-in optical system, mounting the chuck in place of the knife or sliding the binocular back and using a special chuck mount. It is very unwise to attempt to trim a block with the chuck in place on the microtome; the knife edge and trough are so easily damaged, and if the razor blade slips, damage can be done to the microtome, not to mention the fingers. Always remove the chuck first. When fine trimming, two of the edges of the pyramid (those parallel with the knife edge) must be made exactly parallel,

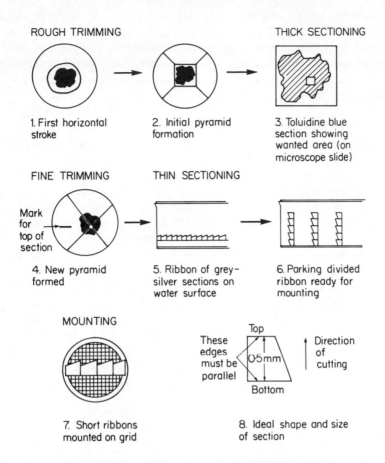

ROUGH TRIMMING

1. First horizontal stroke

2. Initial pyramid formation

THICK SECTIONING

3. Toluidine blue section showing wanted area (on microscope slide)

FINE TRIMMING

Mark for top of section

4. New pyramid formed

THIN SECTIONING

5. Ribbon of grey–silver sections on water surface

6. Parking divided ribbon ready for mounting

MOUNTING

7. Short ribbons mounted on grid

These edges must be parallel

Top

0·5mm

Bottom

Direction of cutting

8. Ideal shape and size of section

Figure 18.15 The stages in obtaining ultrathin sections. Rough block trimming (1,2) is followed by the localization of the required tissue area in a 0·5 μm thick section stained for the light microscope (3). The block is retrimmed to the size and shape shown in (8), and a ribbon of grey sections is cut (5). The ribbon is divided up, using the finely pointed ends of mounted eyelashes, into short lengths which are 'parked' against the side of the trough (6). The grid is then introduced beneath the water surface, and each short length of sections is drawn on to a grid by means of an eyelash. The mounted sections (7) are then stained before examination in the electron microscope

or the ribbon of sections will curve, hit the trough side and crumple up. The smaller the area of the block face, the easier it is to cut really thin (grey) sections. The final trimming must be done with a brand-new oil-free blade, or the inclined surfaces of the pyramid will be ragged, and the thin sections will not part from the block during sectioning, and will be dragged over the knife edge and disappear.

18.29.3 Thin Sectioning

Having fine-trimmed the block, replace the chuck on the microtome arm making sure that the parallel top and bottom edges are exactly horizontal. Mount a new knife. Advance the knife to the block as before, using the very greatest care, as the block face is now much smaller. Centre the block on the best part of the cutting edge. Trim off a single thick section if needed for 'thick-and-thin' techniques, using combined light and electron microscopy, then set the required advance for the ultrathin sections (500 Å). Start the microtome; if automatic, sit back and watch. Do not perturb the microtome thermally by breathing on it, or mechanically by leaning on the bench or standing up and walking around. If hand operated, perform each operation very gently and slowly, keeping the unwanted hand out of the way. If all before has gone well, perfect grey sections should now peel off, adhering together in a ribbon. The interference colours can be clearly seen in the pool of light reflected from the water surface as shown in Fig. 18.14. Actual section cutting is the simplest operation in ultramicrotomy; the preliminary steps are the really difficult ones.

18.30 Mounting

When sufficient thin sections have been cut, stop the microtome. Next, divide the ribbon up using two eyelash manipulators as before into lengths which are a little longer than the width of a grid. Turn these lengths sideways, and 'park' the end section of the short ribbon by sticking it on the adhesive of the trough tape (Fig. 18.15, 5). Now seize a grid by the extreme edge with a pair of locking forceps, bend the edge slightly so that the grid lies flat when the forceps are at a convenient angle for holding, and slip the grid under the water surface well behind the rearmost set of parked sections. Bring the grid up beneath this short ribbon, using an eyelash manipulator to guide them on the exact centre of the grid. Mount them parallel with the meshwork of the grid (Fig. 18.15, 6). A ribbon of 4–10 sections is the normal number to mount on one grid. The greater number, the greater the likelihood of finding exactly the wanted feature unobscured by a grid bar and uncontaminated with dirt or other undesirable features. The sections dry down and adhere tenaciously to the surface of the grid without the use of any adhesive. If the grids used have a dull side and a shiny side, mount the sections on the dull side. After mounting, remove the surplus water from the grid by touching the edge to a piece of filter paper, and place the grids on filter paper in a closed Petri dish to allow them to dry naturally. Blowing hot air on grids generally carries dirt on to the sections. Mark full details on the filter paper with a ball-point pen, not a pencil; graphite scrapings are a common cause of grid dirt.

18.31 Flattening Sections

Fig. 18.12 shows how the plastic block is deformed during the process of sectioning. The exact processes taking place at the molecular boundary of the cutting edge are speculative; remarkably little is known about them. It would seem reasonable to assume, however, that considerable compression takes place in the body of the section as it is torn from the body of the block, which in turn is forced back and recoils elastically. The compressive forces in the section make it thicker. It is also probable that very high temperatures are generated at the actual point of contact due to friction. The thickening of the section by compression shortens it, and causes a characteristic foreshortening of circular objects embedded in the section; circular fat droplets for instance become ellipses with the major axis parallel to the knife edge and perpendicular to the direction of cutting. Many early electron micrographs show this characteristic appearance. The stress in compressed sections can often be relieved by exposing them to the vapour of a solvent such as xylene or ethylene dichloride, applied simply by holding a small brush dipped in the solvent close to the surface of the compressed section as it floats on the water surface. Compressed sections can readily be seen to stretch out to their uncompressed length. However, really well-cut, really thin sections do not show compression, and this flattening process should be unnecessary as a routine procedure.

18.32 Section Staining

18.32.1 For Light Microscopy

The 0·5 µm sections are dried down on to the slivers of cover slips so that they adhere thoroughly without wrinkling. They are dyed with a slightly alkaline solution of toluidine blue (1 %) in borax (1 %). One drop of the dye solution is placed on the end of the cover glass sliver, covering the section. Place the sliver on a microscope slide on an electrically heated hotplate at about 120 °C. Leave the dye for about a minute, until it begins to evaporate at the edges. Wash off the dye with distilled water, and then differentiate with 50 % alcohol until the epoxy resin is colourless. Dry the section on the glass sliver under a current of warm air; do not dry it on the hotplate or the colours will fade. Next place a drop of Depex or other standard mounting medium in the centre of a $3'' \times 1''$ glass slide, and carefully slide the end of the glass sliver bearing the dyed section upwards into the drop so that the section is well covered by the mountant. Drop a $\frac{3}{4}''$ square cover slip down on top, and press gently to expel any air bubbles. Do not use excess of mountant. Examine in the usual way with the light microscope, remembering when using immersion oil that the cover slip will not be adherent to the slide until the mountant has completely dried. The tissue will be strongly stained in various shades of purples, deep reds and blues. The stain binds to the sites of osmium binding, and so the degree of osmium fixation and hence the penetration of the fixative can be estimated. The cytological detail is highly differentiated against a colourless background of unstained resin. Sections thicker than 0·5 μm stain too heavily. The slides are permanent if kept in the dark; sunlight fades them rapidly. The method enables the experienced worker to judge at a glance the adequacy of fixation, penetration of the fixative and the general suitability of the tissue for ultrathin section staining. The exact area required for examination in the electron microscope is then selected for final fine block trimming. Other dyes and mixtures of dyes can be used for staining epoxy resin sections, which are in fact far more easily stained than was earlier thought possible.

18.32.2 For Electron Microscopy

Section staining is used to increase the differential electron scattering power of various components of tissues and cells. The precise mode of action of these electron stains has not been studied in close detail, so all tend to be somewhat non-specific. The staining technique has developed rapidly over the past few years, due to the lower inherent contrast of osmium-fixed, epoxy resin-embedded material. The method is to react the section with a solution of a heavy metal salt such as lead hydroxide by floating the sections mounted on the grid on the surface of the staining solution for a length of time known to result in the desired increase in contrast. The technique is to place a single drop of the staining solution on the surface of dental wax in the bottom of a Petri dish, place the grid with the sections downwards on the surface of the drop, cover the

Petri dish and leave the sections at room temperature or higher for the required time. A few pellets of NaOH surrounding the lead stain drop will absorb atmospheric CO_2 and prevent the deposition of lead carbonate 'dirt' on the section (see below). The grid is then removed, and the stain is flushed off by washing the surface of the grid very gently with a stream of water flowed from a plastic wash bottle. Great care must be taken to remove all the staining solution, or it will dry on the section surface, leaving crystals of contamination. It is especially necessary to take care that staining solution does not become trapped between the tines of the forceps, or it will run out over the sections as the grid is put down. The washed grids are allowed to dry on filter paper in covered Petri dishes. Specimen preparation is now completed, and the grid can be inserted into the electron microscope.

The most commonly used heavy metal for section staining is lead. It has great affinity for many epoxy resin embedded cellular structures, and is especially useful for staining glycogen, which is unstained by osmium fixative. A number of recipes are in common use, designed in an attempt to prevent the formation of basic lead carbonate by contact with atmospheric carbon dioxide, which forms gross contamination on the surface of sections. The most successful appears to be a recipe due to Reynolds. Lead citrate, made from lead nitrate and sodium citrate, is dissolved in sodium hydroxide to give a highly alkaline, concentrated solution (about pH 12). The solution is stable and will keep in tightly stoppered bottles. It 'matures' with age. The staining is especially intense with tissue which has been double-fixed with glutaraldehyde and osmium; a staining time of about 2 minutes is ample. The stain can be differentiated to a certain extent by washing the sections with very dilute sodium hydroxide solution. The stain provides a general, unspecific increase in contrast, especially to PAS-positive substances such as glycogen. It is the most generally useful all-purpose stain.

Uranium is the other commonly used section stain. Uranyl acetate as a 1 % solution in water or alcohol is used. Alcohol appears to give more rapid staining. The disadvantage of this stain is the long staining time— 30 minutes or more. The time may be reduced by warming the stain. It also gives a generalized, non-specific increase in staining of a subtly different type to lead. It stains collagen more than other tissue components. Because of the different staining properties of uranium and lead, it is becoming common practice to 'double-stain', first using warm uranyl acetate for 30 minutes followed by lead for 1–2 minutes.

Tungsten, as a 1 % solution of phosphotungstic acid in water or alcohol, is sometimes used, as it stains collagen fairly specifically, greatly enhancing the banding. Potassium permanganate solution is also used as a general section stain, also as a 1 % aqueous solution. After staining, the grids are rinsed with citric acid solution followed by distilled water.

As time goes on, more specific staining methods will doubtless be devised, which will allow definite histochemical conclusions to be drawn. As yet, there is still much information of a purely morphological nature to be obtained from the present staining methods.

18.33 Faults in Sectioning

Unsatisfactory sections arise either from faulty embedding procedures, or from faulty cutting. Section faults must be clearly distinguished from artifacts of faulty tissue preservation, which are discussed in the following section. Faulty sections are almost invariably due to faulty knives, knife angle, infiltration, polymerization, incorrect matrix hardness or just plain sheer lack of skill on the part of the operator. Modern microtomes are almost incredibly reliable; the likelihood of faulty sections being due to any fault in the microtome is almost negligible. Do not under any circumstances, however tempted, interfere with the mechanical or electrical parts of the microtome, or the result is most likely to be a costly repair bill.

It is very difficult to list and classify difficulties in sectioning, partly because there is insufficient knowledge of the actual physics of the cutting process, and partly because there are so many variables which are interrelated in the most complex fashion. If the microtome simply will not cut a ribbon of ultrathin sections after the block has been faced and two or three $0.5 \ \mu m$ thick sections have been cut, try the following: Is the mechanical advance at the end of the travel? Is the clearance angle enough? Is the water surface touching the edge of the knife? Is the knife just plain blunt? Is the block face wet? Is the clearance face of the knife wet?

Generally, some sort of sections will be cut. It is most infuriating if when cut, they cling to the upper edge of the block and disappear down with it. Cause—pyramid trimmed with blunt razor blade. The pyramid sides must be mirror-smooth, or the ragged surface will cause tiny threads to remain between section and upper edge of block face. This generally happens only if the knife is blunt. Another cause is a wet block face.

Alternate thick and thin sections may be cut. Cause—blunt knife, bad match between block and tissue hardness, poor infiltration, irregular polymerization due to inadequate mixing of resin components. The phenomenon of 'skipping' is an aggravated form of the above fault—a section is cut, then no section, then a section and so on. The cut sections are generally twice the thickness set on the advance. This may be due to too small an advance being set, in which case the alternate sections will be grey. In general, the alternate sections will be purple or thicker. Change the knife, check the clearance angle and try again. Frequently the surface of the water in the trough becomes contaminated with oil or other floating matter. It is essential to use glass distilled water and not 'deionized' water in the trough. The exchange resins of deionizers are frequently greasy. If the water is mixed with acetone or alcohol to reduce the surface tension, make sure the solvents are grease-free. Laboratory reagent acetone generally contains grease. The edge of the knife itself becomes contaminated during sectioning, and can often be cleaned by gently rubbing the broad end of a soft wooden toothpick from side to side along the knife edge; care must be taken not to break the knife-trough seal. Contamination can also arise from the tape adhesive; this must be grease-free and water insoluble. Use plain water only with tape troughs; never add any surface active agent.

Sections when cut may 'shred' into long ribbons which float apart. This is almost invariably due to irregularities in the clearance face of the knife, made when the inch-square piece was cracked off. It is a waste of time to make a knife if the face into which the free break runs is not a perfect mirror finish (Fig. 18.7).

If a ring of embedding material is present around an area of embedded tissue, it sometimes happens that alternate tissue-resin-tissue etc. sections are cut. This always indicates a hardness mismatch, but can generally be overcome by retrimming the block. Try to keep the distribution of material of the section as uniform as possible, avoiding areas of blank resin.

Even if apparently perfect sections have been cut, they may display imperfections invisible to the eye when examined in the electron microscope. The commonest of these is 'chatter', which takes the form of ridges and furrows or alternating lines of thick and thin areas of section, parallel to the cutting edge of the knife. These are caused by vibration of the knife or the block or both during cutting. This can actually be heard as a high-pitched squeak in aggravated cases. The causes are: the block is protruding too far out of the chuck, or is not firmly gripped; the knife is not being properly gripped, the clearance angle is wrong or the cutting speed is too high. The knife may just be blunt, the block too hard or the cut too heavy. As always—change the knife and tighten everything up.

Very few sections are absolutely flawless; most have a slight knife scratch, which may open up into a line of holes or even tear completely apart under electron bombardment. In this case, try to use a different part of the section. Sometimes tears and chatter occur only on particular areas or organelles; lipid droplets are notorious for showing chatter. This is localized vibration due to localized hardness mismatch, and beyond trying another knife to see if it is sharper, little can be done about it. This tissue mismatch can be carried to extremes, where the hard areas are actually forced across the section, leaving holes behind. Again, change the knife and hope for the best. A blunt knife may cause compression in the section, causing circular objects to assume an elliptical shape. If the major axes of the ellipses are all parallel to the knife edge, the section is compressed. The shape of the section when floating on the water should always be that of the block face; any large degree of compression can be noticed by eye. Try flattening the sections with ethylene dichloride vapour; the proper remedy is to use a sharper knife.

18.34 Criteria of 'Good Preservation'

In the early days of biological electron microscopy, the cry of 'artifact' tended to be levelled at practically every new discovery as soon as it was announced. This is understandable, because the process of fixation gives rise among other things to protein precipitation, and the regular, ordered arrays of cytoplasmic and mitochondrial membranes might conceivably have been produced as precipitation artifacts by the physical processes consequent upon fixation. The problem at that time was that there was no way other than the electron microscope of getting information about structures in the size range between 2,000 and 20 Å. Above this size, checks can be made with the light microscope; below this size, the spacing of regular arrays can be measured by X-ray diffraction. The latter method becomes more sensitive and accurate as the spacing becomes smaller, but is only applicable to regularly repeating structures. The spacings of certain tissue components, notably myelin and collagen, could be measured accurately using X-ray diffraction. The results confirmed those found with the electron microscope. Since this time, other methods have been devised for preparing electron micrographs of cellular fine structure, notably the process of 'freeze etching' (see p. 484). All these methods show similar structures for mitochondria, the Golgi apparatus, and the endoplasmic reticulum and all the other cell components made familiar from micrographs of ultrathin sections. Examination of biochemical preparations of particulate fractions using negative staining and shadowing techniques as described in the following chapter also tend to confirm these findings.

There has now been sufficient time (about 20 years) since the first tentative results were obtained from embedded and sectioned tissue for an overwhelming body of evidence to have been amassed which makes it virtually certain beyond any reasonable doubt that the very detailed and aesthetically pleasing micrographs of cells and organelles which are published nowadays by skilled workers do in fact represent a substantially true picture of the cell in life, certainly down to the present resolution generally available in sections (about 20 Å). Whether or not the exact molecular architecture of components such as phospholipid unit membranes is correctly represented is still a matter for dispute. Painstaking analysis of disrupted membrane fractions at the molecular level using freeze-etching and negative and other staining methods is beginning to resolve this particular problem.

How do we recognize 'good preservation' and distinguish it from 'bad preservation'? Fixation is not the only disruptive process suffered by the tissues before they reach the electron microscope. Considerable extraction of lipid occurs during the processes of dehydration and infiltration, and dimensional changes take place during polymerization, since the polymer always occupies less space than the monomer.

A number of criteria are now recognized by most workers, and are looked for in electron micrographs. These indicate whether or not the processing of the tissue has been successful. These can be classified as follows.

1. Absence of empty spaces. The presence of 'empty spaces' completely devoid of any granular or fibrous structural feature, especially between the membranes of the nuclear envelope and between the cristae and matrix of mitochondria, gives rise at once to the suspicion that extraction of material or shrinkage have taken place. In life, all spaces in the tissue are filled with an aqueous solution of protein colloids and salts (tissue fluid), and these should show some kind of precipitated structure. In 'well preserved' micrographs these spaces contain a greyish, granular and fibrous structure and not a startling white, featureless area. This should be looked for particularly in the lumina of capillaries and other small blood and lymph vessels. If preservation of tissue fluid in these larger areas is good, then it is likely to be so at the cell organelle level. An electron micrograph of well-preserved tissue does not have a 'soot-and-whitewash' appearance; it tends to be a 'symphony in greys', which possibly has a rather unattractive, 'muddy' look to the portrait or landscape photographer. The aesthetic criteria of normal photography cannot be applied to electron microscopy, which is concerned with the transfer of the maximum amount of information, and demands the maximum use of half-tone. If there are areas of dead white or dead black, information may be lost.

2. Membrane continuity. There should be no free ends of membranes visible, or signs of sharp breaks. Membranes never seem to have free ends in life.

3. Comparison with life. There should be no sign of any distortion which could be detected with the phase contrast microscope in a living cell. If any gross damage can be seen with the light microscope in the 0·5 μm thick sections stained with toluidine blue, then the block should be discarded, because it is unlikely if the ultrastructural features will be well preserved.

4. Aesthetic considerations. It is apt at this point to end with a quotation from Dr. D. W. Fawcett, one of the acknowledged masters of the art of electron microscope histology: 'Perhaps it is more an article of faith for the morphologist, than a matter of demonstrated fact, that an image which is sharp, coherent, orderly, fine-textured and generally aesthetically pleasing is more likely to be true than one which is coarse, disorderly and indistinct. Like other matters of faith, this may not withstand logical analysis but it has proven to be operationally sound and has been responsible for much of the progress that has been made in descriptive cytology at the electron microscopic level. To accept any other guiding principle is to encourage carelessness and technical ineptitude' (from D. W. Fawcett: 'Histology and Cytology': in *Modern Developments in Electron Microscopy*, ed. Siegel, B. M., Academic Press, New York, 1964).

CHAPTER 19

Particulate Material

19.1 Introduction

All the early specimen preparation techniques had perforce to deal with small particles and the preparation of replicas of surfaces. With the advent of ultrathin sectioning, which became generally available around 1957, interest in particulate specimens waned and the major attention of biologists turned towards obtaining information from sections. There are now signs of a return to particulate material, using refined techniques for obtaining information at the molecular level which thin sectioning cannot in the present state of the art provide. Also, the comparatively new technique of 'freeze etching' requires the preparation of surface replicas.

The particulate materials which the biologist is mainly interested in examining with the electron microscope are the purified fractions such as ribosomes and membrane fragments of cells which are prepared by biochemical ultracentrifugation techniques, and also preparations of the macro-molecules of cells and tissues, such as nucleic acids, fibrous proteins and enzymes. Information can also be obtained about larger particles and organelles such as virus, small bacteria, cilia and flagella, but information can in general only be obtained about the surface structure.

With an instrumental resolving power of 5 Å, it should be possible in theory to detect spherical protein particles of 5 Å radius. In practice, particles as small as this are difficult to resolve, because of the difficulty of distinguishing them against the granular background of the carbon supporting film, which as we have seen (p. 100) is enhanced and magnified by phase contrast effects on either side of true focus. But features of this order of size can be distinguished in particles which are large enough to be recognized, such as virus particles, or long, fibrous macromolecules such as tropocollagen and nucleic acids. Once the particle has been detected and recognized, finer features can be seen.

Particles of biological materials must be mounted on an organic sub-

473

strate film for examination in the electron microscope, and as we have already seen (p. 101), they have no differential electron scattering power over that of the substrate film, and are therefore invisible. They must be differentially stained before they can be seen. There are two basic methods of staining: positive staining and negative staining. Section staining, which we have discussed on p. 465, is positive staining, where particular components of tissue macromolecules combine with heavy metal atoms, and thus give some clue to the unit molecular structure of the macromolecule. The type of staining which is more commonly used in the examination of particulate materials is the opposite of positive staining; hence the term 'negative staining'. Here, the object to be visualized is not stained at all; the background is stained, leaving the electron transparent object unstained. The particle is thus embedded in a very thin film of electron-dense material. This technique allows the shape, size and surface structure of the object to be studied, but little of the internal structure. The method is not novel or peculiar to electron microscopy; it has been used by light microscopists since before the beginning of this century. A somewhat different but closely related form of negative staining is also frequently used; this is called 'metal shadowing', and was developed specifically for electron microscopy by Wyckoff in 1944. Here, the particles to be studied are embedded in a film of metal which is evaporated on to them from a point source at a low angle of incidence. Metal atoms coat the surface features of the particles and also leave electron transparent areas or 'shadows' on the substrate film. The overall effect when suitably printed is a very lifelike effect of light and shadow, which is immediately interpreted by the eye in topographical terms, showing the shape and surface features of the particles very clearly.

The results obtained by the three particle contrasting methods—positive staining, negative staining and metal shadowing—are shown in the case of collagen in Fig. 19.1. Collagen was one of the first biological materials to be studied with the electron microscope. When fragmented in a Waring blendor, collagen forms a suspension of fibrils having diameters around 200–500 Å and which are several micrometres in length. The electron microscope shows these filaments to have a lengthwise banded structure of major periodicity about 700 Å, with many minor subbands. The banding reflects the regular arrangement of the repeating amino acid sub-units along the length of the polypeptide long chain macromolecule called 'tropocollagen', which is the monomer from which collagen is polymerized. The reactive groups of the amino acids bind heavy metals, and thus give rise to positive staining at regularly repeating sites along the collagen fibril. This regularly repeating molecular structure also gives rise to a surface feature: a regular thickening and thinning of the fibril of about 350 Å periodicity.

Fig. 19.1 shows the effect of the three contrasting methods on a short length of collagen fibril supported on a thin carbon membrane. In Fig. 19.1(a), the fibril is positively stained with uranyl acetate, which binds at periodic sites along the filament. The stain is then washed away, and the fibril is examined in the electron microscope. The electron micrograph

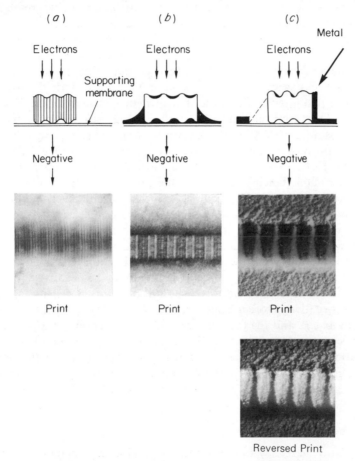

Figure 19.1 The effects produced by (a) positive staining, (b) negative staining and (c) metal shadowing on a fragment of collagen lying on a supporting membrane. The overall effect of banding is clearly shown by all three methods, but each supplies certain information which is lacking in the others. Shadowing shows clearly the granular nature of the carbon support film

shows a complex pattern of regularly repeating lines which are alternately longer and shorter, indicating a regular increase and decrease in diameter. There is a major periodicity of about 700 Å, each period showing at least 10 minor periods.

In Fig. 19.1(b), the fibril is negatively stained by allowing a dilute solution of sodium phosphotungstate (PTNa) to dry around it. The stain dries down in the surface features of the fibril, and also surrounds it so that the organic matter shows as electron-transparent (light) areas against an electron-dense (dark) background. The surface features are now very much more clearly shown. The fine axial filaments which probably represent the actual tropocollagen molecules can now clearly be seen running along the length of the fibril. The major periodicities are now seen to be composed of a shorter, light 'A' band and a longer, dark 'B' band, but the minor periodicities are not so sharply defined as in the case of positive staining. This negatively-stained picture is unlikely to be formed purely by negative staining: PTNa is also known to stain collagen positively, so the picture is formed by a combination of negative and positive staining.

In Fig. 19.1(c), the fibril is metal shadowed by having gold–palladium evaporated at a glancing angle on to the surface. When the plate negative is printed on to paper, as was done with the preceding two methods, the result is perplexing; the electron-dense areas are black and the electron-transparent ones white. But the electron-transparent areas are the 'shadows' where the surface features shielded parts of the specimen and the substrate film from the evaporated metal. The eye is accustomed to black shadows, and cannot immediately evaluate the negative effect of the print. An intermediate negative should therefore be made from the original negative before a print is made. When this is done, the shadows are now black, and the eye can immediately interpret the topographical features of the collagen fibril. It will immediately be apparent that the resolution of the method is very much poorer than that of the two preceding methods. This is because evaporated metals tend to deposit in coarse, granular particles and not as an atomic matrix.

19.2 Vacuum Evaporation

It is now necessary to discuss the technique of vacuum evaporation, since the preparation of metal-shadowed specimens and the making of carbon substrate films requires the use of this technique. A typical vacuum evaporation plant (Edwards 12E1) is shown in Fig. 19.2, and a diagram of the vacuum system is shown in Fig. 19.3. The apparatus consists essentially of a work chamber in the form of a stout glass bell jar 12″ in

Figure 19.2 A typical vacuum evaporation plant
with front cover removed used for metal shadowing
and preparing carbon support films. (Edwards 12E1
fitted with carbon evaporation pulse timer; photograph Michael Turton)

diameter which can be evacuated very rapidly to a vacuum of 10^{-5} torr or better by means of a wide-throated oil diffusion pump backed by a rotary pump. The vacuum system is identical with that of an electron microscope (see Chapter 7), the column being replaced by the bell jar. The backing tank can be used as a vacuum desiccator for drying electron microscope plates. The base plate of the bell jar is drilled for the insertion of vacuum-tight insulated electrical connections capable of carrying very high currents (of the order of 100 A). The diffusion pump must be capable of dealing with a large gas flow, since the heat from the evaporation process causes the evolution of quantities of adsorbed gases and vapours from the surface of the bell jar, and it must be mounted as close to the bell jar as possible. Vacuum evaporators are available with varying degrees of sophistication and automation. Since it is of no great consequence if the bell jar becomes contaminated (it can easily be removed and washed), a simple manually operated apparatus such as is shown is entirely satisfactory. A float valve should be fitted between the rotary pump and the diffusion pump to prevent carry-over of rotary pump oil into the diffusion pump if the rotary pump air admittance valve should accidentally be left closed on shutting the plant down.

The vacuum system must be fitted with an accurate high vacuum Philips gauge, and there should also be a low vacuum gauge provided. A number of pairs of electrical connections (up to 4) are generally provided in the base of the bell jar. A heavy duty, high current rotary switch enables one pair of connections to be selected at a time, so that sequential evaporation processes may be carried out without breaking the vacuum. Electrical power for evaporation is provided by a massive transformer, which can be switched to provide either 30 V at 20 A for carbon evaporation or 10 V at 60 A for metal evaporation. A high-voltage transformer (5,000 V) and special electrodes are optionally available to enable the inside of the bell jar and the apparatus and workpiece therein to be outgassed by ionic bombardment, in order to achieve the highest vacuum possible. The current flowing in the evaporator circuit is controlled by means of a rotary Variac transformer and measured with an A.C. current meter in the primary circuit of the low-voltage transformer.

The principle of vacuum evaporation is shown in Fig. 19.4. A source of metal or carbon is used which approximates as closely to a point source as possible. In the case of metal, a V-shaped length of tungsten wire is clamped into a holder which is bolted to a pair of high-current connectors (Fig. 19.4(b)). A short length of a fine wire of the metal to be evaporated is bent into a U-shape and hung on the tungsten V. Small lumps of metals such as chromium which sublime can be placed in a tungsten wire basket.

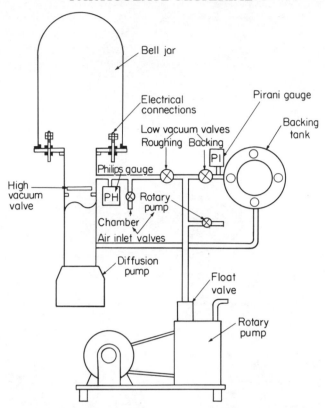

Figure 19.3 The vacuum system of the Edwards 12E1
evaporator. The backing tank can be used for desiccating
photographic material for the electron microscope

The grids to be coated are placed so that the evaporated metal strikes
them at a low angle, determined by the ratio of the distances MO:OG
(Fig. 19.4(a)). This angle, expressed as a ratio, is usually between 1:3 and
1:5. The distance between the grids and the tungsten V is about 25 cm.
The amount of metal evaporated is judged by observation of an 'indicator
plate' as the evaporation proceeds. This plate, I in Fig. 19.4(a), is a small
square of thin white glazed tile on which a small drop of diffusion pump
oil is placed. When evaporated metal strikes the tile, it coats it, and the
colour darkens. The metal which strikes the surface of the transparent oil
drop is removed by surface tension, and so the oil drop effectively shields
part of the area of the white tile, which remains white. Comparison of

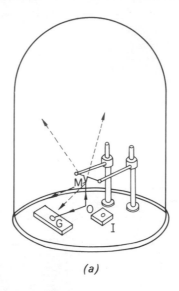

(a)

(b)

(c)

Figure 19.4(*a*)   The principle of metal shadowing in the vacuum bell jar. (*b*) The arrangement of apparatus used for evaporating low melting point metal wires. (*c*) The arrangement used for evaporating carbon. (Photographs by Michael Turton)

the white oil area with the darkened coated area gives a reasonably accurate measure of the amount of metal which has been evaporated.

19.3 Metal Evaporation

The procedure for metal evaporation is as follows. A V-shaped wire is first bent from a suitable length of 1 mm tungsten wire. The wire must not be bent too sharply or it will split. The V is clamped into the connector, and a short length (about 5 mm) of fine (about 30 SWG) wire of the desired metal (platinum, gold, gold–palladium, etc.) is bent into a U and hung on the V. The grids to be shadowed are laid on a 3" × 1" microscope slide, and are positioned on the base plate. A beaker is convenient for adjusting the height O–M. A drop of oil is placed on the indicator plate, the bell jar is carefully positioned, making sure the rubber seal is clean and that there are no pieces of wire, grids, etc. on the base plate to prevent the vacuum seal from being made. The arrangement is shown in Fig. 19.4(b). Before evacuating the bell jar, it is a wise precaution to cover it with a perforated metal cylinder in case the glass implodes during evacuation. Close the chamber air admittance valve and open the roughing valve. When the chamber has pumped to below 1 torr, remove the implosion screen and test the electrical continuity of the evaporation wire. Select the 10 V transformer secondary, select the correct connectors, switch on the current to the transformer primary, and advance the Variac while watching the current meter. Do not make the wire red-hot; simply check that all is well. Switch off the heater current. When the rotary pump has reached its ultimate vacuum (about 0·2 torr in 2 minutes), close the roughing valve, open the backing valve and open the high vacuum valve while watching the low vacuum meter. This should dip and rise again, indicating that the diffusion pump is operating correctly and dealing with the low vacuum. Switch on the high vacuum gauge, and continue to pump until the ultimate vacuum has been reached. The value of this vacuum and the rapidity with which it is reached depends on the design of the vacuum system and the general cleanliness and dryness of the bell jar and the apparatus inside it. A good evaporator in clean condition should reach about 10^{-5} torr in 10–20 minutes. The metal may now be evaporated.

First, shut off the rotary pump to reduce vibration. Shut the backing valve, turn off the pump motor and open the rotary pump air inlet valve. Advance the Variac slowly and carefully until the V-wire is red-hot. The wire will be passing 30–40 A. Continue to advance the Variac until the metal wire begins to melt. This must be done very slowly, or the wire may outgas on melting and 'spit', sputtering large droplets of molten metal around inside the bell jar. When the wire has melted, it should wet the V. Now watch the indicator plate carefully and advance the Variac until the wire is white-hot. Metal will now begin to evaporate, and the indicator plate will darken. When it is a light grey colour, turn the Variac back to zero smartly, or excess metal will be evaporated, resulting in a coarse, low-resolution preparation. The actual evaporation time should be about 1 second or less. Now switch off the heater current, switch off the high vacuum gauge, close the high vacuum valve and carefully open the chamber air admittance valve. Do not let air in with a rush, or the grids may be blown off the microscope slide. When the pressure in the bell jar has equilibrated with the atmosphere, carefully lift off the bell jar and set it down on a clean, lint-free cloth. Remove the grids and place them evaporated side up on filter paper in a covered Petri dish to await examination in the electron microscope.

19.4 Carbon Evaporation

The procedure for evaporating carbon is basically similar to the procedure outlined above for evaporating metal. The source of carbon is a pair of spectroscopically pure carbon rods, usually $\frac{1}{4}''$ dia., each pointed at one end. Since it is not possible to hold two sharp points in contact, one point is flattened to a diameter of 1–2 mm, and the sharp point of the other carbon rests on the flat. Carbons can conveniently be shaped in a rotary pencil sharpener; a special sharpener should be kept for carbons only. The other end is flattened with fine emery paper. The pointed ends are held lightly in contact by means of coil springs wound concentrically around the carbons (Fig. 19.4(c)). When a voltage of about 30 V is applied across the points in vacuum, the point of the sharper rod becomes dazzlingly white-hot and about 2–3 mm of carbon point is evaporated away during the space of about 0·5 second. This amount of carbon is sufficient to coat the indicator plate a light tan colour. A space must not be allowed to develop between the pointed ends during evaporation, or an arc will form and far too much carbon will be evaporated. The very short time necessary for the evaporation process is difficult to reproduce accurately, so some form of relay opened and closed by a fast-acting time delay circuit should be fitted in the primary circuit of the 30 V transformer. Suitable transistor-operated short time delay circuits operating a heavy-duty relay switch are obtainable commercially; few commercial evaporators are sold equipped with them.

The carbon rod holder for carbon evaporation is interchangeable with the V-wire clamp used for metal evaporation. The microscope slide bearing the grids must be placed so that they are face on to the carbon source, not at a glancing angle to it. It is best not to place the grids directly below the carbon points, as carbon powder may fall on them while the bell jar is being evacuated, leading to specimen contamination.

19.5 Preparation of Particulate Specimens

Several simple methods are currently in use for transferring small particles on to the grid substrate film. The simplest of these is to make a suspension of the particles using an ultrasonic disperser or a Waring blendor, place a drop of the suspension on a coated grid, and allow the drop to dry down, leaving the particles adherent to the grid. The grids can be metal shadowed without further preparation. They can be positively stained by floating the grids on staining solutions as for sections, leaving them for a given time to allow the staining reaction to take place to completion, removing the grids, washing them thoroughly in a gentle stream of glass distilled water from a wash bottle or by floating them on distilled water, and then allowing them to dry with the particles uppermost on filter paper in closed Petri dishes.

Negative staining can be done by adding a drop of very dilute negative stain solution (about 0·001 % PTNa) to the film with dried-down particles and allowing this to dry in turn. The staining solution will be drawn up around the particles by surface tension, and will surround the particles as shown in Fig. 19.1(b). The right amount of negative stain to use must be found empirically; an excess will lead to large areas too dense for the electron beam to penetrate, in which case the particles will be completely embedded, obscuring all detail. Too little stain will result in low contrast, and delicate structures may be injured by excessive surface tension.

An even simpler method for negative staining is to mix the negative stain with the particle suspension, and to dry one drop of the mixture down on a grid. If the stain concentration is too dilute, the particles may clump together, obscuring detail. It may be necessary to add spreading agents to improve the dispersion of particles; bovine serum albumin, glycerine and propylene glycol have been recommended. Buffers may also be necessary to preserve some biological structures. These must be of volatile ammonium salts. With particle-stain mixtures, better results may be obtained by spraying the mixture on to grids from a nebulizer. The very tiny droplets dry down rapidly and frequently give an improved appearance to the preparation. Great care must be exercised when spraying pathogenic virus or bacteria; the spray method is best reserved for non-pathogenic materials.

These two methods can be used to distinguish between positive and negative staining effects. The first method, in which a drop of negative stain solution is dried rapidly down on to dried particles, gives the least amount of positive staining, since the stain is in contact with the particles for the minimum time. If, on the other hand, the negative stain is mixed with the particle suspension and allowed to stand overnight, there will be a maximum of positive staining. These effects are particularly noticeable in the case of collagen, and must be borne in mind when interpreting the results.

19.6 Surface Replication

One of the earliest techniques of electron microscopy was to make casts or replicas of surfaces, using evaporated carbon or silica. These replicas are very thin (200 Å or so), and consequently have little contrast. This is added by shadowing the replica with evaporated metal.

The technique is one which demands the very highest skill and patience. The major difficulty is to part the replica from the surface to be studied. In general, this necessitates the destruction of the specimen by dissolving it away from the carbon replica. It may in some cases be possible to separate the two by simple flotation. If for instance carbon is evaporated on to the surface of freshly-cleaved mica, the sheet of mica can then be slid gently under the surface of water and the evaporated carbon film will float off. This is a very convenient method for preparing very thin carbon films for specimen supports directly, without the formation of an intermediate plastic film which must subsequently be dissolved away. In general, the evaporated carbon cast adheres tenaciously to the surface to be replicated. The 'single-stage' replica technique outlined above can be replaced by the 'two-stage' technique, in which a primary cast of the surface is made with a thick layer of plastic, which can then be stripped of the surface simply by pulling. The underside of this primary replica, bearing the imprint of the specimen surface, is then coated with evaporated carbon, shadowed, and the plastic is then dissolved away, leaving both the secondary replica and also the specimen surface intact. A specimen such as an etched metal surface can be replicated time after time in this manner.

19.7 Freeze Etching

The technique of surface replication has recently been revived by the so-called 'freeze etching' technique, which is yielding information on membrane structure and other cellular fine structure, especially of plant cells. The method was introduced by Steers in 1957 and was further developed by Moor and Mühlethaler in Switzerland. A suitable vacuum apparatus is manufactured and marketed by Balzers of Liechtenstein. The principle of the method is basically as follows. A piece of tissue or a suspension of cells (e.g. yeast) in water is held on a specimen support and is cooled as rapidly as possible to below $-100\ ^\circ C$. The specimen support bearing the frozen tissue is then introduced into a vacuum evaporator, where it is rapidly evacuated and maintained at a low temperature. The frozen specimen is then cut across with a cooled knife, exposing a flat surface to the vacuum. This surface is now left for a time, during which ice sublimes from the cut surface, leaving the cell ultrastructure standing in relief above the ice surface. This etches the cut surface by vacuum sublimation. The ice sublimes on to a surface cooled with liquid nitrogen held in close proximity to the cut surface. After sufficient ice has sublimed, the surface is then shadowed with evaporated metal, and is immediately coated with a layer of evaporated carbon. This forms a shadowed replica of the cut, sublimation-etched surface. The specimen is then removed from the vacuum, thawed, and the replica is removed by floating it off on to a water surface. Any adherent particles of organic matter are then removed by floating the replica on a solution of a caustic alkali, after which the replica is washed by transference to several changes of glass distilled water and is then picked up on a grid and examined in the electron microscope. Suitable areas show remarkable detail of cytoplasmic structures, which in general bear out the previous conclusions derived from data obtained from thin sections. Additional information is gained from areas where for instance paired membranes have been pulled apart, disclosing face-on the structure between the paired membranes, information which is difficult or impossible to obtain from sections. The method is a useful auxiliary specimen preparation technique of limited application, but because the approach is quite different from that of orthodox embedding and sectioning, it can yield new information on cellular ultrastructure. Results have been obtained on animal and plant tissues, showing the inner surface of vacuolar, nuclear, plasma and mitochondrial membranes. The resolving power of the method is limited by the coarseness of the granular deposits of evaporated metal, but individual protein molecules can be identified in the lattice structure of virus particles. The useful limit of magnification is at present about $\times 100,000$, but improvements in the resolving power of the method will doubtless enable higher magnification micrographs to be obtained.

Epilogue

The Future of Electron Microscopy

Before attempting any predictions about the way in which a subject might be expected to advance, it is instructive first of all to examine its history. The history of electron microscopy, short as it is, falls into three phases. First, there was the pioneering phase between 1930 and 1940 when the great innovators such as Gabor, Knoll, Ruska, von Ardenne, Marton, le Poole, Zworykin and others too numerous to mention by name laid the technological and theoretical foundations of transmission, reflection and scanning electron microscopy. Second came the feasibility phase. Could these first crude instruments, which could be induced to perform only in the hands of their creators, be made into a practical proposition? Could they be manufactured commercially and could they be applied to practical problems in other fields by non-specialists in electron microscopy?

The great breakthrough in transmission instrumentation came with the introduction of Ruska's Siemens Elmiskop I in 1954. This instrument proved itself to be rugged and reliable and it gave excellent results in the hands of non-technical biologists and metallurgists. Revolutionary new discoveries came tumbling out, and the great grant-giving foundations began to buy these expensive new instruments for laboratory after laboratory as fast as they could be manufactured. This was the golden age of transmission electron microscopy; the plums were ripe for the plucking, and almost every investigation bore fruit that could be published. A similar breakthrough in scanning came with Oatley's scanning instrument, the Stereoscan, by Cambridge Instruments in 1958. These two instruments, transmission and scanning, clearly showed that electron microscopes could be successfully handled by non-physicists. The novel biological techniques devised by Palade, Porter, Sjostrand and others, and which have recently brought the Nobel Prize to George Palade, opened up a completely new world of 'ultrastructure', and brought about a revolution in biology. This era of fantastic productivity slowly petered out as fewer new techniques were devised and fewer new fields were left to conquer.

We are now well into the third phase, the phase of easy-to-use super-performance second-generation instruments. Resolving power has slowly been pushed to the theoretical limit calculated by Ruska in 1932, and, as far as the conventional transmission electron microscope is concerned, no further breakthrough seems to be in sight. The next breakthrough looks as though it will come from the greatly enhanced image contrast and information processing facilities offered by the scanning transmission electron microscope. The imaging of individual atoms, albeit only those of relatively heavy metals, is now becoming a commercially feasible proposition, and the great grant-giving foundations will now doubtless turn their attention to STEM. We may be on the brink of another great outpouring of new information at the molecular level, comparable to the ultrastructural advances of the '50s. The conventional TEM is now regarded in the same light as the common light microscope—a workhorse which individual laboratories are expected to provide for themselves. The present inflated cost of a high-performance TEM puts it out of the reach of unassisted university and hospital laboratories. It is therefore likely that the TEM manufacturers will be forced by economic circumstances to concentrate on smaller, simpler routine instruments costing little more than some of the highly complex automated light microscopes at present on the market. New techniques in the miniaturization of magnetic lenses and modern integrated solid-state circuits should reduce the cost of medium-performance instruments. Electron microscopes such as these, having adequate performance, low initial cost and minimal maintenance problems, should appeal to the hospital pathology laboratory, where the routine use of the TEM in the diagnosis of leukaemias, for example, is now becoming accepted. The general slow increase of high-performance TEMs in industry leads to a demand for trained operators. Small, cheap instruments might be within the reach of schools, which could then provide a basic training in theoretical and practical electron microscopy.

Even though the conventional TEM seems to be approaching the limit of its development, it is by no means approaching the limit of its usefulness. It may shortly be supplanted as the king of research microscopes by the STEM, but its place in future research involving the use of microscopes is as assured as that of the light microscope.

Appendix I

The H.T. accelerating voltages given in the specifications are those which are immediately switchable during operation. Lens currents are automatically adjusted during the switching operation to maintain focus and magnification. 'Plate' cameras can all be used with cut sheet film of the appropriate format, although special clips or holders may be necessary for the cassettes. Abbreviations: ES = electrostatic; EM = electro-magnetic; PM = permanent magnet; SP = single phase.

1. Basic Instruments

JEM 50B 'Superscope'

Performance

Claimed RP:	less than 100 Å (depends on grain of film emulsion)
Magnification:	$1,000 \times$, $1,500 \times$, $2,000 \times$

Illuminating system

HT voltage:	50 kV
Gun:	triode tungsten hairpin, self-biased
Emission control:	variable, 5-step, beam current meter
Lenses:	none
Alignment:	mechanical
Apertures:	3, externally centrable, clickstop

Imaging system

Lenses:	2, EM, air-cooled
Objective:	top entry stage
Aperture:	one, fixed, not externally centrable
Stigmator:	none
Projector:	excitation coupled to mag. control
Focusing:	variable objective current
Alignment:	factory pre-aligned

Specimen facilities

Stage:	top, single grid cartridge, side entry
Airlock:	none

Anticontaminator:	optional extra
Tilting:	$\pm 15°$ on special diffraction holder only
Special holders:	diffraction, mounted above projector
Diffraction:	208 mm camera length, (optional extra)

Viewing facilities

Window:	transmission screen viewed direct
Screen:	10 cm dia.
Focusing aids:	2 × magnifier lens, hood, 10 × binoculars: (optional extras)

Recording facilities

Camera:	35 mm film, 36 exp., manual, auto. wind-on and reset
Airlock:	none
Shutter:	mechanical in camera body, reset by wind-on

Vacuum system

Pumps:	1 rotary + 1 air-cooled oil diffusion
Gauge:	Pirani
Operation:	manual, linked single lever

Electronics	transistorized, built into console
Units	console only; 70 × 90 × 116 cm
Weight	230 kg
Utilities	100 V 50/60c/s SP; 1 kVA no water
Accessories	viewing hood, magnifier and binocular; objective aperture centring jig; diffraction holder; anticontaminator
Manufacturer	Japan Electron Optics Lab. (JEOL) Ltd., Tokyo 100, Japan.

Siemens Elmiskop 51 (Figure 9.1)

Performance

| Claimed RP | 25 to 40 Å (depends on grain of film emulsion) |
| Magnifications: | 1,250 × , 2,500 × , 5,000 × , 12,500 × |

Illuminating system

HT voltage	50 kV
Gun:	triode tugsten hairpin, self-biased, filament voltmeter
Emission control:	continuous 0–8 µA, 2-range Wehnelt shield, beam current meter
Lenses:	none
Alignment:	mechanical
Aperture:	fixed, pre-centred

Imaging system

Lenses:	2, PM series double-lens
Objective:	PM, fixed excitation
Apertures:	2, externally centrable
Stigmator:	none
Projector:	PM, revolving turret of 4 polepieces changes magnification
Focusing:	variable HT voltage $\pm 3·5$ kV
Alignment:	factory pre-set

Specimen facilities

Stage	Top; 15 position revolving specimen table; 2 × 2 mm translation
Airlock:	none; up to 15 specimens per loading
Anticontaminator:	none
Tilting:	none
Special holders:	none
Diffraction:	none

Viewing facilities

Window:	single front, rectangular 115 × 54 mm
Screen:	8 cm dia.
Focusing aids:	9 × binocular standard

Recording facilities

Camera:	6·5 × 9 cm, 30 exp., 16 × enlargement for max. resolution
Airlock:	none
Shutter:	manual raising of screen

Vacuum system

Pumps:	1 rotary + 1 air-cooled oil diffusion
Gauge:	discharge tube
Operation:	manual; non-linked valving
Desiccator:	built into console

Electronics　　built into console

Units　　console only, 140 × 70 × 166 cm

Weight　　230 kg

Utilities　　220 V 50/60 Hz SP 1·3 kVA; no water

Accessories　　none

Manufacturer　　Siemens AG, Wernerwerk für Messtechnik, Karlsruhe, W. Germany.

2. Medium Performance Instruments

AEI 'Corinth' (Figure 9.2)

Models	3: 'Clinical'—low mag. range
	'275'—medium resolution
	'500'—high resolution

Performance

Claimed RP:	'Clinical', 10 Å; '275', 8 Å; '500', 5 Å
Magnification:	Scan: 500 ×, push-button selection
	Main: 600–100,000 × in 12 steps
	Clinical model approx. half above values

Illuminating system

HT voltages:	60 kV, Clinical and 275
	20–40–60–80 kV, 500
Gun	triode tungsten hairpin, self-biased, DC heating

Emission control: continuously variable to max. 150 μA, beam current meter
Lenses double EM condenser, water-cooled, independent excita-
tation, 5μm min spot dia.
Alignment: anode, C1 and C2 mechanically centrable. X–Y ES
beamshift
Apertures: 3 in C2, clickstop, externally centrable
Beam tilt: $\pm 2°$ optional with focus wobbler

Imaging system
Lenses: 3, EM, watercooled
Objective: $f_0 = 2·1$ mm
Apertures: 3, clickstop, externally centrable
Stigmator: ES 8-pole, amplitude/azimuth, electrical alignment
Intermediate: excitation coupled to mag. control and HT
Projector: excitation coupled to mag. control and HT
Focusing: beam wobbler (optional extra)
Alignment: mechanical

Specimen facilities
Stage: side entry; 4-grid holder; $2·5 \times 2·5$ mm translation
Airlock: pre-pumped; automatic; pump-down 25 sec
Anticontaminator: cold finger (optional extra)
High-angle tilt: $\pm 20°$ (2·3 mm grid) or $\pm 30°$ (1·5 mm grid) orthogonal,
not eucentric, full resolution, motor-driven, meter readout,
foot pedal controls (optional extra)
Special holders: double tilt; serial section 2·3 mm grid (optional extras)
Dark field: EM tilt in gun combined with focus wobbler (optional
extra)
Diffraction: selected area, camera length 80 cm, 3 apertures (optional
extra)

Viewing facilities
Window: direct viewing of tilted aluminized transmission screen
Screen: 16·5 cm square (275 cm^2)
Focusing aids: $1·5 \times$ full area lens; $7 \times$ binocular (optional extra)

Recording facilities
Camera: 70 mm 50 exp. roll film; manual transport linked to
shutter
Airlock: none
Shutter: electrical; 0·5–8 sec $+ T$; photometric exposure meter,
manual setting; automatic enumerator; digital readout
count-down

Vacuum system
Pumps: 1 rotary + backing tank + 1 water-cooled oil diffusion
Operation: fully automatic, push-button, fail-safe, signal lights
Gauge: Penning + Pirani (switchable)

Electronics all solid state; interchangeable modular boards in console

Units console only; $127 \times 82 \times 102$ cm

Weight 500 kg

Utilities	230/250 V 50 Hz or 110 V, 60 Hz, SP, 3 kVA. Water 4·5 l/min 15 p.s.i.
Accessories	tilt stage, serial-section stage, anticontaminator, selected area diffraction, beam tilt and wobbler, standardize mag. circuit, binocular magnifier, film desiccator, $3\frac{1}{4}''$ square 18 exp. plate camera, off-screen TV system
Special feature	the inverted column construction is unique, reducing overall height and facilitating servicing
Manufacturer	AEI Scientific Apparatus Ltd., Urmston, Manchester M31 2LD England.

Hitachi–Perkin–Elmer HS–9

Performance
Claimed RP:	4·5 Å
Magnification:	Scan: 200 ×, push-button
	Main: 500–100,000 × in 12 steps

Illuminating system
HT voltages:	50 and 75 kV
Gun:	triode tungsten hairpin, self-biased
Emission control:	variable, beam current meter
Lenses:	double EM condenser, water-cooled
Alignment:	EM lateral and tilt
Apertures:	4, on externally centrable strip

Imaging system
Lenses:	3, EM, water-cooled
Objective:	$f_0 = 2·0$ mm
Apertures:	4 on externally centrable strip
Stigmator:	EM 8-pole, X–Y system
Intermediate:	EM, excitation coupled to mag. selector and HT
Projector:	EM, fixed excitation
Focusing:	beam wobbler
Alignment:	objective mechanical tilt; rest pre-aligned

Specimen facilities
Stage:	top insertion, single grid cartridge, 2·0 × 2·0 mm translation
Airlock:	pre-pumped, manual
Anticontaminator:	cold finger (optional extra)
High-angle tilt:	none
Special holders:	none
Dark field:	beam tilt in gun
Diffraction:	selected area, camera length 50 cm, 3 apertures

Viewing facilities
Window:	1 at front, rectangular, 120 × 125 mm
Screen:	120 mm dia.
Focusing aids:	7 × binocular on tilted main screen

Recording facilities

Camera:	55 × 80 mm or 80 × 95 mm cut film/plate, 18 exp., motorized auto. wind-on
Airlock:	manual
Shutter:	electrical, coupled to exposure meter and wind-on, auto. enumerator

Vacuum system

Pumps:	2 separate rotary + buffer tank + oil diffusion
Operation:	fully auto. push-button fail-safe
Gauge:	signal lights

Electronics   plug-in solid state modules in console

Units   2; console plus separate rotary pumps; 140 × 827 × 230 cm

Weight   console 500 kg, pumps 55 kg

Utilities   100–120 V 50/60 Hz SP 2 kVA. Water 2 l/min

Accessories   high-resolution diffraction holder, stereo stage, image intensifier, TV system

Manufacturer   Hitachi Ltd., Tokyo, Japan

Philips EM–201 (Figure 9.3)

Models   2: EM–201S standard high-resolution stage
                     EM–201G tilt/goniometer stage

Performance

Claimed RP:	Standard, 5 Å; goniometer 15 Å
Magnification:	Scan: 200 × on mag. selector
	Main: 1,500–200,000 × standard
	600–80,000 × goniometer

Illuminating system

HT voltages:	40–60–80–100 kV
Gun:	triode tungsten hairpin, self-biased
Emission control:	variable in 6 steps to max. 100 μA, beam current meter
Lenses:	single high-excitation condenser, 6 μm min spot dia.
Alignment:	gun mechanical; X–Y EM beamshift
Apertures:	3, clickstop, externally centrable

Imaging system

Lenses:	3, EM water-cooled
Objective:	$f_0 = 1·6$ mm standard, 4·1 mm goniometer; excitation coupled to mag.
Apertures:	3, clickstop, externally centrable
Stigmator:	EM 8-pole, electrical alignment with modulator, X–Y system
Intermediate:	excitation coupled to mag. and HT
Projector:	excitation coupled to mag. and HT
Focusing:	objective excitation coupled to mag. ('proximity focusing') plus beam wobbler

Alignment: mechanical; pre-programmed alignment settings; lens current modulator

Specimen facilities

Stage: side entry, single grid holder, 2·4 mm × 2·4 mm translation

Airlock: pre-pumped; 8 sec pump-down time

Anticontaminator: cold finger (optional extra)

High-angle tilt: ±60° single eucentric tilt plus 360° rotation on EM 201G only involving loss of RP and mag. and increased specimen change time

Special holders: ±45°/±30° non-eucentric double tilt with full resolution, rotation, straining, heating, cooling, stereo, reflection diffraction

Dark field: EM tilt in gun

Diffraction: selected area, camera length 36 cm, 3 apertures

Viewing facilities

Window: single circular front, 140 cm dia.

Screen: 120 mm dia.

Focusing screen: 2 cm dia., removable from beam

Focusing aids: 9 × binocular on small screen

Recording facilities

Cameras: 3 available: 35 mm 40 exp. roll film; 6·5 × 9 cm or 3¼″ × 4″ plate/cut film; 70 mm 50 exp. roll film. Choice of one, others optional extra; all 3 can be used together in any sequence. TV camera (optional extra) fits in place of 70 mm camera; all auto. motorized wind-on

Airlocks: on all 3 cameras; pump-down time 3½ min

Shutter: electrical, 1/8–64 sec + T. Beam current exposure meter, manual setting (fully auto. optional extra), auto. enumerator (optional extra)

Vacuum system

Pumps: 1 rotary + backing tank + water-cooled oil diffusion

Gauge: Penning

Operation: fully auto., push-button, fail-safe, signal lights

Electronics   all solid state; interchangeable plug-in boards in console

Units   console only; 124 × 87 × 230 cm

Weight   750 kg

Utilities   220 V 50 Hz SP 3 kVA. Water 1·5 l/min, 2 kg/cm2

Accessories   2 interchangeable stages, 8 different specimen holders, anticontaminator, high-resolution diffraction unit, lens or fibre optics TV system

Manufacturer   N.V. Philips Gloeilampenfabrieken, Eindhoven, Netherlands

RCA EMU–4C

Performance
Claimed RP: 5 Å
Magnification: scan 400 × (variable)
                       main 1,400–240,000 × in 15 steps

Illuminating system
HT voltages: 25–50–75–100 kV
Gun: triode tungsten hairpin, self-biased, DC excitation
Emission control: continuously variable, beam current meter
Lenses: one EM standard; double EM condenser optional extra
Alignment: mechanical; X–Y ES beamshift, HT alignment modulator
Stigmator: ES on C2
Apertures: 3 externally centrable in C1 and C2

Imaging system
Lenses: 3, EM water-cooled
Objective: $f_0 = 3{\cdot}1$ mm standard; special holder $3{\cdot}1$–$6{\cdot}0$ mm (optional)
Apertures: single, permanent, heated to 275°C, centrable externally
Stigmator: ES 8-pole
Intermediate: excitation coupled to mag. control
Projector: low mag. polepiece (main range × 0·35) optional
Alignment: mechanical on obj. and int.; HT modulator
Focusing: beam wobbler (optional extra)

Specimen facilities
Stage: top entry, single grid cartridge, 2 mm × 2 mm traverse
Airlock: pre-pumped, pump-down time 10 sec
Anticontaminator: cold finger (optional extra)
Special holders: cold, hot, variable magnetic field, high-resolution diffraction chamber (replaces intermediate lens)
Diffraction: selected area, 4 apertures. HR camera length 50–175 cm camera length

Viewing facilities
Window: single front, rectangular 24 cm × 12 cm
Screen: 12 cm × 12 cm, tilts to focus
Focusing aids: 10 × binocular

Recording facilities
Cameras: 3 available: 2″ × 10″ glass plates, 6 per loading, up to 40 exp. per plate with manual masking. $3\tfrac{1}{4}″$ × 4″, 18 exp. 70 mm film (optional extra); all motor-driven, auto. wind-on
Airlock: on all cameras; motor-driven, auto.
Shutter: electrical, 0·4–6 sec by tenths, manual setting; exposure meter coupled to shutter

Vacuum system
Pumps: 1 rotary (separate) + ballast tank + water-cooled oil diffusion

| Gauges: | Penning + thermocouple |
| Operation: | fully auto., push-button, fail-safe, signal lights |

Electronics   solid state plug-in modules in console and HT cabinet

Units   3: console + HT cabinet + rotary pump. Console 158 × 95 × 244 cm

Weight   1040 kg total

Utilities   108–125 V 50 or 60 Hz SP 4 kVA. Water 1 l/min

Accessories   double condenser, diffraction chamber, low mag. projector polepiece, various specimen holders, focus beam wobbler and darkfield beam tilt, 70 mm camera, image intensifier, TV, scan generator for SEM and STEM

Manufacturer   Radio Corporation of America, Scientific Instruments Division, Camden, New Jersey 08102, U.S.A.

Zeiss EM–10

Performance
Claimed RP:	better than 5 Å
Magnification:	scan: 100 ×
	main: 1,000–200,000 × in 25 steps; digital display

Illuminating system
HT voltages:	40–60–80–100 kV
Gun:	triode tungsten hairpin, self-biased, DC heating
Emission control:	8 steps to max.60 μA; filament volts and beam current meters
Lenses:	double EM condenser, water-cooled, independent excitation, 1 μm min spot dia.
Alignment:	EM X–Y beamshift, both lenses mechanically adjustable
Apertures:	3, clickstop, centrable in C2
Beam tilt	±2° optional with focus wobbler

Imaging system
Lenses:	3, EM, watercooled
Objective:	$f_0 = 2.6$ mm
Apetures:	3, clickstop, centrable
Stigmator:	EM, electrically centrable
Intermediate:	excitation coupled to mag. control; EM stigmator
Projector:	excitation coupled to mag. control
Focusing:	square wave wobbler (optional extra)
Alignment:	mechanical

Specimen facilities
Stage:	side entry, single grid, 2 mm × 2 mm translation; optional 3-grid holder
Airlock:	pre-pumped; automatic; 3 sec pump-down time
Anticontaminator:	cold finger (optional extra)
Tilting:	double tilt and double tilt/rotation (optional extras)
Special holders:	double tilt, stereo, cooling, heating, straining (extras)

Dark field:	gun tilt optional extra with focus wobbler
Diffraction:	selected area and high resolution (optional extras)

Viewing facilities

Window:	one front, 22·5 × 12 cm
Screen:	14·5 cm dia.
Focusing aids:	permanent tilted focusing screen in centre of main screen 9 × binocular

Recording facilities

Cameras:	3 available: 8 cm × 10 cm 30 exp. plates; 70 mm 75 exp. roll film; 35 mm film; motorized transport
Airlock:	manual
Shutter:	auto., linked to exposure meter, 0·2–100 sec, digital enumeration of negative, mag., HT and year

Vacuum system

Pumps:	2 separate rotary + water-cooled oil diffusion pump + backing tank
Operation:	fully auto. fail-safe push-button
Gauge:	discharge tube + signal lights

Electronics	lens power supply in console, HT separate; transistorized
Units	3: console 160 × 76 × 235 cm + HT cabinet + pump cabinet
Weight	800 kg
Utilities	220 V 50/60 Hz SP 3·5 kVA. Water 21/min 30 p.s.i.
Accessories	focus wobbler and beam tilt; anticontaminator; 7 special specimen cartridges; high-resolution diffraction chamber
Manufacturer	Carl Zeiss AG, 7082 Oberkochen, W. Germany

3. High Performance Instruments

Hitachi–Perkin–Elmer H-500

Performance

Claimed RP:	2 Å (line), 3 Å (p/p)
Magnification:	Upper specimen chamber: scan 250 ×, main 100–800,000 in 35 steps or continuous 'zoom' control: digital readout Lower (diffraction) specimen chamber 50–3,000 ×

Illuminating system

HT voltage:	10–25–50–75–100–125 kV
Gun:	triode tungsten hairpin, self-biased, DC heating; automatically adjusted Wehnelt-anode distance with HT; gun airlock; pointed filaments available
Emission control:	continuously variable
Lenses:	EM double condenser, water-cooled, min spot size 2 μm
Alignment:	EM X–Y spot deflector
Stigmator:	EM in C2, electrically centrable, X–Y system, single knob

Beam tilt: EM in gun for darkfield
Apertures: 4, clickstop, in C2
Meters: beam current, HT

Imaging system
Lenses: 4, EM, water-cooled
Objective: $f_0 = 2 \cdot 0$ mm
Focusing: beam wobbler; coupled to magnification and HT ('zoom focusing')
Apertures: 4, clickstop, thin foil, in objective
Stigmator: EM, electrically centrable, X–Y system, single knob
Intermediate: used as low-mag. objective when lower specimen chamber is in use; excitation coupled to mag. control
Projector 1:   coupled to mag. control
Projector 2:   fixed excitation for mag.; coupled to camera length
Alignment: mechanical centration on all lenses; 'block lens' construction

Specimen facilities
Specimen chambers: 2: high mag./high resolution above objective; low mag./high contrast/diffraction above intermediate
Specimen holders: upper chamber: top entry, rotates, 6 cartridges/load; lower chamber: side entry, 6 specimen clickstop (optional extra)
Airlocks: auto. pre-pumped
Anticontaminators: cold fingers in upper chamber and in vacuum manifold
Tilting: $\pm 10°$ (stereo) or $\pm 35°$ (high angle), both with 360° rotation, non-eucentric (optional extras)
Special holders: 7 different available (optional extras)
Diffraction: selected area in upper chamber, high resolution in lower chamber. 50 cm–60 m camera length
Darkfield: EM beam tilt in gun

Viewing facilities
Windows: 3, 13 cm × 10 cm
Main screen: 19 cm dia.
Focusing screen: 4 cm dia., removable from beam
Focusing aids: 8 × binocular on small screen, beam wobbler

Recording facilities
Camera: one, $3\frac{1}{4}'' \times 4''$, exp, fully auto., motorized
Airlock: fully auto., motorized
Shutter: 0·5–32 sec in 7 steps; auto., coupled to exposure meter; screen current readout; auto. enumerator records neg. number, mag., specimen chamber, HT voltage

Vacuum system
Pumps: 2 separate rotary + buffer tank + oil diffusion pump
Gauges: Penning + 2 Piranis + signal lights
Operation: fully auto. push-button fail-safe, signal lights
Desiccator: built into console

Electronics   fully transistorized, plug-in modules in console and power supply cabinet

Units	4: console, power supply cabinet, measuring resistor, rotary pumps
Weight	1,480 kg
Space requirement	5 m × 4 m × 2·6 m high minimum
Utilities	100/240 V, 50/60 Hz, SP, 3·5 kVA. Water 4 l/min, 0·5 kg/cm²
Accessories	7 specimen holders, high-resolution diffraction, TV, image intensifier, SEM/STEM scanner, EXMA
Manufacturer	Hitachi Ltd., Tokyo, Japan

JEOL JEM-100CX

Performance

Claimed RP: 1·4 Å (line) 3 Å (p/p) HR stage
3·4 Å (line) 5 Å (p/p) top entry goniometer (TEG)
4·5 Å (line) 7 Å (p/p) side entry goniometer (SEG)

Magnification: Scan range: 90, 250, 500, 750 ×
Main range: 1050–500,000 × HR stage
660–300,000 × TEG
330–165,000 × SEG and wide field holder
all in 22 steps; digital readout

Illuminating system

HT voltage: 20–40–60–80–100–120 kV
Gun: triode tungsten hairpin (pointed filaments available) self-biased, DC heating; electrical tilt and traverse; pneumatic gun lift; auto. airlock; field emission gun with UHV pumping system optional extra
Emission control: continuously variable
Lenses: EM double condenser, water-cooled
Alignment: auto. centration; EM X–Y spot deflector
Stigmator: EM in C2, electrically centred
Beam tilt: ±6° for darkfield
Apertires: 4, clickstop, in C2, thin foil
Meter: beam current

Imaging system

Lenses: 4, EM, water-cooled
Objective: f_0 = 1·6 mm HR, 3·1 mm TEG, 5·0 mm SEG
Focusing: beam wobbler, square wave
Apertures: 4, clickstop, thin foil
Stigmator: EM, coupled to mag., electrically centrable, X–Y system
Intermediate 1:   excitation coupled to mag. selector
Intermediate 2:   excitation coupled to mag. selector
Projector: excitation coupled to mag. selector
Alignment: C1, C2 *en bloc*, mechanically centrable, P mechanically centrable, obj. mechanical tilt plus EM alignment

Specimen facilities

Stages: 3, interchangeable: 1. standard HR, 6 specimen, top entry; 2. top entry gonio. $\pm 30°$ double tilt, 6 specimen; 3. $\pm 60°$ single tilt + rotation, 2 specimen side entry goniometer

Airlock: auto. pre-pumped, pump-down time 10 sec

Anticontaminator: cold finger with heater

Tilting: on goniometer stages only

Special holders: 9 available

Diffraction: selected area, 4 apertures, camera length 20–120 cm; high-resolution chamber optional extra

Darkfield: $\pm 6°$ beam tilt in gun

Viewing facilities

Windows: 3: front $13 \cdot 4$ cm $\times$ $23 \cdot 6$ cm, 2 side 9 cm. $\times$ 8 cm (optional extra)

Screen: 160 mm dia., centre tilts for focusing

Focusing aids: beam wobbler; 10 $\times$ binocular

Recording facilities

Camera: one; fully auto. motor driven; either 25 cassettes $6 \cdot 5$ $\times$ 9 cm or $3\frac{1}{4}'' \times 4''$, or 70 mm 50 exp roll film

Airlock: fully auto.

Shutter: fully auto., coupled to exposure meter; auto. enumeration and mag. recording, digital readout

Exposure meter: screen brightness readout

Vacuum system

Pumps: 2 rotary + buffer tank + 2 water-cooled cascade oil diffusion pumps; 10^{-7} torr claimed

Gauge: Penning + Pirani + signal lights

Operation: pneumatic valves operated by air compressor, fully auto. fail-safe, push-button, signal lights

Desiccator: built into console

Electronics fully transistorized, plug-in circuit modules in console and power supply cabinet

Units 4: console, power supply cabinet, rotary pumps, air compressor

Weight 1,655 kg

Space requirement $2 \cdot 8 \times 3$ m $\times 2 \cdot 5$ m high minimum

Utilities 220/240 V 50/60 Hz SP $4 \cdot 5$ kVA; water 4 l/min 1–5 kg/cm^2

Accessories SEM/STEM; image intensifier, EXMA, HR diffraction, 35 mm camera

Manufacturer Japan Electron Optical Laboratories (JEOL) Ltd., 1419 Nagami Akishima, Tokyo 196, Japan

PHILIPS EM 400

Stages	1. Standard high resolution (HM) out
	2. High tilt ($\pm 60°$) eucentric goniometer (HTG)
	3. High mag. ($\pm 30°$) eucentric goniometer (HMG)

Performance

Claimed RP:	HM stage: 1·4 Å line, 3·0 Å p/p
	HTG stage: 3·4 Å line, 7·0 Å p/p
	HMG stage: 2·0 Å line, 5·0 Å p/p
Magnification:	HM stage: 50–800,000 × in 39 steps
	HTG stage: 50–300,000 × in 35 steps
	HMG stage: 50–440,000 × in 37 steps

Illuminating system

HT voltages:	20–40–60–80–100–120 kV
Gun:	triode tungsten hairpin (pointed filaments and FE gun available), self-biased, DC heating, externally adjustable shield-anode distance, gun airlock, EM alignment
Emission control:	2·5–100 μA coupled to filament heating (automatic-filament saturation)
Lenses:	EM double condenser, min. spot size 0·2 μm, coupled to magnification (automatic brightness control)
Alignment:	EM X–Y spot deflector
Stigmator:	EM 8-pole in C2, X–Y system, electrically centrable with 2c/s alignment modulator
Beam tilt:	$\pm 4°$ EM in gun for darkfield
Apertures:	3, clickstop in C2
Meter:	beam current

Imaging system

Lenses:	5, EM, water cooled
Objective:	$f_0 = 1\cdot7$ mm HM, $4\cdot3$ mm HTG, $3\cdot0$ mm HMG
Focusing:	coupled to mag. control and HT ('zoom focusing'), square wave beam wobbler, no image rotation, focus maintained automatically to within 800 Å
Apertures:	3 in objective, automatic insertion and withdrawal -
Stigmator:	EM 8-pole, X–Y system, electrically centrable with 2 c/s modulator
Diffraction lens:	low mag. and selected area diffraction ranges, excitation coupled to mag. control
Intermediate lens:	high mag. range, coupled to mag. control
Projector lens:	2 separate lenses (zero image rotation)
Alignment:	factory pre-aligned with EM fine adjustments

Specimen facilities

Stages:	8-pole specimen chamber takes 3 interchangeable stages
Specimen holders:	7 different types available; all single grid, side entry
Diffraction:	selected area 9 cm–10 m camera length

Viewing facilities
Windows: 3, circular; 16·5 cm dia front, 15·4 cm dia sides
Main screen: 16 cm dia
Focusing screen: 2 cm dia, removable from beam
Focusing aids: 12·5 × binocular on small screen; beam wobbler

Recording facilities
Cameras: 2, fully automatic, compressed air driven, can be mounted simultaneously and used in any order:
1. Plate/film 6·5 × 9 cm or 3¼″ × 4″, 36 exp.
2. 35 mm roll film, 40 exp. One camera supplied
Printout on plate: Neg. no., mag., HT voltage, operator code no.
Shutter: 1/8–32 sec. in 9 steps, fully automatic
Airlocks: separate on each camera, auto. pumping

Vacuum system
Pumps: single rotary + buffer tank + cascade mercury/oil-pump + ion getter pump on gun and specimen chamber; 10^{-7} torr claimed; anticontaminator unnecessary
Operation: fully automatic push-button fail-safe with signal lights and digital readouts
Desiccator: separate, optional extra
Electronics: all transistorized, pre-programmed plug-in modules
Units: console + external desiccator
Weight: 1,500 kg
Space requirement: 3 × 3 × 2·55 m high minimum
Utilities: 190–440 V SP 50/60 Hz, 5–8 kVA, water 3 litres/min
Accessories: SEM/STEM unit; EDAX-EXMA unit; fibre optics TV display system
Manufacturer: NV Philips Gloeilampenfabrieken, Eindhoven, Netherlands

Siemens Elmiskop 102

Performance
Claimed RP: 2 Å (line) 3 Å (p/p)
Magnification: 200–500,000 × in 33 steps; digital readout

Illuminating system
HT voltage: 20–40–60–80–100–125 kV
Gun: triode tungsten hairpin (pointed filaments available) self-biased, DC heating, mech. tilt and centration
Emission control: continuously variable 0–60 μA; auto. brightness control linked to mag.
Lenses: EM double condenser, water-cooled, min. spot size 0·5 μm
Alignment: EM X–Y spot deflector
Stigmator: EM in C2, electrically centrable, X–Y system
Beam tilt: ±2·5° EM in gun for darkfield
Apertures: 3, clickstop, in C2
Meters: Filament volts; beam current

Imaging system

Lenses:	4, EM, water-cooled
Objective:	$f_0 = 2 \cdot 1$ mm
Focusing:	beam wobbler
Apertures:	3, clickstop in objective
Stigmator:	EM 8-pole, electrically centrable, X–Y system
Intermediate 1:	coupled to mag. control. EM stigmator
Intermediate 2:	coupled to mag. control
Projector:	fixed excitation
Alignment:	all lenses mechanicaly centrable; modulator on HT

Specimen facilities

Stage:	top entry, single grid cartridge, 2 mm × 2 mm translation, viewing window in stage chamber
Airlock:	auto. pre-pumping; pump-down time 3 sec
Anticontaminator:	cryopump in stage chamber
Tilting:	$\pm 45°$ X–Y cartridge (optional extra)
Special holders:	12 available
Diffraction:	selected area, camera lengths 30–300 cm, digital readout
Darkfield:	$\pm 2 \cdot 5°$ beam tilt in gun

Viewing facilities

Windows:	3, 95 × 85 mm
Screens:	2: main outer 20 cm dia., central (tiltable) 11 cm dia. Also window above projector with 7 cm dia. screen
Focusing aids:	beam wobbler; 8 × binocular; $1 \cdot 5$ × lens available

Recording facilities

Camera:	one, with 2 exchangeable cassettes: $6 \cdot 5$ cm × 9 cm, 12 exp. or 70 mm roll film, 40 exp.
Airlock:	manual control, auto. pumping
Shutter:	$0 \cdot 4–100$ sec in 33 steps. Semi-auto. with auto. brightness control; auto. negative enumerator, digital readout

Vacuum system

Pumps:	2-stage rotary + buffer tank + cascade water-cooled mercury diffusion/oil diffusion
Gauge:	Penning + Pirani + signal lights
Operation:	fully auto., fail-safe, push-button, signal lights
Desiccator:	built into console

Electronics	fully transistorized, plug-in circuit modules, in 3 separate cabinets
Units	4: console, HT cabinet, power supply cabinet, measuring resistance
Weight	1,430 kg
Space requirement	floor area of 18 m², $2 \cdot 6$ m high
Utilities	220 V 50 Hz SP 4 kVA. Water 3 l/min, $2 \cdot 5$ kg/cm²
Manufacturer	Siemens AG, Wernerwerk für Messtechnik, Karlsruhe, W. Germany

Appendix II

List of Suppliers

The following list is of general suppliers of materials and accessories for biological electron microscopy. Specialist manufacturers of single items, e.g. grids, resins etc. have not been included.

Great Britain

Agar Aids,
127a Rye Street,
Bishop's Stortford,
Herts.

EM Aids,
6 Lime Trees,
Christian Malford,
Chippenham, Wilts.

Emscope Laboratories,
99 North Street,
London SW4 OHQ

Graticules Ltd.,
Sovereign Way,
Tonbridge, Kent

LKB Instruments Ltd.,
232 Addington Road,
South Croydon,
Surrey CR2 8YD

Nortec Products,
16 Front Street,
Sherburn Hill,
Durham

Polaron Equipment Ltd.,
60/62 Greenhill Crescent,
Holywell Industrial Estate,
Watford, Herts.

TAAB Laboratories,
52 Kidmore End Road,
Emmer Green,
Reading

United States

BEEM Inc.,
PO Box 132,
Jerome Avenue Station,
Bronx N.Y. 10468

Ernest F. Fullam Inc.,
PO Box 444,
Schenectady N.Y. 12301

Extech International Corp.,
177 State Street,
Boston Mass. 02109

Ladd Industries,
PO Box 901,
Burlington Vt. 05401

LKB Instruments Inc.,
12221 Parklawn Drive,
Rockville Md. 20852

Polysciences Inc.,
Paul Valley Industrial Park,
Warrington Pa. 18976

Ted Pella Co.,
PO Box 510,
Tustin Calif. 92680

Europe

Balzers Union AG,
Postfach 75,
FL 9496 Fürstentum,
Liechtenstein

LKB-Produckter AB,
S–161 25
Bromma 1,
Sweden

Touzart et Matignon,
3 Rue Amyot,
75 Paris 5,
France

VECO b.v.,
Karel van Gelreweg 22,
Eerbeek (GLD),
Netherlands

Bibliography

It is not the intention in this book to cite extensive lists of references to original papers published in the various Journals. Many of the works of reference and reviews cited contain particularly full lists of references; these are denoted by (R).

The number of papers and articles which have been published on electron microscopy, preparative procedures and biological results must by now run into many tens of thousands. The Electron Microscope Society of America (EMSA) made a valiant attempt to keep up with the literature, but the task became so monumental that it was abandoned in 1962. The EMSA references were issued on punched cards for needle sorting. These cards are a very valuable source of references to the early literature of electron microscopy. The great majority of these references will be found in the major text-books on the subject cited below.

Symposia

The progress of electron microscopy is reported in great detail in the Proceedings of the major Symposia, which are held biennially in various parts of the world. These are now alternately International and European Symposia, and are known generally by the name of the place where they were held. In the following list, the latest is cited first. The publisher is given in parenthesis.

1974 Eighth International, Canberra, Australia (Australian Academy of Science).
1972 Fifth European, Manchester, England (EMCON '72, Institute of Physics).
1970 Seventh International, Grenoble, France (S.F.M.E., Paris).
1968 Fourth European, Rome (Tipographia Poliglotta Vaticana).
1966 Sixth International, Kyoto (Maruzen, Tokyo, Japan).
1964 Third European, Prague (Royal Microscopical Society, Oxford).
1962 Fifth International, Philadelphia (A.P.).
1960 Second European, Delft (N. Holland Pub. Co.).
1958 Fourth International, Berlin (Springer-Verlag).
1956 Third International, Stockholm (Alunqvist and Wiksell).
1955 First European, Toulouse (Comptes Rendus).
1952 Second International, Washington (Electron Physics Circular No. 527).
1949 First International, Delft.

Several other minor conferences have been held, notably in 1950 in Paris; 1954 in Ghent; 1956 in Cambridge; and 1964 in London. The next major conferences are (provisionally): Sixth European, Israel 1976; Ninth International, Toronto 1978; Seventh European, Holland 1980.

Journals

Instrumentation and techniques:

Journal of Microscopy (Blackwell, Oxford; English).
Journal de Microscopie (CNRS, Paris; French, English).
Journal of Electronmicroscopy (Jap. Soc. E.M.; Japanese, English).

Biological results:

Journal of Cell Biology (Rockefeller Press N.Y.; English).
Journal of Ultrastructure Research (A.P., N.Y.; English, French, German).
Journal of Cell Science (Cambridge Univ. Press; English).
Experimental Cell Research (A.P., N.Y. and London; English, French, German).
Tissue and Cell (Oliver and Boyd, Edinburgh; English).

Annual Reviews

Advances in Optical and Electron Microscopy, eds. R. Barer and V. E. Cosslett, A.P., London and New York. (R)

The first four volumes (1966, 1968, 1969 and 1971) contain a great deal of information and references pertinent to Parts I and II.

General Source Books and Basic works: Electron Optics and Instruments

In English

Cosslett, V. E. (1946). *Introduction to Electron Optics*, 2nd edn., Oxford.
Cosslett, V. E. (1951). *Practical Electron Microscopy*, Butterworths, London. (R)
Grivet, P. (1965). *Electron Optics*, Pergamon, Oxford. (R)
Haine, M. (1961). *The Electron Microscope—the Present State of the Art*, Spon, London.
Hall, C. E. (1966). *Introduction to Electron Microscopy*, 2nd edn., McGraw-Hill, New York. (R)
Jacob, L. (1950). *An Introduction to Electron Optics*, Methuen, London.
Klemperer, O. (1953). *Electron Optics*, Cambridge. (R)
Zworykin, V. K., G. A. Morton, E. G. Ramberg, J. Hillier and A. W. Vance (1945). *Electron Optics and the Electron Microscope*, Wiley, New York. (R)

In Other Languages

Ardenne, M. von (1940). *Elektronen Übermikroskopie*, Springer, Berlin.
Borries, B. von (1949). *Die Übermikroskopie*, Langer, Berlin.

Broglie, L. de (1950). *Optique Électronique et Corpusculaire*, Herrmann, Paris.
Dupouy, G. (1952). *Éléments d'Optique Électronique*, Colin, Paris.
Glaser, W. (1952). *Grundlagen der Elektronenoptik*, Springer, Berlin.
Magnan, C. (ed.) (1961). *Traité de Microscopie Électronique* (2 vols.), Herrmann, Paris.

Elementary Books

Grimstone, A. V. (1968). *The Electron Microscope in Biology*, Arnold, London.
Gabor, D. (1945). *The Electron Microscope*, Hutton, London.
Burton, E. F. and W. H. Kohl (1946). *The Electron Microscope*, 2nd edn., Reinhold, New York.
Weakley, B. S. (1972). *A Beginner's Handbook in Electron Microscopy*, Churchill Livingstone, Edinburgh.

Chapter 1

General

Bragg, Sir William (1933). *The Universe of Light*, Bell, London.
Minnaert, M. (1954). *The Nature of Light and Colour*, Dover, New York.
Llowarch, W. (1961). *Ripple Tank Studies of Wave Motion*, Oxford.

Microscopy

Barer, R. (1968). *Lecture Notes on the Use of the Microscope*, 3rd edn., Blackwell, Oxford.
Barron, A. L. E. (1965). *Using the Microscope*, Chapman and Hall, London.
Cosslett, V. E. (1966). *Modern Microscopy*, Bell, London.
Hartley, W. G. (1962). *Teach Yourself Microscopy*, EUP, London.
Martin, L. C. and B. K. Johnson (1958). *Practical Microscopy*, 3rd edn., Blackie, London.

Advanced

Martin, L. C. (1966). *The Theory of the Microscope*, Blackie, London. (R)
Curry, C. (1957). *Wave Optics*, Arnold, London.
Jenkins, F. A. and H. E. White (1954). *Fundamentals of Optics*, McGraw-Hill, New York. (R)

Chapter 2

Historical

Mulvey, T. (1967). The history of the electron microscope. *Proc. R. microsc. Soc.*, **2**, 201–227.
Freundlich, M. M. (1963). Origin of the electron microscope. *Science*, **142**, 185–188.

Early Methods and Instruments

Ardenne, M. von (1940). *Elektronen-Übermikroskopie*, Springer, Berlin.
Borries, B. von (1949). *Die Übermikroskopie*, Langer, Berlin.
Cosslett, V. E. (1951). *Practical Electron Microscopy*, Butterworth, London. (R)

Drummond, D. G. (ed.) (1950). The practice of electron microscopy, *Jl R. microsc. Soc.*, **70**, 1–158. (R)

Gabor, D. (1945). *The Electron Microscope*, Hutton, London.

Wyckoff, R. W. G. (1949). *Electron Microscopy: Technique and Applications*, Interscience, New York.

Zworykin, V. K. and others (1945). *Electron Optics and the Electron Microscope*, Wiley, New York. (R)

Chapter 3

Classification

Cosslett, V. E. (1966). *Modern Microscopy*, Bell, London.

X-ray microscopy

Cosslett, V. E. and W. C. Nixon (1960). *X-ray Microscopy*, Cambridge. (R)

Chapters 4–16

Electron Microscopy, General

Agar, A. W., R. H. Alderson and D. Chescoe (1974). Principles and Practice of Electron Microscope Operation. Vol. 2 of Glauert, A. (ed.) *Practical Methods in Electron Microscopy*, North-Holland–Elsevier, Amsterdam and New York. (R)

Haine, M. (1961). *The Electron Microscope—the Present State of the Art*, Spon, London.

Cosslett, V. E. (1946). *Introduction to Electron Optics*, 2nd edn., Oxford.

Grivet, P. (1965). *Electron Optics*, Pergamon, Oxford. (R)

Magnan, C. (ed.) (1961). *Traité de Microscopie Éléctronique*, 2 vols., Herrmann, Paris. (R)

Hall, C. E. (1966). *Introduction to Electron Microscopy*, 2nd edn., McGraw-Hill, New York. (R)

Heidenreich, R. D. (1964). *Fundamentals of Transmission Electron Microscopy*, Interscience, New York. (R)

High vacuum technology

Dushman, S. (ed. Lafferty, J. M., 2nd edn.) (1962). *Scientific Foundations of Vacuum Technique*, Wiley, New York. (R)

Guthrie, A. (1963). *Vacuum Technology*, Wiley, London.

Turnbull, A. H., R. S. Barton and J. C. Riviere (1962). *An Introduction to Vacuum Technique*, Newnes, London.

Ward, L. and J. P. Bunn (1964). *Introduction to High Vacuum Technology*, Butterworth, London.

Yarwood, J. (1955). *High Vacuum Technique*, 3rd edn., Chapman and Hall, London.

Electronics and stabilizers

Scroggie, M. G. (1965). *Radio Laboratory Handbook*, Iliffe, London.

Patchett, G. N. (1964). *Automatic Voltage Regulators and Stabilizers*, Pitman, London. (R)

Benson, F. A. (1965). *Voltage Stabilization*, Macdonald, London. (R)

Shaw, D. F. (1962). *An Introduction to Electronics*, Longmans, London.

Photography

Horder, A. (ed.) (1958). *The Ilford Manual of Photography*, Ilford, Essex.

Engel, C. E. (ed.) (1968). *Photography for the Scientist*, Academic Press, London. (R)

de Mare, E. (1965). *Photography*, Penguin, Harmonsworth.

Jacobson, C. I. (1954). *Developing—the Technique of the Negative*, Focal Press, New York.

Jacobson, C. I. (1954). *Enlarging—the Technique of the Positive*, Focal Press, New York.

Valentine, R. C. (1966). Response of photographic emulsions to electrons. In R. Barer and V. E. Cosslett (eds.), *Advances in Optical and Electron Microscopy*, Vol. 1, Academic Press, New York. (R)

Mees, C. E. K. (ed.) (1966). *The Theory of the Photographic Process*, 3rd edn., Macmillan, London. (R)

High-voltage Electron Microscopy

Swann, P. R., C. J. Humphries and M. J. Goringe (eds) (1974). *Proceedings of the Third International Conference on High Voltage Electron Microscopy*, A.P., London. (R)

Cosslett, V. E. (1974). Current developments in high voltage electron microscopy. *J. Microscopy*, **100**, 233. (R)

Swann, P. R. (ed.) (1973). Proceedings of EMCON–72 HVEM Symposium, *J. Microscopy*, **97**, 1–269. (R)

Stereo operation

Willis, R.A. and Beeston, B. E. P. (1973). Proceedings of the Symposium on the Interpretation of Electron Scattering. *J. Microscopy*, **98**, 379 and 402. (R)

Chapter 17

Special Methods

Goodhew, P. J., B. E. P. Beeston, R. W. Horne and R. Markham (1972). Electron Diffraction and Optical Diffraction Techniques. Vol. 1 of Glauert, A. (ed.) *Practical Methods in Electron Microscopy*, North-Holland–Elsevier, Amsterdam and New York.

Chapters 18 and 19

Biological Specimen Preparation Techniques

Elementary books

Grimstone, A. V. (1968). *The Electron Microscope in Biology*, Arnold, London.

Mercer, E. H. and M. S. C. Birbeck (1961). *Electron Microscopy—a Handbook for Biologists*, Blackwell, Oxford.

Burton, E. F. and W. H. Kohl (1946). *The Electron Microscope*, 2nd edn., Reinhold, New York.

Skeist, I. (1958). *Epoxy Resins*, Reinhold, New York.

Weakley, B. S. (1972). *A Beginner's Handbook in Electron Microscopy,* Churchill Livingstone, Edinburgh. (R)

Advanced books

In English

Glauert, A. M. (1974). Fixation, Dehydration and Embedding of Biological Specimens, and Reid, N. (1974) Ultramicrotomy. Vol. 3 of Glauert, A. (ed.) *Practical Methods in Electron Microscopy*, North-Holland–Elsevier, Amsterdam and New York.(R)

Hayat, M. A. (1974). *Principles and Techniques of Electron Microscopy*, Vol. 1, Van Nostrand–Reinhold, New York and London. (R)

Kay, D. H. (ed.) (1965). *Techniques for Electron Microscopy*, 2nd edn., Blackwell, Oxford. (R)

Pease, D. C. (1964). *Histological Techniques for Electron Microscopy*, 2nd edn., Academic Press, New York. (R)

Siegel, B. M. (ed.) (1964). *Modern Developments in Electron Microscopy*, Academic Press, New York. (R)

Sjostrand, F. S. (1967). *Electron Microscopy of Cells and Tissues*, Vol. 1, Academic Press, New York. (R)

In German

Reimer, L. (1967). *Elektronenmikroskopische Untersuchungs- und Praparationsmethoden*, 2nd edn., Springer, Berlin. (R)

Index